Design of Marine Facilities
for the Berthing, Mooring, and Repair of Vessels

JOHN W. GAYTHWAITE, P.E.
Consulting Engineer

VAN NOSTRAND REINHOLD
New York

Copyright © 1990 by Van Nostrand Reinhold
Library of Congress Catalog Card Number 90-12230
ISBN 0-442-22900-3

All rights reserved.
No part of this work covered by the copyright
hereon may be reproduced or used in any form or by any means—graphic,
electronic, or mechanical, including photocopying, recording, taping,
or information storage and retrieval systems—without written permission
of the publisher.

Printed in the United States of America

Van Nostrand Reinhold
115 Fifth Avenue
New York, New York 10003

Van Nostrand Reinhold International Company Limited
11 New Fetter Lane
London EC4P 4EE, England

Van Nostrand Reinhold
480 La Trobe Street
Melbourne, Victoria 3000, Australia

Nelson Canada
1120 Birchmount Road
Scarborough, Ontario M1K 5G4, Canada

16 15 14 13 12 11 10 9 8 7 6 5 4 3 2 1

Library of Congress Cataloging-in-Publication Data

Gaythwaite, John.
 Design of marine facilities for the berthing, mooring, and repair
of vessels/by John W. Gaythwaite.
 p. cm.
 Includes bibliographical references.
ISBN 0-442-22900-3
 1. Harbors—Design and construction. 2. Marine terminals—Design
and construction. I. Title
TC205.G29 1990
627'.22—dc20 90-12230
 CIP

TO
Mrs. Margaret Ruth Rich
AND
The Memory of Mr. Charles Allen Rich

Contents

Foreword	ix
Preface	xiii
List of Abbreviations	xv
1. Introduction	**1**
1.1 Scope and Purpose	1
1.2 Marine Civil Engineering	3
1.3 Port and Harbor Facilities	5
1.4 The Design Process	12
References	15
2. Vessel Characteristics and Dimensions	**17**
2.1 Definitions	17
2.2 Hull Form	20
2.3 Hull Construction	23
2.4 Vessel Types and Dimensions	25
References	41
3. General Design Considerations	**44**
3.1 Design Criteria and General Considerations	44
3.2 Site Selection and Layout	48
3.3 Facility Type Requirements	56
3.4 Environmental Conditions	60
3.5 Materials Selection	82
References	92
4 Operational and Environmental Loads	**97**
4.1 Cargo Loads	97
4.2 Vehicular Loads	102
4.3 Rail-Mounted and Material Handling Equipment	109
4.4 Port Buildings and Superstructures	115
4.5 Environmental Loads	116
4.6 Other Load Sources	128
References	130

5 Berthing Loads and Fender System Design — 134
- 5.1 Introduction — 134
- 5.2 The Berthing Ship — 137
- 5.3 The Concept of Added Mass — 141
- 5.4 Berthing Impact Energy Determination — 145
- 5.5 Fender System Types and Selection — 152
- 5.6 Fender System Design — 160
- References — 173

6 Mooring Loads and Design Principles — 176
- 6.1 Mooring System Principles — 176
- 6.2 Mooring Lines, Hardware and Equipment — 188
- 6.3 Wind Forces — 196
- 6.4 Current Forces — 204
- 6.5 Standoff Forces and Passing Vessel Effects — 209
- 6.6 Wave Loads and Vessel Motions — 211
- 6.7 Analytical Treatment and Modeling — 221
- References — 229

7 Design of Fixed Structures — 233
- 7.1 Structure Types and Configurations — 233
- 7.2 Selection of Optimum Structure Type — 238
- 7.3 Pier and Wharf Structural Design — 240
- 7.4 Dolphins and Platforms — 269
- 7.5 Ancillary and Access Structures — 276
- 7.6 Crane Trackage — 277
- 7.7 Miscellaneous Design Features — 280
- 7.8 Ship Services and Utility Systems — 285
- References — 287

8 Geotechnical Design Considerations — 291
by David R. Carchedi, Ph.D., P.E. and Russell J. Morgan, P.E.
- 8.1 Marine Site Investigations — 292
- 8.2 Soil Mechanics and Properties — 301
- 8.3 Marine Foundations — 311
- 8.4 Slopes and Slope Protection — 313
- 8.5 Bulkheads and Retaining Structures — 317
- 8.6 Solid Fill Structures — 326
- 8.7 Pile Foundations — 328
- 8.8 Dry Docks — 341
- 8.9 Site Improvement, Methods, and Materials — 344
- References — 352

9 Design of Floating Structures	**358**
9.1 Structure Types and Applications	358
9.2 Hydrostatics and Stability	366
9.3 Motion Response	369
9.4 Structural Design	375
9.5 Ballast Control and Ancillary Features	382
9.6 Access Bridges and Gangways	386
9.7 Mooring System Design	388
9.8 Marinas and Small Craft Facilities	401
References	405
10 Introduction to Dry Docks	**409**
by Paul M. Becht, P.E. and James R. Hetherman, P.E.	
10.1 General Characteristics, Features, and Factors Affecting Dry Dock Selection	409
10.2 Basin Dry Docks	415
10.3 Marine Railways	438
10.4 Floating Dry Docks	452
10.5 Vertical Lifts	469
10.6 Straddle Lifts and Boatyard Equipment	477
References	481
11 Rehabilitation, Maintenance and Repair	**484**
11.1 Rehabilitation, Reconstruction, and Maintenance	484
11.2 General Modes of Deterioration	489
11.3 General Repair Methods	497
11.4 Pile and Bulkhead Restoration Methods	503
11.5 Protective Coatings and Claddings	506
11.6 Cathodic Protection Systems	509
References	511
Appendix 1 Conversion Factors	**515**
Appendix 2 Selected Information Sources: Journals, Periodicals, and Conference Proceedings	**516**
Index	**519**

Foreword

This book is a useful addition to the lore of the marine engineer who designs shoreside or offshore facilities for ships. It brings together in a single source a synopsis of the vast number of detailed considerations that face the designer who must fashion a connection between the moving object, the ship, in its flotation medium of waves, currents, and winds, and the fixed structures at ports and docks.

The author has made a shrewd selection of facts, physical theory, and accepted engineering practice in order to give as complete a description as possible of the main tasks, without entangling the designer in a web of uncertainties—and the uncertainties in marine engineering begin with our imperfect understanding of the behavior of water, air, and harbor mud, and how they interact with fixed and floating structures.

Marine engineering is a branch of civil engineering, which in turn started with military needs several thousand years ago. It was built on empirical knowledge, ingenuity, and, little by little, science whenever it was helpful. Ingenuity is still a vital ingredient (consider the recent development of the tension leg mooring for offshore oil platforms) and the science is still growing at an exponential rate.

When the author needs to provide more detailed information, or where the reader wants to know more about the background of a particular theory or code of practice, there is at the end of each chapter an extensive list of references. Incorporated in the text are many formulae and tables taken from these references. The figures are plentiful—some 250 or more, both photographic and graphic—to illustrate and simplify understanding of the text.

The focus of the book is clearly on facilities, and how to go about designing them. The book deals with ports, but it is not all about ports. Instead, it is about the structures that are needed at a port in order to make a ship useful. There is just enough information about oceanography, port siting and planning, economics, navigation, channels, breakwaters, dredging, shore processes and cargo handling to make it clear how facility designs fit into the marine industry, without diverting away from the designer's tasks. But in going about those tasks, the way is made clear.

One of the difficulties in marine engineering is the very breadth of the subject, and the diversity of disciplines one must encompass to understand the variables

and the choices that are possible. The menu includes geography, geology, bathymetry, geotechnics, soil mechanics, strength of materials, corrosion, fluid mechanics, dynamic interaction, naval architecture, hydraulics . . . in almost endless array. To find the appropriate data to suit a particular site and a specific design often requires extensive research. This book succeeds very well in shortening the search. It organizes information in a logical sequence, and points out where to get what is missing.

Many aspects of marine facilities change so rapidly that a text can only be current if it is frequently modified. Part of the transitory nature of the science and art of marine design is the result of advances in materials and knowledge; but most of it is caused by changes in the needs of the industry. For example, drilling for oil and gas under water has changed in forty years—from marshes, where water is a few feet deep, and dredging is required to provide flotation for a small drilling barge, to 1700-foot depth for a recent mammoth tension leg platform in the Gulf of Mexico. Finger piers built 50 years ago for general cargo have become obsolete, and have been abandoned in favor of marginal wharves, each with 100-acre upland parking areas for containers, or for imported automobiles, sorted for transport on 18-wheelers and double-decked trains. Military piers built for the heavy loads of 16-inch gun turrets 47 years ago stand unused, and are being converted for bulk material conveyors. Marinas replace cargo sheds. Dockominiums, pleasure boats and apartments stand where railroad carfloats used to berth. Timber piling, which used to be protected from marine borers in many harbors by the polluted water, is again under attack as the waters become cleaner.

One of the changes produced by time is the language in common usage. For example, Chapter 8 in the book is headed "Geotechnical Design Considerations." The key word, *geotechnical*, is quite new. Only a few years ago, we had "soil mechanics" and "foundation engineering," as introduced into the United States by Karl Terzaghi. Today's techniques, using computer programs to model pile-soil systems and slope stability forces, have brought a new perspective. But the judgment factor in assigning coefficients is still essential, and this is recognized by the author.

Many books about the marine industry, or about ports and harbors, try to cover the entire field of subjects related to the sea, from oceanography, waves, storms, beaches, erosion, sedimentation, navigation, and on into breakwaters, channels, ship behavior, commerce, rivers, estuaries, locks, canals, dams, and so on. This is too diverse a subject matter to explain in any detail in one volume and still remain concise enough to be useful.

The increasing specialization of scientific and engineering knowledge has scattered the sources of information and the findings of research. It is a worldwide explosion of data. The designer of marine and waterfront structures needs to synthesize these sources from many disparate disciplines in dealing with his

craft. There is no international codification, but there are international associations. The author has referenced their findings, and has given us a single volume that brings together the options a marine designer faces, and provides the essential guides to accepted practice and to recent developments.

The book should be as valuable a tool for a designer as a dictionary is for a writer.

EUGENE H. HARLOW
Vice President
Soros Associates

Preface

Written to fill a niche in the general literature on port and harbor engineering, this book is intended to give the practicing civil/structural engineer background information needed to design port and harbor structures for the berthing, mooring, and repair of vessels. It provides a thorough introduction and, as such, should also prove useful to port authority engineers, marine terminal operators, marine contractors, port planners, and others with an adequate technical background and the need to understand the principles of marine structures design.

The book is comprehensive enough to serve also as a text or supplementary text for a basic course in port and harbor structures design, with a scope and purpose that are further defined in Chapter 1. There is an emphasis throughout on basic principles combined with practical design applications, which hopefully will contribute to the reader's intuitive understanding of the behavior of moored ships and marine structures and can be applied to offshore terminals and small craft facilities alike.

The book also attempts to serve as a link between theoretical research and design practice. Detailed explanations of all the subject matter of this book would increase its size many times; so frequent references are made to sources that provide further design guidance and development of the theoretical background. As there are no strict step-by-step design codes or standards in the field, the marine structures engineer must be familiar with a wide range of information sources and be prepared to consult the literature relevant to a particular project. Accordingly, prominent international standards are cited throughout.

This book is intended both to complement the existing body of general textbook literature on the subject and to assemble important references to research and design applications. It is a companion volume to my earlier work *The Marine Environment and Structural Design*. Organizing a book that draws from various disciplines and covers a broad area of study is a difficult undertaking, as one must provide explanations or definitions that in turn depend upon subsequent discussions. Therefore, there is considerable cross-referencing to chapter sections, intended to help the reader grasp the interrelation of all the subject matter. The book is organized to flow somewhat as the design process would—that is, from description of design vessels, to establishment of general design criteria, to evaluation of loadings, to structural design, and to functional design requirements as learned from long-term maintenance considerations.

Two chapters were contributed by specialists in the field. The chapter on

geotechnical design considerations is intended to emphasize the critical importance of marine foundations and soil–structure interaction, and in particular to illustrate aspects of marine geotechnical problems as they may differ from those of traditional land-based structures. The chapter on dry docks is more descriptive than the other chapters because of the peculiarities of dry dock structures, as well as the fact that their design remains highly specialized and usually is carried out by firms with long records of experience in the field. Dry docks are prominent port structures, and all port and harbor engineers should be familiar with their design and characteristics.

The construction of important port and harbor works often entails major civil engineering efforts, requiring the varied expertise and concentrated efforts of many experienced individuals. Likewise this book owes its genesis to many contributors to the large and growing body of literature on the subject, such as those cited in its reference sections. I am particularly indebted to many individuals who have contributed materials and comments directly to this project. I gratefully acknowledge my contributing authors: Dr. David R. Carchedi, P.E.; Mr. Russell J. Morgan, P.E.; Mr. Paul M. Becht, P.E.; and Mr. James R. Hetherman, P.E. Their contributions have greatly enhanced the content of this book. Thanks are also due to the staffs of Goldberg Zoino Associates, Inc. and Crandall Dry Dock Engineers, Inc. and to Mr. Paul S. Crandall, P.E., for supporting these efforts and reviewing the contributed chapters. Special thanks go to: Mr. John R. Headland, P.E., of the Naval Facilities Engineering Command, for contributing material on mooring loads and single point moorings and for valuable comments on the overall text; Mr. Peter J. Piaseckyj, P.E., of Duxbury International, Inc., for reviewing the chapter on berthing loads; and Mr. Klaus Schoellner, P.E., of the Maguire Group, Inc., for reviewing the chapter on maintenance and repairs—and to all of them for their assistance in providing additional reference materials. I thank Mr. John T. Muha for his assistance in preparing some of the artwork, and the following individuals for their help in obtaining photographs and/or reference materials: Dr. Gerardus van Oortmerssen, of the Netherlands Ship Model Basin; Mr. Herman Bomze, P.E., of Frederic R. Harris, Inc.; Mr. Leonard Sugin, P.E., of Soros Associates, Inc.; Mr. Robert H. Gruy, P.E., of Sofec, Inc.; and Mr. Richard A. Russell, of W. B. Arnold Co., Inc. I further express my sincere gratitude to all those authors and publishers who have granted me reprint permissions and to those individuals and organizations that have supplied photographs as credited in the text. Finally, special thanks are due to Mrs. Margaret R. Rich for her invaluable assistance in preparing the manuscript copy, and to Michele R. Gaythwaite for her assistance and for typing the final manuscript.

JOHN W. GAYTHWAITE
Manchester, Massachusetts

Lists of Abbreviations

NOTATIONS

The following symbols and abbreviations occur most often in the text. Notations that are not repeated in the text are defined where introduced and may not appear below. Certain characters are used frequently with different subscripts, which are defined where first introduced and are denoted by an asterisk below.

A_*:	Area as defined in text
a:	Acceleration
a_*:	Motion amplitude as defined in text
B:	Vessel beam or width
b:	Net buoyancy force
bhp:	Brake horsepower
BWL:	Vessel beam at waterline
C:	Wave celerity
C_*:	Coefficient as defined in text
c.b.:	Center of buoyancy
c.f.:	Cubic feet
c.g.:	Center of gravity
c.m.:	Cubic meters
D:	Vessel draft; pile diameter or width normal to flow
D_*:	Diameter or dimension as defined in text
D':	Vessel projected width normal to flow
\underline{D}:	Characteristic pile embedment length
d:	Water depth
DT:	Displacement tonnage
DWL:	Design water level or design waterline length of vessel
DWT:	Deadweight tonnage
E_*:	Energy or encounter probability as defined in text
e:	Exponential (base of Naperian logarithms)
F_*:	Force as defined in text
f:	Frequency $= 1/T$
f_*:	Compressive or bending strength as defined in text
FB:	Vessel or floating structure freeboard
fps:	Feet per second
GRT:	Gross registered tonnage
g:	Acceleration of gravity (usually taken as 32.16 fps^2)
gpm:	Gallons per minute

LIST OF ABBREVIATIONS

H_*: Wave height as defined in text
h_*: Height or vertical distance as defined in text
HAT: Highest astronomical tide
K_*: Force factor, spring constant, or coefficient as defined in text
k: Wave number $= 2\pi/L$
k_*: Constant as defined in text
L_*: Wavelength or structure length as defined in text
l_*: Length of line, cable, or pile as defined in text
LAT: Lowest astronomical tide
LBP: Vessel length between perpendiculars
LOA: Vessel length overall
l.t.: Long ton ($= 2240$ pounds)
LWL: Vessel length on waterline
LWT: Vessel light displacement (tonnage)
M_*: Mass or moment as defined in text
M': Virtual mass
m_*: Mass factor as defined in text
MHW: Mean high water
MHWS: Mean high water on spring tide
MLLW: Mean lower low water
MLW: Mean low water
MLWS: Mean low on spring tide
MSL: Mean sea level
m.t.: Metric ton ($= 2205$ pounds), often denoted as "tonne"
MTI: Moment to trim one inch
MTL: Mean tide level
N_*: Nondimensional scale number as defined in text
n: Number or quantity
n_h: Soil subgrade modulus
NRT: Net registered tonnage
P_*: Pressure, force per unit area, or member axial load as defined in text
$P_{(*)}$: Probability of occurrence of * as defined in text
pcf: Pounds per cubic foot
plf: Pounds per linear foot
pH: Negative log of hydrogen ion concentration (measure of acidity)
PPI: Pounds per inch immersion
psf: Pounds per square foot
Q: Volume flow rate
R_*: Reaction force as defined in text
r: Radius or dimension as defined in text
RAO: Response amplitude operator
RMS: Root mean square value
R_T: Return period
S: Dimension as defined in text
S_c: Anchor line scope
$S(\omega)$: Wave height spectral density

s:	Length along curve or catenary
shp:	Shaft horsepower
SWL:	Still water level
T_*:	Wave period or structure period as defined in text
t:	Time or duration
TEU:	Twenty-foot equivalent unit (vessel containers)
U_*:	Current or water particle velocity as defined in text
u_*:	Velocity component as defined in text
V_*:	Wind or vessel velocity as defined in text
v_*:	Velocity component as defined in text
W_*:	Weight as defined in text
w_*:	Velocity component or weight per unit length as defined in text
x, y, z:	Coordinate directions or distance along x, y, z direction
α_*:	Angle or angular acceleration as defined in text
β:	Characteristic length of elastic beam or angle as defined in text
γ:	Unit weight or specific weight
Δ:	Vessel displacement in weight units (inverted ∇ indicates vessel displacement in unit volume)
δ_*:	Displacement or deflection as defined in text
ϵ:	Horizontal excursion of moored vessel or structure
η_c:	Wave crest elevation above SWL
θ:	Angle as defined in text
μ:	Coefficient of friction
ν:	Kinematic viscosity
ρ:	Mass density ($= \gamma/g$)
σ_*:	Soil stress as defined in text
τ:	Shear or torsional stress as defined in text
ϕ:	Angle as defined in text
ψ:	Angle as defined in text
ω:	Circular frequency $= 2\pi/T$ or angular acceleration

ACRONYMS FOR ORGANIZATIONS

AAPA:	American Association of Port Authorities
AASHTO:	American Association of State Highway and Transportation Officials
ABS:	American Bureau of Shipping
ACI:	American Concrete Institute
AISC:	American Institute of Steel Construction
AISI:	American Iron and Steel Institution
ANSI:	American National Standards Institute
API:	American Petroleum Institute
AREA:	American Railroad Engineering Association
ASCE:	American Society of Civil Engineers
ASTM:	American Society for Testing Materials
AWPA:	American Wood Preservers Association
BSI:	British Standards Institution

CERC:	Coastal Engineering Research Center (USACE)
CRREL:	Cold Regions Research and Engineering Laboratory (USACE)
EAU:	Empfehlungen des Arbeitsausschusses' Ufereinfassungen (German Societies of Harbor and Foundation Engineering)
IADC:	International Association of Dredging Companies
IAHR:	International Association for Hydraulic Research
IALA:	International Association of Lighthouse Authorities
IAPH:	International Association of Ports and Harbors
ICE:	Institute of Civil Engineers
ICHCA:	International Cargo Handling and Coordination Association
ISO:	International Standards Organization
NACE:	National Association of Corrosion Engineers
NFPA:	National Forest Products Association
NAVFAC:	U.S. Navy Facilities Engineering Command
NAVSEA:	U.S. Navy Sea Systems Command
NOAA/NOS:	National Oceanic and Atmospheric Association/National Ocean Survey
OCIMF:	Oil Companies International Marine Forum
OTC:	Offshore Technology Conference
PHRI:	Port and Harbour Research Institute (Japanese Ministry of Transport)
PIANC:	Permanent International Association of Navigation Congresses
SNAME:	Society of Naval Architects and Marine Engineers
SSPC:	Steel Structures Painting Council
UNCTAD:	United Nations Commission on Trade and Development
USACE:	U.S. Army Corps of Engineers
USCGS:	U.S. Coast and Geodetic Survey
USNCEL:	U.S. Navy Civil Engineering Laboratory
WODA:	World Organization of Dredging Associations

1
Introduction

The construction of port and harbor works was among the earliest major undertakings of civilized human beings. The ancients often displayed a great intuitive understanding of nature in their port works, which unfortunately have been lost in the decline of empires and changing coastlines. The timber and stone harbor works of less than one hundred years ago, gradually are being replaced with concrete and steel structures, which have extended offshore port facilities into deeper water at exposed locations. Even so, port and harbor engineers depend heavily upon the collection of past experiences to temper their analysis and contemporary design practices.

This book, which reflects rapid progress made in design and construction in the past few decades, focuses on the structural design of marine facilities for the berthing, mooring, and repair of vessels. This first chapter delineates the book's scope and purpose, providing brief overviews of marine civil engineering and the development of port and harbor facilities, as well as literature on the subject. The general approach to the design process is discussed with regard to state-of-the-art practice.

1.1 SCOPE AND PURPOSE

This book primarily is concerned with the structural design of marine facilities for the berthing, mooring, and repair of vessels. It is intended to provide the civil/structural engineer with background information and general design requirements for port and harbor docking structures. It is not meant to be a step-by-step design guide or to cover basic structural design principles, but instead provides structural design criteria and addresses design problems peculiar to marine structures as they differ from traditional land-based construction. In this regard, the determination of design loads and environmental effects and the selection and proper application of suitable materials are emphasized. Because of the probabilistic nature of loads in marine structures, their magnitude, directions, and frequency of occurrence over the lifetime of the structure cannot be so accurately predicted, compared to land-based structures. The harsh marine environment contributes to more rapid deterioration; so the general design approach also differs. The book is organized to follow the general design process, with each chapter providing a general subject background, discussion of contemporary practices, and references for follow-up.

2 DESIGN OF MARINE FACILITIES

As the purpose of a marine facility is to service vessels, Chapter 2 presents a description of vessel design considerations, and their types and characteristics. General design criteria are reviewed, including site selection and layout, operational requirements, environmental conditions, and materials selection. Chapters 4 through 6 are concerned with the determination of design loads. Structure loadings due to cargo storage and handling, vehicles and equipment, superstructures and environmental loads acting directly on the structure, and other load sources are reviewed. Chapters 5 and 6 deal with berthing and mooring loads imposed by vessel motions, which often govern structural design. The design of fixed structures such as piers, wharves, dolphins, platforms, and ancillary structures is presented in Chapter 7. Chapter 8, a contributed chapter, is devoted to geotechnical considerations, emphasizing the importance of marine foundations and soil–structure interaction. Principles of the design of floating structures and their applications in marine facilities are covered in Chapter 9. An introduction to the design of dry docks, contributed by engineering specialists in the field, follows in Chapter 10. All port and harbor engineers should be familiar with the general dry dock types and their principles of operation. Finally, aspects of rehabilitation, maintenance, and repair are reviewed in Chapter 11, as an understanding of structure maintenance problems and the need to consider future expansion or conversion provides the design engineer with important insights that can be incorporated into new designs. The book attempts throughout to emphasize basic principles applicable to vessels of all sizes, from yacht harbors to mammoth tanker berths and from simple near-shore waterfront structures to offshore deepwater terminals.

The design of marine terminal and dry-docking facilities is only one aspect of the broader field of port and harbor engineering. Important related subjects not covered herein include: port planning and economics; cargo handling and transportation systems; dredging and layout of navigation channels, breakwaters, shore processes; and physical oceanography as it relates to environmental design conditions. These subjects are reasonably well covered in the general literature on port and harbor and coastal engineering, such as references (1) through (19). This book is intended to complement, update, and further contribute to the general body of textbook literature. An effort has been made not to duplicate here material that already has been well covered, but rather to fill in some gaps and to provide guidance on the structural design process as well as contemporary design standards and other important references.

A brief note about the usage of mathematical units is warranted. No attempt has been made to use entirely consistent units, such as the SI system; instead the information is conveyed in the units in which it is usually encountered in practice in the United States, and for data reproduced from other sources the units of the original presentation have been retained. In addition to metric and English units, marine work utilizes some units of its own—nautical miles, knots,

fathoms, and so on—and the port engineer should become accustomed to thinking in terns of the various units. Some important conversion factors are included in Appendix 1.

1.2 MARINE CIVIL ENGINEERING

Marine civil engineering encompasses the planning, design, and construction of fixed and stationary moored floating structures along ocean coastlines and the Great Lakes, and in the offshore realm. Traditionally, civil engineering in the oceans could be categorized as *coastal engineering*, dealing principally with shoreline processes and protection, and *port and harbor engineering*, dealing with the berthing, servicing, and navigation requirements of vessels. Today, another broad category, *offshore engineering*, concerned with the recovery of natural resources—namely, oil and gas—can be added.

Table 1.1 summarizes some of the types of civil engineering works involved in marine civil engineering. Contemporary uses for coastline and offshore areas, such as waterfront development and ocean energy conversion respectively, have been added to Table 1.1 as tentative subdisciplines. Another related field is design for inland waterways, including canals, locks, and flood protection structures. Inland waterways often provide a critical transportation link with major seaports. Although each of the disciplines noted serves a different purpose, all the practitioners must cope with the same harsh environment and draw experience from each other. For example, developments in the study of wave forces and deepwater construction techniques conducted by the offshore oil industry have contributed to the technology base of traditional harbor works and aided the development of marine terminals at deepwater exposed locations.

Historically, coastal and harbor works in the form of breakwaters and quays date back to antiquity. Inman (20) described ancient harbor engineering feats and their lessons for posterity, and Johnson (21) reviewed significant academic progress of more recent times. This book is concerned with certain aspects of harbor engineering that began to be studied as an applied science around the middle to late nineteenth century. It was during this period of industrialization that many of the early wave theories were developed, and early texts on harbor engineering were published (22, 23). As the era of canal building faded during the late nineteenth century, the first international conferences on vessel navigation in coastal ports began to focus on port development technology (24). From this time on into the early twentieth century, harbor structures were built largely of timber or stone or composites of both. Vessel sizes began to grow, and shore transportation networks were well developed. The introduction of reinforced concrete and iron and steel into waterfront construction began what could be considered the golden age of harbor engineering works, in the early to middle twentieth century. References (24) through (27) provide insight into

4 DESIGN OF MARINE FACILITIES

Table 1.1. Marine civil engineering disciplines.

	Coastal	Port and harbor	Offshore
Purpose	Shore protection	Navigation, berthing, and servicing of vessels	Recovery of natural resources (oil and gas)
Project types	Breakwaters	Dredging: channels and mooring basins	Fixed platforms
	Jetties and groins		Mobile drilling units
	Seawalls and revetments	*Terminals and port structures	Moorings
	Beach nourishment	*Shipyards and dry docks	Pipelines
	Shoreline stabilization	*Small craft harbors	Offshore terminals
	Flood control	*Moorings	
	Outfalls and pollution control		
Related and sub-disciplines	Near-shore monitoring	Nav-aids/VTS	Exploration
	Hydro-surveying	Inland waterways	Offshore buoys and monitoring
	Waterfront development	Industrial waterfront facilities	Ocean energy
		Waterfront development	
		*Offshore terminals	

*Subjects covered in this text.

the early development of harbor engineering during the past century. The general trend to larger and deeper-draft vessels and more sophisticated/faster material handling and transportation systems has continued, with some slowing in recent times.

Port development and expansion kept pace with ship and material handling technology for a while, perhaps into the late 1960s or early 1970s; but in the United States and other developed countries many port facilities now are in need of upgrading, rehabilitation, or replacement due to deterioration with age and obsolescence. Port development is strongly affected by political factors such as world trade routes, regulation, lack or inaccessibility of suitable virgin sites, and environmental protection concerns. The waterfront has become an increasingly valuable resource, and conflicting uses such as urban/residential development have placed further pressures and limitations on suitable marine terminal sites. These issues are beyond the scope of this text, but the harbor engineer must be attuned to the way in which they affect his or her work.

Marine civil engineering draws heavily from other disciplines and often relies more on the application of engineering principles than on following strict building codes, compared with the design of most land-based facilities. The art and science of harbor engineering has been greatly affected by technology advances and activity in the following fields:

- Ship design and construction.
- Cargo handling and transportation systems.
- Dredging and construction equipment technology.
- Materials science.
- Applied oceanography.
- Mathematical and physical modeling techniques.
- Surveying and electronic measurement.

Many professional and trade organizations publish periodicals and sponsor occasional and regular conferences to help the harbor engineer keep abreast of technical developments and applications. A listing of some of the journals, periodicals, and conference proceedings is included in Appendix 2.

1.3 PORT AND HARBOR FACILITIES

Contemporary marine facilities tend to be highly specialized in their operations, which focus upon the type of cargo being handled. Marine terminals are active berthing facilities that serve as transfer points for the vessel's cargo. Major facility types include:

- General (break bulk) cargo.
- Container.

6 DESIGN OF MARINE FACILITIES

- Roll on/roll off (Ro/Ro) (trailer ship).
- Ferry.
- Passenger/Cruise ship.
- Barge carrier (LASH).
- Dry bulk.
- Liquid bulk and liquefied compressed gases such as LNG.

Permanent berthing facilities ("home ports") and shipyards are generally considered less active, in that vessels typically remain in berth for long periods. Such facilities include:

- Military and coast guard bases.
- Shipyards (construction and repair).
- Fishing and commercial small craft.
- Research vessels and workboats.
- Marinas (yachts and recreational craft).

Permanent berth facilities and most marine terminals are built near shore and are shore-connected in relatively well-protected harbors. Liquid and dry bulk facilities, however, may be constructed offshore in relatively unprotected deep water. Their cargoes may be transferred to shore via pipelines or conveyor systems or may be transferred to small vessels, in a process termed *lightering*, for distribution to smaller coastal ports. Offshore ports may consist solely of large single point moorings, SPMs, equipped with hose connections for transferring liquid cargoes through underwater pipelines.

Figure 1.1 shows an offshore "sea-island" berth for mammoth tankers of too deep draft and too large size to enter most conventional harbors. Such facilities must be designed to survive severe environmental conditions associated with long return periods and the berthing and mooring forces of mammoth vessels under maximum operating conditions. Petroleum products and liquefied gases are considered hazardous cargoes and are subject to strict regulations and fire protection requirements. Such facilities typically are isolated as much as possible from other waterfront activities.

All facility types have particular design requirements. For example, container terminals, such as that shown in Figure 1.2, generally require considerable backland area for the marshaling and storage of containers and wharf structures designed to accommodate large cranes and heavy deck loads. The trend today seems to be toward expansion of the largest container facilities to serve as load centers for the larger ocean-going vessels, with smaller vessels providing local feeder service and/or direct intermodal connection via highway and rail lines.

Current thinking in the planning and design of new and major marine terminals is to consider multipurpose use. The model terminal (28) of the future

Figure 1.1. An offshore "sea-island" type mammoth tanker terminal off the Kyukyu Islands, Okinawa. The vessel is the *Universe Kuwait* of 326, 000 DWT. (Photo courtesy of Frederic R. Harris, Inc.)

8 DESIGN OF MARINE FACILITIES

Figure 1.2. A contemporary small to medium-size container terminal with dual container cranes. Note stacking area and control tower in background. (Photo by Paul UBl, courtesy of Massachusetts Port Authority.)

would be suitable for handling a mix of containers, general cargo, and Ro/Ro, as well as being designed to handle the largest loads of today. Considering the long lifetimes of such infrastructure works and planners' limited ability to see very far into the future, this seems a prudent approach, especially for developing nations.

Bulk terminals such as that shown in Figure 1.3 often feature elaborate material handling equipment and conveyor systems. Space for stockpiling bulk materials must support very heavy loads. Ferry terminals, as in Figure 1.4, require direct highway linkage and marine terminal structures with substantial fender systems and efficient transfer bridge and passenger boarding systems, because of the rapid-turnaround nature of ferry service. Shipyard facilities, by contrast, require close alongside berthing space with substantial utility and ship service systems and revolving crane service. A central feature of ship repair yards, in particular, is dry docking capability. Figure 1.5 illustrates a ship repair pier with portal crane and full utility systems, which provides vehicular access to a large floating dry dock. The dry dock in turn requires deep water for submergence while docking and undocking vessels, and is moored in position by two large mooring platforms.

Facility type requirements are discussed in Section 3.3. The design of all

Figure 1.3. A bulk cargo terminal facility equipped with a catenary continuous unloader multipurpose Portainer® Crane. (Photo courtesy of Paceco, Inc.)

facility types in general must consider: vessel berthing requirements, which often focus on one or more "design vessels"; navigation channels and water depths; backland space requirements and shore transportation connections; site exposure and subsurface soil conditions; and other factors peculiar to the facility type. Port directories such as references (29) and (30) list over two thousand port facilities worldwide giving data on piers, wharves, services, water depths, climatic conditions, and so on.

Marine facilities typically consist of piers, wharves, quays, bulkheads, fills, dolphins and platform structures, trestles and access bridges or catwalks and buildings, and backland civil/site works. The design of buildings and civil engineering site works is the same as for any other land-based construction (provided the possibility of flooding is considered) and thus is not covered herein. The distinction between piers, wharves, quays, and so on, and the description of other fixed structural types is given in Section 7.1. Port facilities also may consist of floating structures such as piers, floating dry docks, mooring buoys, floating plant and equipment, and so on, which are further described in Section 9.1. Figure 1.6 shows a floating liquefied propane gas (LPG) storage facility permanently moored to an SPM, which serves as an isolated offshore port for the loading of LPG vessels.

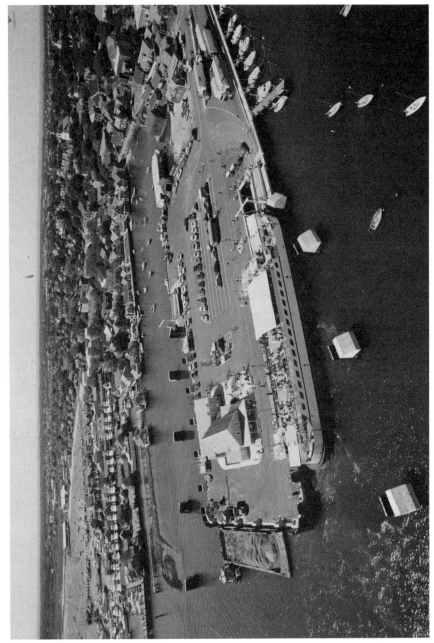

Figure 1.4. A passenger and vehicle ferry terminal with two ferry slips at Nantucket Island, Massachusetts. Note guide-in dolphins and wharf corner fenders for warping of vessel. Also note queuing areas for cars and tractor trailers. (Photo courtesy of The Maguire Group, Inc.)

Figure 1.5. A ship repair facility with 600-foot-long repair pier fitted with 65-ton-capacity revolving crane and providing access to an 81,000-ton-capacity floating dry dock. (Photo courtesy of The Maguire Group, Inc.)

12 DESIGN OF MARINE FACILITIES

Figure 1.6. A floating LPG terminal consisting of a 461 × 136-foot prestressed concrete hull with a gross displacement of 66,000 tons for LPG storage and moored to an SPM with yoke type connection. (Photo courtesy of Concrete Technology Corporation.)

1.4 THE DESIGN PROCESS

The structural design process must consider all available materials and construction techniques with regard to both local site and environmental conditions and operational and vessel berthing requirements. The design procedure can be divided into key activities:

- *Establishment of design criteria:* Considerations include environmental and operational loadings and their probability and frequency of occurrence, as well as structural reliability as measured by factors of safety and allowable stresses and movements.

- *Calculation of forces for design conditions:* These forces include combinations and short-term normal loads and long-term extreme loads.
- *Determination of structural response:* This usually involves analysis of preliminary structural elements and arrangements on a trial basis until an optimum structure is found. Dynamic loads and structural response often form a feedback loop.
- *Comparison of structural responses with structural reliability criteria:* This is done to meet minimum safety requirements and to ensure cost-effectiveness.

Unlike traditional land-based structures, which are designed for loads and allowable stress criteria established by building codes, marine structures must be designed to loading and performance criteria established by the designer in concert with facility owner/operator requirements. A probabilistic approach is followed, considering probability of occurrence, usually in terms of statistical return periods of specific events, and the structure's level of reliability under the specified conditions. For example, it often is assumed that large vessels will not remain alongside at their berth when winds or wave conditions exceed a given threshold velocity or height, respectively. Cranes usually cease operation, and/or loading arms are disconnected at even lower threshold values. Berthing and mooring loads usually govern the lateral load capacity of structures considered herein, and vessel dimensions and berthing maneuvers affect the facility layout and structure configuration. Therefore, a maximum design vessel under various conditions of ballast and water levels usually will govern structural loads, whereas consideration of smaller or alternative vessel types will dictate overall layout, number and locations of fenders, and so forth. Design criteria also must consider: long-term effects such as scour and erosion, general deterioration, fatigue and wear, construction techniques and weather windows, and the possibility of future expansion. The design life of a structure affects the material and durability design criteria. However, it should be noted that for most marine facilities the physical life often exceeds the functional life because of obsolescence and changing socioeconomic conditions.

Although there are no definitive, binding building codes or standards that control the design of marine facilities, there are several guideline codes of practice to which the designer can refer for general design and for certain specific requirements (31–35). In addition, specific criteria that have been established by international organizations relate to specific facility type requirements. These standards are introduced in Chapter 3 along with a more detailed discussion of general design criteria.

The evaluation of loads and forces acting on most near-shore waterfront structures usually can be carried out with simple closed-end-solution equations. In deeper water and at sites exposed to significant wave action, more sophisti-

cated analytical techniques may be required. Mathematical/numerical modeling techniques and/or physical hydraulic model studies may be employed. In some instances the results of such studies may be extrapolated and applied with caution to the design of similar structures under similar vessel and load conditions. Most waterfront structures can be analyzed using static methods, but offshore structures should be subjected to dynamic analysis. Under dynamic analysis the forces and structural response become a function of each other and often lead to very interesting results when compared with static analyses. Dynamic analysis, however, requires relatively detailed information about subsurface soil conditions, the expected sea conditions, and vessel motion characteristics, as well as the elastic properties of the structure.

Reliability criteria are most simply expressed in terms of a factor of safety, F.S., where:

$$\text{F.S.} = \frac{\text{Structure resistance}}{\text{Applied load}}$$

The factor of safety may vary greatly according to material type, structure type, degree of confidence in and probability of exceedance of specified load conditions, and consequences of failure in terms of potential loss of life and/or economic loss. The F.S. typically will be lower for long-term extreme events than for normal operation conditions. It is most important for the designer to consider all possible failure modes and examine the safety margin. Factors of safety cannot be applied directly to fatigue loading conditions, for example.

Marine structures should always possess ductility and some form of redundancy or alternate load path to preclude catastrophic failure under extreme conditions and local yielding of structural elements. Probabilistic approaches such as load and resistance factor design (LRFD) methods are the most rational means of dealing with the uncertainties of marine structure design. However, at the time of this writing, unified LRFD criteria suited to the design of typical waterfront and offshore structures have not yet been devised.

Waterfront structures traditionally have been designed with a high degree of conservatism, largely because of unknowns in possible loadings but also because of relatively rapid deterioration rates of structures in the sea. Today, with economic pressures, a better understanding of environmental loadings, and a statistical approach to determining design criteria, and with high speed computers to aid their analysis, as well as better quality control of materials and protective coatings and systems, designers have the ability to produce more economical designs on a rational basis. Engineers must be cautious when establishing design criteria, however, as the knowledge of marine loads and structure interaction is still incomplete, and the ability to accurately forecast future events remains limited.

REFERENCES

Port and Harbor Engineering

1. AAPA (1973). *Port Planning, Design and Construction*. American Association of Port Authorities, Washington, D.C.
2. Agerschou, H., Lundgren, H., and Sorensen, T. (1983). *Planning and Design of Ports and Marine Terminals*. Wiley, Chichester, U.K.
3. Bruun, P. (1981). *Port Engineering*, 3rd ed. Gulf Publishing, Houston, Tex.
4. Cornick, H. (1962). *Dock and Harbour Engineering*, 4 vols. Griffen, London.
5. Du-Plat-Taylor, F. M. (1968). *The Design, Construction and Maintenance of Docks, Wharves and Piers*. Ernest Benn, London.
6. Quinn, A. D. F. (1972). *Design and Construction of Ports and Marine Structures*. McGraw-Hill, New York.
7. Thoresen, C. A. (1988). *Port Design—Guidelines and Recommendations*. Tapir Publishers, Trondheim, Norway.

Port Planning and Finance

8. Frankel, E. (1987). *Port Planning and Development*. Wiley, New York.

Coastal Engineering and Marine Environment

9. Gaythwaite, J. W. (1981). *The Marine Environment and Structural Design*. Van Nostrand Reinhold, New York.
10. Goda, Y. (1985). *Random Seas and Design of Maritime Structures*. University of Tokyo Press, Tokyo.
11. Horikawa, K. (1970). *Coastal Engineering—An Introduction to Ocean Engineering*. University of Tokyo Press, Tokyo.
12. Ippen, A. T., ed. (1966). *Estuary and Coastline Hydrodynamics*. McGraw-Hill, New York.
13. Minikin, R. R. (1963). *Wind, Waves and Maritime Structures*. Griffen, London.
14. Silvester, R. (1974). *Coastal Engineering*, 2 vols. Elsevier Scientific, New York.
15. Sorenson, R. (1978). *Basic Coastal Engineering*. Wiley, New York.
16. Wiegel, R. (1964). *Oceanographical Engineering*. Prentice-Hall, Englewood Cliffs, N.J.

Dredging

17. Bray, R. N. (1979). *Dredging—A Handbook for Engineers*. Edward Arnold. London.
18. Herbich, J. G. (1975). *Coastal and Deep Ocean Dredging*. Gulf Publishing, Houston, Tex.
19. Houston, J. (1970). *Hydraulic Dredging*. Cornell Maritime Press, Cambridge, Md.

Historical

20. Inman, D. L. (1974). "Ancient and Modern Harbors: A Repeating Phylogeny," *Proc.* of the 14th Conf. on Coastal Engineering, ASCE, New York.
21. Johnson, J. W. (1974). "History of Some Aspects of Modern Coastal Engineering," *Proc.* of the 14th Conf. on Coastal Engineering, ASCE, New York.
22. Stevenson, T. (1864). *On the Construction of Harbours*. England.

23. Vernon-Harcourt, L. F., (1885). *Harbours and Docks*. Clarendon Press, Oxford.
24. PIANC (1985). *Centenary of the Permanent International Association of Navigation Congresses: 1885 to 1985*, Centenary Yearbook. PIANC, Brussels.
25. Clearwater, J. L. (May 1985). "Port Construction Since 1885: Evolving to Meet a Changing World," *The Dock and Harbour Authority*, Special Feature, Vol. LXVI, No. 769, London.
26. Greene, C. (1917). *Wharves and Piers*, McGraw-Hill, New York.
27. Small, E. W. (1941). *Early Wharf Building*. Eastern National Parks and Monuments Association, Salem, Mass. (Rev. 1970).

General

28. PIANC (1987). "Development of Modern Marine Terminals," Report of the Working Group of the Permanent Tech. Comm. II, PIANC, Supp. to Bull. No. 56, Brussels.
29. —— (Annual). *Lloyd's Ports of the World 1988*. Lloyd's of London Press, New York.
30. —— (Biannual). *Fairplay World Ports Directory*. Fairplay Publications, London.

Codes and Standard Practices

31. BSI (1984). *British Standard Code of Practice for Maritime Structures*, Part I: "General Criteria," British Standards Institution, BS6349, London. Additional Parts: Part 2: "Design of Quay Walls, Jetties and Dolphins"; Part 3: "Design of Dry Docks, Locks, Slipways and Shipbuilding Berths . . ."; Part 4: "Design of Fendering and Mooring Systems" (1985).
32. EAU (1985). *Recommendations of the Committee for Waterfront Structures, EAU 1985*, Comm. for Waterfront Structures Soc. for Harbor Engineering and The German Soc. for Soil Mechanics and Foundation Engineering, 4th English trans. Wilhelm Ernst and Sohn, Berlin–Munich, 1986.
33. PHRI (1980). *Technical Standards for Port and Harbor Facilities in Japan*, The Overseas Coastal Area Dev. Inst. of Japan, Bureau of Ports and Harbors, Ministry of Transport, Port and Harbor Research Institute, Tokyo.
34. NAVFAC (1980). *Piers and Wharves*, Design Manual 25.1. Dept. of the Navy, Naval Facilities Engineering Command, NAVFAC DM-25.1, Alexandria, Va. Additional parts: DM 25.2, *Dockside Utilities for Ship Service* (1981); DM 25.3, *Cargo Handling Facilities* (1981); DM 25.4, *Seawalls, Bulkheads and Quay Walls* (1981); DM 25.5, *Ferry Terminals and Small Craft Berthing Facilities* (1981); DM 25.6, *General Criteria for Waterfront Construction* (1981).
35. NAVFAC (1981). *Harbors*, Design Manual 26.1. Dept. of the Navy, Naval Facilities Engineering Command, NAVFAC DM-26.1, Alexandria, Va. Additional parts: DM 26.2, *Coastal Protection* (1982); DM 26.3, *Coastal Sedimentation and Dredging* (1981); DM 26.4, *Fixed Moorings* (1986); DM 26.5, *Fleet Moorings* (1985); DM 26.6, *Mooring Design Physical and Empirical Data* (1986).

2
Vessel Characteristics and Dimensions

The principal purpose of most marine facilities is the servicing of vessels, so it is essential that the harbor engineer have a basic understanding of vessel design, construction, operating requirements, and, in particular, overall dimensions and form. Ship development and port development are interrelated, each at times influencing the characteristics of the other. This chapter first presents some important definitions and terms, followed by brief discussions of hull form and construction and a broad review of vessel types and dimensions, giving salient features of many of the vessels now in use.

2.1 DEFINITIONS

A vessel's principal dimensions are illustrated in Figure 2.1. The vessel's overall length is designated *LOA*. The length along its waterline, which may vary slightly with load condition, is designated *LWL*. The length between perpendiculars (*LBP*) is the length between the forward and aft vertical structure of the ship's hull where it intersects the designed waterline (*DWL*). The *DWL* is the nominal waterline at which a vessel is designed to operate, usually the full load condition. For all practical port design purposes, the *LBP* and *LWL* are very nearly the same, and are sometimes used interchangeably. Many contemporary vessels have bulbous bows that project forward at the forward perpendicular below the waterline, in some cases for a considerable distance; so caution should be exercised in allowing for such projections. The beam (*B*) is the vessel's maximum width, which usually occurs at or near the vessel's transverse centerline or midships section, designated by the symbol ⊕. The depth of a vessel's hull (D_s) usually is measured at midships from the bottom of the keel to the top of the main deck (strength deck). The draft (*D*) is the distance from the vessel's waterline to the bottom of its keel or baseline.

Many vessels draw more water aft than forward; so one must distinguish between draft forward, draft aft, and mean draft. For hydrostatic calculations the mean draft must be used, and for navigation requirements the maximum draft must be known. All U.S. vessels of 150 tons (GRT, defined below) and over are given a load line assignment by the American Bureau of Shipping (ABS) as required by the U.S. Coast Guard (USCG). The load line is indicated by a plimsol mark displayed on the side of the hull, which indicates various maximum safe drafts to which a vessel can be loaded for various ocean areas

18 DESIGN OF MARINE FACILITIES

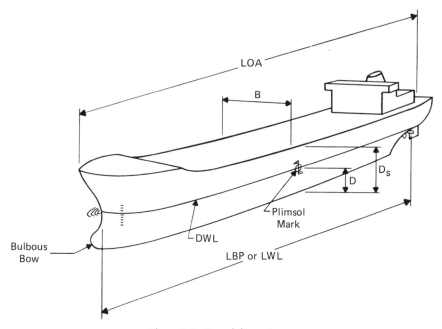

Figure 2.1. Vessel dimensions.

and seasons. A vessel's draft will increase approximately 2% to 3% in fresh water compared to seawater. Trim refers to the difference in drafts fore and aft, and list refers to a difference in drafts side to side.

When a tanker or cargo ship has discharged its cargo, it will often take on ballast by flooding internal tanks with seawater in order to reduce windage and alter the vessel's trim. Maximum ballast capacities of most vessels are on the order of one-half of their full load displacement, and a tanker "in ballast" will ordinarily be at from 30% to 50% of its full load displacement, depending upon weather conditions. Therefore, a vessel usually will be somewhere between lightship weight (LWT) and full load condition for most berthing and mooring situations. For dry-docking, however, there usually is an attempt to get the vessel into LWT condition. Freeboard (FB) refers to the height of the vessel's deck above the waterline; that is, $FB = D_s - D$.

A vessel's total weight, termed its displacement (Δ), varies with its load condition from fully loaded, which is the displacement or displacement tonnage (DT) (the figure usually given, as in the tables that follow in this text), to the weight at the lightship condition (LWT), which is the weight of the empty vessel, sometimes including minimal stores and partial fuel. Vessel displacement normally is given in long tons (l.t.) of 2240 pounds, each equivalent to approximately 35 cubic feet (c.f.) of seawater (36 c.f. fresh water). As a vessel

displaces a volume of water equal to its own weight (Archimedes principle), it is convenient to use 35 c.f./l.t. for converting from volume to weight. A vessel's displacement in cubic feet sometimes is designated by the inverted delta, ∇. Note that the metric ton (m.t.), which is used for most foreign vessels, is very nearly equal to a long ton (1 m.t. = 2205 pounds).

The stowage factor refers to the relative density of the vessel's cargo, usually expressed in c.f./l.t. Tankers, for example, carry relatively high-density cargoes in the range of 39 to 46 c.f./l.t. and float low in the water, with the ship's draft to hull depth ratio (D/D_s) usually around 0.75. Containerships and Ro/Ro vessels, by comparison, carry relatively low-density cargoes with storage factors in excess of 100 c.f./l.t. and D/D_s around 0.5, depending upon other specific design characteristics. Merchant vessels usually are referred to in terms of their cargo-carrying capacity, or dead weight (DWT), which is essentially the vessel's loaded DT minus its LWT. Note that the actual cargo capacity equals the DWT minus the weight of the ship's crew, stores, and fuel. For purposes of figuring port duties, shipping costs, and so on, registered tons or tonnage is employed. A registered ton is figured as 100 c.f. (2.83 c.m.) of internal space, with a gross registered ton (GRT) equal to the ship's total internal volume and a net registered ton (NRT) being the total internal volume available for cargo (i.e., GRT minus living, machinery, and fuel storage spaces, etc.). For most contemporary tankers and cargo ships, the LWT is on the order of 25% to 35% of the loaded displacement. The facility's designer is primarily interested in the vessel's actual displacement, and often needs to know both the LWT and the DT.

According to reference (1), the following approximations can be applied for preliminary design purposes when specific vessel data are lacking, in order to obtain vessel displacement tonnage, DT, from GRT or DWT:

General cargo ships: DT = 2.0 × GRT, or DT = 1.4 to 1.6 × DWT.
Container ships: DT = 1.4 × DWT.
Bulk carriers (and tankers): DT = 1.2 to 1.3 × DWT.
Passenger liners: DT = 1.1 GRT.
Fishing vessels:
 Small: DT = 2.0 to 2.5 × GRT.
 Large: DT = 1.5 to 2.0 × GRT.

The lower end of the ranges given above generally applies to the smaller vessels of the given type. As a general rule, the ratio of DT/DWT decreases with increasing vessel size.

The vessel form and dimensions are of obvious importance to port engineers, and are discussed further in following sections of this chapter. Knowledge of a vessel's hull structure is important in fender design applications and dry-dock-

ing. Other vessel characteristics of interest are speed and maneuverability, which are addressed in the general literature on port engineering, navigation channel layout and design, and vessel projected areas to wind and currents, which are discussed in Chapter 6.

2.2 HULL FORM

A ship's overall shape is defined on a lines drawing, such as that illustrated in Figure 2.2. The vessel's design and construction use molded dimensions, which may be measured to the inside or the outside of the vessel's shell plating, as specified in the plans. The vessel's shape is defined by various form coefficients, the most important of which is the *block coefficient* (C_B), which can be applied to the vessel's overall dimensions to give the vessel's displacement:

$$\Delta = \frac{C_B \times LWL \times B \times D}{35} = \frac{\nabla}{35} \tag{2-1}$$

for Δ in l.t. in seawater. For freshwater, ∇ is divided by 36 instead of 35. Values of C_B normally range from 0.36 for fine high-speed vessels to 0.92 for slow full ships such as Great Lakes ore carriers (2). Table 2.1 gives the usual range of C_B for selected types of merchant ships.

Other form coefficients of interest are: the *waterplane coefficient* (C_W), which gives the area of the vessel's waterplane (A_W):

$$C_W = \frac{A_W}{LWL \times B} \tag{2-2}$$

and is in the range of 0.75 to 0.85 for most commercial ships; the *midship section coefficient* (C_M), which gives the cross-sectional area of the midship section (A_M):

$$C_\text{M} = \frac{A_\text{M}}{B \times D} \tag{2-3}$$

and *the prismatic coefficient* (C_P) which is a measure of the fineness of a vessel's ends and is given by:

$$C_P = \frac{\nabla}{LWL \times A_\text{M}} \tag{2-4}$$

where $\nabla = LWL \times B \times D \times C_B$.

VESSEL CHARACTERISTICS AND DIMENSIONS 21

Figure 2.2. Example of a vessel lines drawing. (Courtesy of Crandall Dry Dock Engineers.)

Table 2.1. Representative block coefficients (C_B)* for selected vessel types.

Vessel type	C_B
Great Lakes ore carrier	0.88–0.92
Tanker	0.82–0.87
Bulk carrier	0.72–0.84
General cargo	0.55–0.78
Containership	0.54–0.64
Ro/Ro	0.52–0.66
Barge carrier	0.58±
Passenger liner	0.55–0.65
Auto ferry	0.45–0.50
LNG	0.72

*Typical range for vessels in loaded condition.

Note:

$$\text{Displacement: } \Delta = \frac{LWL \times B \times D}{35} \times C_B$$

for Δ in long tons.

Note that the above coefficients are usually given for the condition of the vessel on its *DWL*, and all of them vary with change in draft or trim. The length of parallel midbody, or the distance along a vessel's length over which the sides are essentially flat and vertical, is often of interest in berthing and mooring calculations. Ships with long parallel midbodies, say 50% or more of the *LOA*, have relatively higher values of C_B and C_P and also have large cargo capacities.

The volumetric coefficient, C_V, is defined as the displacement volume, ∇, divided by the cube of one-tenth of its length, or:

$$C_V = \nabla/(L/10)^3 \qquad (2\text{-}5)$$

Values of C_V normally range from near 1.0 for long fine ships to 15.0 for short heavy ships such as trawlers. The C_V is similar to the displacement/length ratio, which is an important parameter used in comparing a vessel's resistance and powering requirements. The amount a vessel will sink or emerge when weight is added or removed is given in terms of tons per inch immersion (TPI), and the moment to trim one inch (MTI) gives the change in trim when weights are shifted fore or aft.

Means of calculating TPI and MTI and other hydrostatic and stability parameters are presented in Section. 9.2.

A vessel's general proportions can be described in terms of certain dimensional ratios of interest to harbor engineers, such as the vessel length to beam ratio (L/B), beam to draft ratio (B/D), and length to draft ratio (L/D). Al-

most all vessels lie within the following approximate limits (2): $L/B = 3.5$ to 10; $B/D = 1.8$ to 5; and $L/D = 10$ to 30.

In the above ratios, L is usually taken as the *LBP*. Most commercial merchant vessels have L/B ratios in the range of 5.5 to 7.0 and B/D ratios on the order of 2.7 to 4.0. Certain vessels such as ocean liners and Great Lakes ore carriers have a relatively high L/B on the order of 9.0 to 10.0, whereas for work boats, fishing boats, and certain automobile ferries L/B is often in the range of 3.5 to 4.5. A vessel's proportions are a direct consequence of its function. For example, compared to container ships, break bulk cargo ships are slower and have lower L/B ratios; and because they are generally smaller and have fewer stability problems due to containers on deck, they have lower B/D ratios as well. Ro/Ro vessels have lower L/D ratios than container ships, as they need the extra hull depth to accommodate the trailer chassis. Tankers vary considerably in their dimensional ratios, but the larger vessels tend to have a moderate L/B in the range of 5.5 to 6.5, and being slower they also have relatively high block coefficients.

Other factors affecting vessel proportions include physical constraints on the vessel's trade route, such as water depths and canal or waterway widths. The maximum beam for a vessel that can transit the Panama Canal is 105.75 feet, resulting in the so-called Panamax-size vessel, for example. Vessels transiting the Suez Canal are limited to a 128-foot beam, with additional limitations on length and draft depending upon other navigability requirements. Environmental regulations also may impact a vessel's form, such as double bottom and segregated ballast requirements for tankers. A more in-depth treatment of vessel form and proportions can be found in the general textbook literature on naval architecture, such as references (2) through (4).

2.3 HULL CONSTRUCTION

The vast majority of vessels of interest herein are constructed of structural steel, although smaller vessels usually may be constructed of aluminum, wood, fiberglass reinforced plastics (GRP), or even reinforced concrete or ferro-cement. This discussion will be confined to steel hull construction. The vessel's hull can be thought of as a box girder made up of stiffened plate elements that form the top deck (also called the weather deck or strength deck), which is the girder's top flange, plus bottom shell plating (bottom flange) and side shell plating. Today's larger tankers and some bulk carriers have double bottoms, which include an inner watertight bottom plate above the bottom shell plating. There are two distinct types of overall framing arrangements: transverse and longitudinal framing. Both types employ transverse frames and longitudinal beams and stringers. The difference between them is that in the transverse system the frames are continuous and closely spaced, with longitudinal girders that are deeper and

Figure 2.3. Ship hull framing schematics.

spaced farther apart than in the longitudinal system, which has closely spaced but shallower longitudinals with widely spaced deep web frames. The construction of given ships, military vessels especially, may lie anywhere between these two extremes. Figure 2.3 illustrates the two basic types of framing and defines some basic structural terms.

The term "scantlings" refers to the dimensions of a ship's structural members. The American Bureau of Shipping's *Rules for Building and Classing Steel Vessels* (5) gives minimum scantling requirements for ship hull strength.

The most important stresses that the hull girder must resist arise from the longitudinal bending moment induced by waves. When a ship is supported on a wave with a length nearly equal to the vessel's *LWL*, the vessel is in a sagging condition when the wave crests support its ends and in a hogging condition when supported amidships by a wave crest. The longitudinal strength and stiffness of floating structures are discussed further in Section 9.4 and in references (2) through (5). Locally the ship's hull plating must resist hydrostatic and dynamic pressures plus the combined stresses induced through its overall girder action. The ability of the hull to distribute and resist localized loads is very important in designing fender systems, and is discussed further in Section 5.6.

2.4 VESSEL TYPES AND DIMENSIONS

Table 2.2 gives a statistical breakdown of the number and tonnage of the world fleet by vessel type. Merchant ship statistics are published annually (6, 7), and

Table 2.2. Number and tonnage of world fleet by vessel type.*

	World fleet	
	Number	Tonnage
Oil tankers	8.0	30.7
Gas/chemical tankers	0.7	1.1
Chemical tankers	1.1	0.8
Liquefied gas tankers	1.0	2.4
Bulk/oil carriers	0.5	5.3
Ore and bulk carriers	6.5	27.5
Miscellaneous tankers	0.2	0.1
General cargo ships	26.9	17.9
Container ships	1.4	4.8
Vehicle carriers	0.4	1.1
Ferriers/passenger ships	5.2	2.2
Passenger/cargo ships	0.2	0.2
Supply ships and tenders	2.9	0.3
Tugs	10.5	0.7
Fishing ships and fish factories	29.0	3.3
Miscellaneous	5.5	1.6
	100%	100%
Total gross reg. tonnage		404.9 mill.
Total number of ships	75,266	

*Data from Lloyd's Register (1986) (7).
For vessels over 100 GRT.

a detailed analysis of vessel size, type, tonnage, age, and so on, by country of register can be found in reference (8).

In the United States, the Merchant Marine Act of 1936 began the construction of the "C" series cargo ships and "T" series tankers; and some of these vessels, built in the post–World War II boom, are still active. Dillon et al. (9) present a detailed listing of vessels built under the Maritime Administration's (MarAd) Merchant Marine Act of 1936, which includes dimensional and other data of interest for vessels built under the act up until 1976. In the mid-1950s, containerized cargos were introduced, and they became prominent in international trade during the 1960s. In the late 1960s and early in 1970s, roll on/roll off (Ro/Ro) vessels came into vogue, especially on the shorter-voyage routes. Lighter aboard ship (LASH) barge-carrying type vessels also came into being along ocean routes that service inland waterways. Tankers grew from the World War II vintage T-2 class of 16,700 DWT and 524 feet *LOA* to over 500,000 DWT and 1300 feet *LOA* today (see Figure 2.4). The historical development of the U.S. Merchant Marine has been described by Whitehurst (10).

Vessels cover a wide range of sizes and shapes, from the smallest yachts and harbor craft to supertankers and aircraft carriers. Trade routes, port development, commodities, and cargo-handling methods may have a profound effect on ship configuration. There are too many specialized types of craft to describe in a text such as this; instead, a review of the most common types is presented, along with follow-up references. Although representative data for various ship types are useful for instruction and comparison, the designer always must gather specific information on the particular vessels to be served at a given facility. Vessels can be broadly categorized with regard to their mission as: commercial vessels, consisting primarily of merchant ships engaged in trade and the transportation of cargoes and/or passengers; industrial vessels, engaged in specific tasks or operations such as drill ships, research vessels, dredgers, fishing vessels, and so on; service vessels, such as tugs, pilot boats, fireboats, and so forth; and military vessels. As a rule, commercial-type vessels are of the greatest general interest to port and harbor engineers, and the following discussion focuses on them. References (3), (9), (11), and (12) contain useful data on ship design characteristics for various commercial vessel types.

Tankers are liquid bulk products carriers, with crude oil by far the most common cargo, although chemical and other specialized tankers are also active. Virtually all tankers over 100,000 DWT (supertankers) are crude oil carriers. Tankers have established the state of the art in ship size, with vessels of 550,000 DWT and over 1300 feet *LOA* in service. Local feeder lines may operate coastal tankers of 2000 DWT or less, although smaller shipments are usually transported by barge. Tanker capacities, especially for smaller vessels, are sometimes given in barrels (bbl), where 1 bbl = 42.5 U.S. gallons. The crude capacity of a ship can be approximated by multiplying its DWT by 6.5 barrels per ton. Tankers can be classed by size for convenience, as follows:

Figure 2.4. Two tankers engaged in lightering illustrate the dramatic increase in tanker sizes from the World War II vintage T-2 to the contemporary VLCC. (Photo by Lloyd Koenig, courtesy of Seaward International, Inc.)

1. *Conventional tankers* of generally less than 100,000 DWT and 40 feet or less draft. Such vessels usually are operated in domestic/coastal trade routes and can enter almost any major seaport. Tankers over approximately 25,000 DWT to 50,000 DWT up to around 150,000 DWT are generally referred to as supertankers and have loaded drafts in excess of 40 feet, but they may enter certain U.S. ports partially loaded. Tankers of less than approximately 25,000 DWT are usually coasters or general-purpose product carriers, including specialized parcel tank ships that carry liquid chemicals in special segregated tanks or containers.
2. *VLCCs*, or very large crude carriers, generally between 150,000 DWT and 300,000 DWT with loaded drafts from 60 to 80 feet. This size is the workhorse of the general overseas trade, with 200,000 DWT to 250,000 DWT being especially common. These vessels may enter some deep draft ports but more commonly will load/offload at offshore terminals and/or by lightering, transferring their cargo to smaller vessels.
3. *ULCCs*, or ultralarge crude carriers, of over 300,000 DWT and with drafts of 80 feet to 100 feet. Although vessels of 1,000,000 DWT technically are feasible and have been forecast by the year 2000 (13), the practical economic limit is probably somewhere around 800,000 DWT, and it is possible that the largest of these already have been built. These vessels are strictly relegated to offshore terminals or mooring systems.

Note that there is no strict technical definition of VLCC, ULCC, or supertanker; the above descriptions represent a rather loose classification.

Many tankers over 70,000 DWT have double bottoms as a result of the 1973 Intergovernmental Maritime Consultive Organization (IMCO) pollution control regulations and the requirement that such vessels have segregated ballast tanks. The general effect of this has been to increase the ship's hull depth, D_s, and increase freeboard rather than draft. Large tankers typically are longitudinally framed, to better cope with large sea bending moments due to their long lengths.

Figure 2.5 shows the range of *LOA*, *B*, and *D* versus DWT for tankers. Contemporary tankers of 100,000 DWT and above usually have L/B ratios within the 5.5 to 6.5 range, compared with around 7.5 for smaller, earlier tankers. Block coefficients are in the 0.80 to 0.85 range, compared with 0.75 for earlier vessels (3). For additional data on tankers see references (3), (11), and (14).

LNG and LPG, liquefied natural gas and liquefied propane gas carriers, are a distinct specialized class of vessels. Their hulls cradle heavily insulated cryogenic tanks that contain liquefied gas at very low temperatures at atmospheric pressure. LNG must be cooled to $-285°F$, at which point it liquefies and occupies a volume 1/600 of its gaseous volume. The capacity of such vessels usually is given in cubic meters (c.m.). Most contemporary ocean-crossing ves-

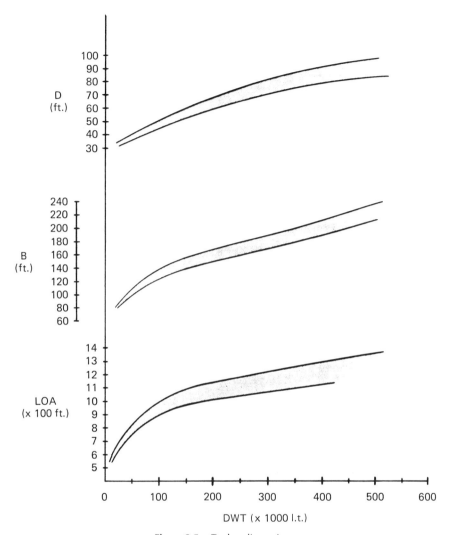

Figure 2.5. Tanker dimensions.

sels are in the 125,000 c.m. range and around 900-plus feet *LOA*. Up until the early 1970s, vessels in the 40,000 c.m. to 75,000 c.m. range were most common. Table 2.3 gives general dimensions and characteristics of selected LNG and LPG carriers. Such vessels usually are dedicated to runs between particular supply and receiving terminals. Their movements within harbors are controlled by very stringent safety requirements, and terminal facilities are of a very specialized nature. Further information on LNG vessels is found in reference (15).

Table 2.3. Principal dimensions of selected LNG and LPG carriers.

Vessel name/type	LOA (ft.)	Beam (ft.)	Depth (ft.)	Draft (ft.)	DWT (l.t.)	Capacity (c.m.)
Aquarius/LNG	936.0	143.5	82.0	36.0	63,600	125,000
Staffordshire/LNG	742.3	112.2	70.8	42.6	55,800	75,000
Sener/LNG	334.6 (LBP)	60.7	34.5	18.9	4,100	5,000
Pioneer Louise/LPG	747.8	120.0	72.2	37.6	55,843	77,500
Donau/LPG	600.2	98.4	56.1	38.9	32,339	30,207
Becquer/LPG	277.5	47.6	28.9	21.6	—	3,250

Combination bulk carriers are specially configured to carry both liquid and dry bulk cargoes, the most common type being the OBO (ore/bulk/oil). Other adaptations include ore/oil, bulk/oil, and ore/slurry/oil. There vessels usually are relatively large, typically in the 50,000 DWT to 200,000 DWT range. Their relative proportions are similar to those shown in Figure 2.5 for tankers. These vessels have the flexibility required to engage in different trades and/or haul one cargo one way and a different cargo on the return voyage. Nearly all new vessels below 130,000 DWT are OBOs with drafts generally less than 50 feet. Dorman (16) has provided detailed information on combination carrier characteristics.

Dry bulk carriers may be engaged in carrying a wide range of dry bulk cargoes such as ore, coal, grain, and so on, and may also be adapted for combination cargoes such as bulk and containers, vehicles, wood chips, and other specialized bulk products. Dry bulk cargoes can vary greatly in nature, ranging in density from 10 c.f./l.t. for iron ore to 100 c.f./l.t. for certain grains, compared with the usual range of 36 c.f./l.t. to 48 c.f./l.t. for petroleum products. Bulk carriers may be outfitted with their own crane or conveyor systems for self-unloading. Almost all dry bulk carriers are below 150,000 DWT, with general-purpose bulk carriers being dominant and in the 20,000 DWT to 60,000 DWT range. Panamax-size vessels are limited to about 80,000 DWT. Bulk carriers tend to have full hulls with long parallel midbodies similar to those of tankers and combination carriers. Bulk carriers differ from tankers in having deck hatches that typically cover from 50% to 75% of the midbody deck area. The size, number, and extent of hatch openings is of obvious importance in laying out berths and cargo-handling equipment. The principal dimensions and other characteristics of selected bulk carriers are given in Figure 2.6 and references (16) through (18).

General cargo ships may carry a wide variety of cargoes, stowed in palletized, baled, crated, containerized, or other form. Break-bulk cargo refers to individually wrapped, packaged, crated, and so forth, items that are stowed individually in the ship's hold, as compared with unitized cargoes that may be

D.W.T.	L.O.A.	BEAM	DEPTH	DRAFT	END/END HATCHES	HATCH WIDTH
40,000	207	27.5	16.5	11.5	135	13.0
60,000	235	32.0	18.0	12.8	155	14.0
150,000	305	43.5	25.0	18.0	220	20.0
200,000	327	52.0	27.0	19.1	230	22.0
250,000	344	54.0	28.5	21.0	245	24.0
350,000	390	61.0	34.5	25.5	300	25.0

NOTE: ALL DIMENSIONS IN METERS

Figure 2.6. Bulk carrier dimensions. (Courtesy of Soros Associates Consulting Engineers.)

bundled or palletized together or containerized into larger units. General cargo ships may be multipurpose breakbulk types or dedicated to specific commodities such as refrigerated cargoes (reefers), lumber carriers, and so on. General cargo ships are typically self-unloading, equipped with their own cargo-handling gear ranging from approximately 20 tons to 100 tons (heavy lift) capacity. Vessels engaged in regular scheduled trade routes are often referred to as cargoliners, as compared with general-purpose tramp ships in which the cargo and the destination vary with consignments. General cargo ships are typically within the 12,000 DWT to 25,000 DWT capacity range, and within 450- to 550-foot *LBP* with drafts on the order of 30 feet, allowing them to enter virtually all major U.S. ports. Small vessels may be employed on shorter and coastal trade routes. Table 2.4 gives principal dimensions of representative cargo vessels. See also references (3) and (11).

Containerships are the current greyhounds of the sea, featuring faster turn-around times at port and a greater cubic capacity per DWT as well as their high speed en route. Containerships have become prominent in recent years despite the fact that intermodal transport requires specialized port facilities and access to land transportation systems. Containership capacity is most often expressed in TEUs, or twenty-foot equivalent units, referring to the number of equivalent 20-foot-long standard-size containers carried. A vessel's capacity in TEU can be found by dividing its deadweight capacity (DWT) by 17. The vast majority of containerships have beams of less than the Panama Canal limit of 105.75 feet and drafts less than 40 feet. The largest vessels are around 950 feet long (Panamax length limit) with 4300 TEU capacity. The greatest number of vessels are in the 1000 TEU to 1500 TEU range, however. Vessels of up to 10,000 TEU and 1400 feet *LOA* have been anticipated by the year 2000 (13). Table 2.5 gives principal dimensions and other characteristics for selected vessels. Dimensions and weights of standard containers are given in Section 4.1. For further information on containerships, see references (19) through (21).

Table 2.4. Principal dimensions of selected general cargo ships.

Class/type	LOA (ft.)	LBP (ft.)	Beam (ft.)	Depth (ft.)	Draft (ft.)	DWT (l.t.)	DT (l.t.)
C-3 class	492.0	465.0	69.5	33.5	28.6	12,000	18,215
Cargo liner/M.P.	498.5	475.6	75.2	43.6	31.3	16,698	—
PD-159	504	470	74.0	43.0	28.3	11,850	17,900
Mariner class	563.7	528.5	76.0	35.5	29.9	12,910	21,093
Kennebec class/ M.P.	570	541	75.0	45.0	33.5	21,920	29,300
Merrimac class/ Cont.-break bulk	592	550	86.0	54.8	34.0	15,900	26,400
Alaskan Mail	605.0	582.6	82.0	46.5	35.1	22,208	31,995

VESSEL CHARACTERISTICS AND DIMENSIONS 33

Table 2.5. Principal dimensions of selected containerships.

Class or identification	LOA (ft.)	LPB (ft.)	Beam (ft.)	Depth (ft.)	Draft (ft.)	Capacity TEU	DT (l.t.)
U.S. Lines	950	—	105.7	—	35	4,258	—
PD-161	941.0	862.0	105.0	64.0	32.0	2,294	44,600
C-10	902	—	129	—	41	4,300	53,648
President Lincoln	860	810	105.7	66	35	2,500	49,500
C8-5-85	813.3	769.0	90.0	53.0	33.0	1,708	42,350
PD-160	758	706.0	101.0	55.0	28.5	1,672	32,500
C6-5-85	669.3	625.0	90.0	53.0	33.0	1,186	30,490

Ro/Ro, or roll on/roll off, vessels can accommodate virtually any over-the-road vehicle. They are essentially large oceangoing ferries originally envisioned as pure trailer ships. The major drawback to such a system in competing with other means of cargo handling is its loss of cubic capacity due to the parking space required by vehicles. Ro/Ro vessels carry their own vehicle ramps and thus are self-unloading. They are most adaptable to shorter ocean routes, usually with a 1000- to 2000-mile range, because of the economics of loading/unloading containerized cargoes versus driving on and off. Most Ro/Ro vessels are less than about 20,000 DWT capacity, although in the early 1980s larger third-generation vessels (22) approaching 40,000 DWT and having combined container and Ro/Ro capability were introduced into the transoceanic trade routes. Table 2.6 gives principal characteristics of selected Ro/Ro vessels, and references (23) and (24) contain valuable information on Ro/Ro design particulars.

Vehicle carriers are distinguished from Ro/Ro type vessels in that they carry only the vehicle (usually autos) and not loaded trailers with variable cargoes. Vehicle carriers have very large volumetric capacities and are easy to identify with their typically very large, wall-sided superstructures. Vehicle carriers capable of carrying on the order of 5000 automobiles, corresponding to approximately 15,000 DWT, are in service. Also, vehicles often are carried as an alternate cargo on combination carriers. Because of their relatively low-density cargoes, Ro/Ro's and vehicle carriers typically have higher freeboards, lower drafts, and slightly larger overall dimensions than general cargo vessels of the same DWT capacity.

Barge carriers generally are built according to one of three patented systems: the LASH (lighter aboard ship), SEABEE, or BARCAT system. These vessels have square open sterns with either gantry cranes (LASH) or elevators (SEABEE) for lifting barges aboard. Relatively small barges are brought to the vessel's stern by tugs or towboats and lifted aboard. Alternatively, the barges may be floated aboard by partially submerging the vessel. This method generally is restricted to smaller feeder line vessels. This cargo-handling system is compet-

Table 2.6. Principal dimensions of selected large roll on/roll off vessels.

Name/type	LOA (ft.)	LBP (ft.)	Beam (ft.)	Depth (ft.)	Draft (ft.)	DT (l.t.)	Capacity (as noted)
Barber Blue Sea	859.4	808.2	105.8	68.9	38.4	—	43,300 DWT (630 cars + 2464 TEU)
ACL G-3	820.2	761.2	105.8	66.4	36.1	—	37,000 DWT (600 cars + 2130 TEU)
—*	684.0	640.0	102.0	69.5	34.0	33,765	18,989 DWT
Car carrier/ Ro/Ro	639.6	—	105.0	53.8	36.0	—	20,600 DWT (5500 cars *or* 921 TEU)
Atlantic Saga	—	600.4	90.0	63.3	29.5	27,531	15,725 DWT (1200 cars + 500 TEU)
Australian Enterprise	—	551.2	82.0	53.8	29.5	22,173	14,082 DWT
USNS Comet	—	465.0	78.0	48.9	27.1	18,286	10,111 DWT

*A representative contemporary design with slewing quarter ramp and container carrying capability.

Table 2.7. Principal dimensions of selected barge carriers.

Class or type	LOA (ft.)	LBP (ft.)	Beam (ft.)	Depth (ft.)	Draft (ft.)	Capacity* (No. of barges)	DT (l.t.)
LASH	893.3	797.3	100.0	60.0	31.0	89	62,314
SEABEE	875.9	740.0	105.7	74.8	39.0	38	57,290
LASH/ PD-162	872	771	105.0	60.0	35.0	82	50,300
LASH	820	724	100	60	35	61	32,650
BACO	669.4	—	93.5	45.9	20.0	12	—

*Barge dimensions vary with system type; see Table 2.8.

itive where river or canal systems exist at each terminus, especially underdeveloped areas, and where a limited number of port calls are involved. These vessels generally are of moderate to large size, the largest vessels being on the order of 50,000 DWT and 900 feet LOA. Principal dimensions and data for selected barge carriers are given in Table 2.7, and standard barge dimensions are given in Table 2.8.

Ferry boats, which are employed principally to transport highway traffic vehicles and passengers, form a bridge or link between overland roadways. They vary in size from small river-crossing commuter ferries carrying fewer than a dozen automobiles to large oceangoing ferries carrying over 200 vehicles on overnight runs. Double-ended ferries, with propellers at both ends, frequently are used on short commuter runs, whereas single-ended ferries are more economical and preferred for exposed water crossings. Ferry boats typically have broad beams to maximize the number of cars carried and provide stability for the relatively high center of gravity of the cargo and superstructures. Ferry boats also typically have high B/D ratios, in the 4 to 5 range and higher for side loaders. They also are typically fine-lined below the waterline to reduce wavemaking resistance, with a C_B in the 0.4 to 0.5 range. Table 2.9 presents some typical ferry boat characteristics.

Passenger vessels, sometimes referred to as liners, have been in decline since their pre- and post-World War II heyday, but have in recent years made a strong comeback in the form of cruise ships. The great transatlantic liners once were among the fastest and most majestic vessels ever built. Today's liners are not so much involved with transporting people as with providing on board vacations in the form of short cruises to no specific destination or to local ports.

Table 2.8. Barge carrier systems—barge dimensions.

System	Length × Beam × Draft	Tonnage (gross)
LASH	61.5' × 31' × 9.5'	450
SEABEE	97.5' × 35' × 10.5'	835
BACO	79' × 31' × 13.5'	800

Table 2.9. Principal dimensions of selected ferries.

Type	LOA (ft.)	LWL (ft.)	BOA (ft.)	BWL (ft.)	Depth (ft.)	Draft (ft.)	DT (l.t.)	Capacity (as noted)
Baltic Sea Ferry	545.3	—	93.2	—	—	31.3	(4,500) (DWT)	1440 pass./283 autos
Alaskan single ender	418.0	385.0	85.1	72.0	24.0	16.0	6700	1000 max. pass./326 autos
Puget Sound double ender	328	317.5	78.5	62.8	23.0	15.5	2930	300 max. pass./100 autos
Staten Island double ender	294.0	285.0	60.8	50.9	20.6	13.5	2360	300 max. pass./175 autos
Nantucket Sound single ender	239.5	—	60.1	—	17	11	1565	980 max. pass./50 vehicle units
Nantucket Sound double ender	200.5	191.7	60	58	17	13	1106	780 max. pass./45 vehicle units
Casco Bay single ender	133.0	—	36.3	—	—	7.8	255	10 autos/pass. not available

VESSEL CHARACTERISTICS AND DIMENSIONS 37

The domestic cruise ship industry and its vessels have been well described by Leeper and Boylston (25). The short-vacation-cruise business has brought about an essentially new class of small passenger vessel (26). Principal dimensions and data for selected passenger type vessels are given in Table 2.10.

Barges are a common sight in most ports, used to transport a wide variety of cargoes, some of which call for specially adapted hulls. Types include flat deck, covered hopper, bulk carrier, cement bulk, rail (sometimes referred to as car floats), chemical, wood chip, log, tanker, and other special-purpose barges (27). Barge sizes range upward to over 900 feet *LOA* and around 50,000 DWT. Figure 2.7 shows the usual range of barge *LOA*, *B*, and *D* versus DWT. A barge's proportions and dimensional ratios may vary greatly, depending upon its purpose, although values of C_B typically remain at around 0.95 or greater. Barges also may be used as platforms for floating equipment, especially lifting equipment, such as stiff-leg derricks or revolving cranes.

An important adaptation of barges in marine transportation is the integrated tug/barge system (ITB). In this system, the tug and barge are rigidly or semi-rigidly linked, with the tug secured to the barge's stern. One specially adapted barge type has a notched stern, into which the bow of a specially designed tug fits and is secured. The bow of the barge is rounded or spoon-shaped and when linked together the ITB combination somewhat resembles a bulk carrier vessel in its proportions. Most ITBs are engaged in dry bulk or liquid bulk transport. This system of transport has certain advantages when the loading time is long compared to the transit time, the cargo movement is intermittent or seasonal, and the cargo is stockpiled or warehoused. Further information on ITBs can be found in references (28) through (30).

Workboats and harbor craft include tugs, towboats, offshore supply vessels, firefighting vessels, pilot boats, and various other utility craft. Tugboats and towboats are distinguished from one another in that tugs normally pull their loads, whereas towboats normally push. Towboats are most common on inland waterways where they transport rafted barge tows. Towboats normally have heavy bow knees for engaging their tow, their superstructures usually extend nearly the full length of their hulls, and often they have elevated pilot houses so that operators can see over their cargo. Tugs, on the other hand, have lower superstructures that terminate well forward from their sterns so that the tow rope pull will act close to the vessel's turning center, allowing better control of the tow. Tugs usually are engaged in assisting larger vessels in berthing, general towing, salvage and rescue, and so on, and generally can be categorized by size as harbor, coastal, and oceangoing.

As their primary use is in moving loads, tugs and towboats generally are referred to in terms of their pulling power or engine (brake) horsepower (bhp), or often the power delivered to the shaft or shaft horsepower (shp). Harbor tugs usually range from 50 to 100 feet *LOA* and 200 to 1000 shp, coastal tugs from

38 DESIGN OF MARINE FACILITIES

Table 2.10. Principal dimensions of selected passenger/cruise ships.

Name	LOA (ft.)	LBP (ft.)	Beam (ft.)	Draft (ft.)	DT (l.t.)	GRT	Capacity (pass./crew)
Queen Elizabeth II	963.0	887.1	105.0	32.5	48,886	—	—
Michelangelo	904.8	800.4	101.7	30.6	41,328	—	—
Sovereign of the Seas	874.1	774.1	105.6	24.8	—	74,000	2276/750
Royal Princess	761.2	—	95.8	25.6	—	40,000	1200/—
Tropicale	669.3	580.7	86.3	23.0	—	36,600	1396/491
Sea Venture	553.7	475.7	80.7	24.6	15,685	19,903	650/301
Cunard Adventurer	448.0	413.0	70.5	19.2	9,774	14,555	750/300
Sea Goddess	343.8	—	47.9	14.0	—	4,000	120/70
Majestic Explorer	152	—	31	8	—	99.7	44 pass. staterooms

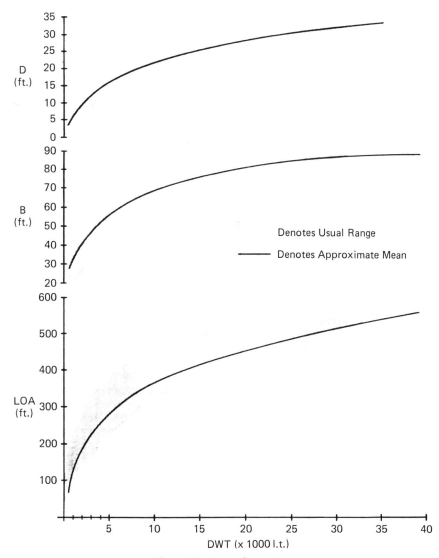

Figure 2.7. Barge dimensions.

100 to 120 feet *LOA* and 1000 to 3000 shp, and oceangoing tugs from 150 to over 200 feet *LOA* and up to 23,000 shp. Traditional harbor and coastal tugs typically have drafts within the range of 8 feet to generally less than 15 feet and are characterized by a relatively low L/B in the 3 to 4 range and a moderate to high B/D in the 2.2 to 3.2 range. They have moderately fine lines below the waterline, with C_B values in the range of 0.52 to 0.58. Displacements typically

range from 100 l.t. to less than 1000 l.t. An important result of the tug's power and propulsion system is its towline pull, or bollard pull, which is usually given at zero speed (or 100% propeller slip). The bollard pull of a tug with diesel electric power plant can be roughly estimated at 25 pounds per bhp for conventional propellers and 30 to 35 pounds per bhp for screw tugs equipped with Kort nozzles.

The largest tugs generally used for docking vessels usually are within the 3000 to 4000 bhp range, and must operate cautiously to prevent damage to the berthing vessel's hull plating (3). The tug's characteristics that are most important to docking operations include its bollard pull capacity and effective use of its power, maneuverability, and stability while thrusting against a vessel. Certain propulsion systems offer greater maneuverability and control than the traditional screw-type propeller; they include the Voith-Schnieder vertical axis cycloidal type propeller and the water-tractor propulsion system consisting of steerable twin unit drives mounted near the vessel's midships. Both of these systems have the potential disadvantages of deeper draft and slower speed. The stern drive tug, which has steerable drive units mounted aft below the vessel's counter, is similar to the conventional tug, but it lacks some of the maneuverability of the midships drive systems. Criteria for matching tugboat specifications to vessel berthing requirements can be found in references (31) and (32). Treatment of the general design of tugs and typical dimensions and characteristics can be found in references (3) and (33) through (35).

Another prominent workboat seen today is the offshore supply vessel (OSV), with a limited superstructure well forward and plenty of large open deck space aft. Large OSVs equipped with towing machinery, which are called anchor handling tugs, are used to tow and deploy offshore rigs and moorings. OSVs have evolved from the early supply boats of generally less than 150 feet *LOA* and 1000 shp to over 200 feet *LOA* and up to 10,000 shp for contemporary offshore tug/supply vessels. Many of the earlier generation of OSVs have found their way into various commercial transportation uses and services outside of the offshore oil industry. Like traditional harbor tugs, they are characterized by a generally low L/B but an even higher B/D, in the range of 3 to 4, reflecting their relatively shallow draft and wide open deck capacity. They also have slightly higher values of C_B, in the range of 0.62 to 0.75 when fully loaded. The larger contemporary OSVs tend to have more moderate proportions with an L/B in the 4 to 5 range and a B/D in the 2.5 to 3.0 range. A detailed description and dimensional data for these vessels can be found in references (36) through (38).

Numerous *specialized vessel types* fall under the industrial and miscellaneous service vessel categories. Fishing vessels are perhaps the most common of vessel types worldwide, ranging in size from the smallest skiff to tuna catchers of 250 feet *LOA* or more to factory/mother ships of up to around 30,000 DWT. Their sizes and arrangements vary greatly with the type of fishing carried out.

Most fishing vessels, however, have in common relatively high displacement ratios or cubic coefficients because of their need to maximize catch capacity. Descriptions of fishing vessels and their gear can be found in references (39), (40), and (3). Another fairly common industrial-type vessel is the oceanographic research vessel. These vessels may be specially built for their given purpose or converted from some other vessel type; so their characteristics vary widely, as seen in references (41) and (42). Other important industrial vessel types include dredges, pipelaying and ocean construction platforms, cable ships, ocean drilling and mining vessels, and various floating plants such as incinerator vessels.

Other miscellaneous vessels of interest include certain advanced vehicle types such as surface effect ships (SES), hydrofoils, and various catamaran (twin hull) configurations that are in limited commercial use, in particular for rapid passenger transport. An overview of advanced vehicle types is provided by Miller in reference (35). Information on yachts and recreational small craft can be found in the general small craft references given in Section 3.3.

Military vessels include a wide variety of vessel types and sizes, ranging from the smallest patrol craft to the largest aircraft carriers. Warships tend to be fine-lined, and many contemporary warships have relatively fragile hulls requiring special care in the berthing and dry-docking of these vessels. The reader should see references (43) and (44) for particulars of U.S. Navy and Coast Guard vessels and reference (45) for general dimensions and particulars of the world's warship fleet.

Additional tabulated and graphical data on various vessel types can be found in the cited literature, especially the article by Michel in reference (3), and the general references on port and harbor engineering given in Chapter 1. Several marine engineering and shipping industry professional journals and trade magazines regularly publish detailed data on vessels of all types (see Appendix 2).

In summary, it is essential that the harbor engineer seek out specific vessel data for those vessels that will use and/or are likely to visit the facility being designed. The dimensional data presented here, on representative vessels of the types most commonly encountered by the port engineer, are included for general information and instructional purposes. However, it is most important that the port and harbor engineer understand a ship's basic hull form and proportions as they relate to the vessel's mission and berthing requirements. Subsequent chapters will illustrate how a vessel's shape and proportions relate to berthing and mooring loads and how its hull structure relates to berthing and dry-docking problems.

REFERENCES

1. BSI (1984). *British Standard Code of Practice for Maritime Structures*, Part I, BS-6349. London.

2. Lewis, E. V., ed. (1988). *Principles of Naval Architecture*, 3 vols. SNAME, Jersey City, N.J.
3. Taggert, R., ed., (1980). *Ship Design and Construction*. SNAME, New York.
4. Gillmer, T. C. (1970). *Modern Ship Design*. Naval Institute Press, Annapolis, Md.
5. ABS (1988). *Rules for Building and Classing Steel Vessels*. American Bureau of Shipping, Paramus, N.J. (Issued Annually).
6. ——— (1988). *Shipping Statistics Yearbook*. Inst. of Shipping Economics and Logistics, Bremen and Lloyd's Shipping Economist, Lloyd's of London Press, Essex, England. (Annual.)
7. ——— (1988). *Register of Ships*. Lloyd's Register of Shipping, London. (Published annually.)
8. MarAd (Jan. 1984). *A Statistical Analysis of the World's Merchant Fleets*. U.S. Dept. of Transportation, Maritime Administration.
9. Dillon, E. S. et al., (1976). "Forty Years of Ship Designs Under the Merchant Marine Act; 1936-1976," *SNAME, Trans.*, Vol. 84, New York.
10. Whitehurst, C. H., Jr. (1983). *The U.S. Merchant Marine: In Search of an Enduring Maritime Policy*. Naval Institute Press, Annapolis, Md.
11. Johnson, R. P. and Rumble, H. P. (Apr. 1965). "Weight, Cost and Design Characteristics of Tankers and Dry Cargo Ships," *Marine Technology*, Vol. 2, No. 2, SNAME.
12. Kiss, R. K. and Garvey, J. J. (Oct. 1970). "CMX—Designs—Merchant Ships for the 70's," *Marine Technology*, Vol. 7, No. 4, SNAME.
13. AAPA (June 1969). *Merchant Vessel Size in U.S. Offshore Trades by the Year 2000*, Am. Assoc. of Port Authorities, Comm. on Ship Channels and Harbors, New York.
14. Gannet, E. (1969). *Tanker Performance and Cost; Measurement, Analysis and Management*, Cornell Maritime Press, Cambridge, Md.
15. Thomas, W. and Schwendtner, A. (1971). "LNG Carriers, the Current State of the Art," *SNAME, Trans.*, Vol. 79, New York.
16. Dorman, W. J. (Oct. 1966). "Combination Bulk Carriers," *Marine Technology*, Vol. 3, No. 4, SNAME.
17. Jacobs, C. L. (Jan. 1983). "Development of the Specialized Dry Bulk Carrier," *Marine Technology*, Vol. 20, No. 1, SNAME.
18. Roseman, D. P. et al. (1974). "Characteristics of Bulk Products Carriers for Restricted Draft Service," *SNAME, Trans.*, Vol. 82, New York.
19. Dorman, W. J. and deKoff, D. J. (Oct. 1971). "Characteristics of Recent Large Containership Designs," *Marine Technology*, Vol. 8, No. 4, SNAME.
20. Sartor, T. J. and Gibbon, R. P. (Jan. 1979). "Farrell Lines '85'-Class Container Ships," *Marine Technology*, Vol. 16, No. 1, SNAME.
21. Boylston, J. W. et al. (1974). "SL-7 Containerships: Design, Construction and Operational Experience," *SNAME, Trans.*, Vol. 82, New York.
22. ——— (May 1984). "First of ACL's Third Generation Ro/Ro Ships—Atlantic Companion," *Marine Engineering/Log*, Vol. 89, No. 5.
23. Argyriadis, D. A. et al. (1979). "Design and Construction of Modern Roll on/Roll off and Container Carriers," *SNAME, Trans.*, Vol. 87.
24. Taylor, G. R. (Oct. 1976). "Ro/Ro Ships—State of the Art in Australasia," *Marine Technology*, Vol. 13, No. 4, SNAME.
25. Leeper, J. H. and Boylston, J. W. (Jan. 1987). "The Emerging Domestic Cruise Industry," *Marine Technology*, Vol. 24, No. 1, SNAME.
26. Cashman, R. M. (Oct. 1985). "The Small Passenger Vessel," *Marine Technology*, Vol. 22, No. 4, SNAME.
27. Ward, T. C. (Jan. 1980). "The Ubiquitous Barge: Some Trends in Barge Design and Construction in British Columbia," *Marine Technology*, Vol. 17, No. 1, SNAME.
28. Waller, D. B. (Feb. 1972). "Integrated Tug/Barge Units for Ocean Transportation," SNAME, Gulf Section.

29. Giblon, R. P. and Tapscott, R. J. (July 1973). "Design and Economics of Integrated Tug/Barge Systems," *Marine Technology*, Vol. 10, No. 3, SNAME.
30. Seymour, D. J. and Kossa, M. (Jan. 1981). "Integrated Tug/Barge Valerie F: Design and Bulk Cargo Handling Operations," *Marine Technology*, Vol. 18, No. 1, SNAME.
31. —— (Sept. 1977). "The Use of Tugs for Maneuvering Large Vessels in Ports—A Preliminary Study," National Ports Council, Dept. of Industry, General Council of British Shipping.
32. Allen, R. G. and Woodward, P. (July 1988). "Ship Handling in the Port of Churchill—The 'Bear' Essentials," *Marine Technology*, Vol. 25, No. 3, SNAME.
33. Argyriadis, D. A. (1957). "Modern Tug Design," *SNAME, Trans.*, Vol. 65, New York.
34. Spaulding, P. F. (1973). "Ocean Tug Boat Design," *Proc.* First American Tug Convention, Ship and Boat International.
35. Miller, R. T. (1969). "Ships," in *Handbook of Ocean and Underwater Engineering*, Myers, Holm, and McAllister, eds. McGraw-Hill, New York.
36. Mok, Y. and Hill, R. C. (July 1970). "On the Design of Offshore Supply Vessels," *Marine Technology*, Vol. 7, No. 3, SNAME.
37. Raj, A. and White, C. N. (May 1979). "Trends in Offshore Towing and Supply Vessel Designs," *Proc.* OTC, No. 3388, Houston.
38. Patten, L. M. (July 1983). "The Offshore Supply Vessel," *Marine Technology*, Vol. 20, No. 3, SNAME.
39. Blair, X. and Ansel X. (1968). *A Guide to Fishing Boats and Their Gear*. Cornell Maritime Press, Cambridge, Md.
40. Cole, J. A. (Apr. 1988). "Fishing Vessels of the Pacific Northwest: The Past 30 Years," *Marine Technology*, Vol. 25, No. 2, SNAME.
41. Daidola, J. C. and Griffin, J. J. (1986). "Development in the Design of Oceanographic Ships," *SNAME, Trans.*, Vol. 94, New York.
42. Derrickson, J. K. et al. (Jan. 1974). "A New Class of Coastal Research Vessels," *Marine Technology*, Vol. 11, No. 1, SNAME.
43. NAVFAC (1986). *Mooring Design Physical and Empirical Data*, DM-26.6. Dept. of the Navy, Alexandria, Va.
44. Polmar, N. (1984). *The Ships and Aircraft of the U.S. Fleet*. Naval Institute Press, Annapolis, Md.
45. Moore, J. E. (1978). *Jane's Fighting Ships*," Franklin Watts, New York.

3

General Design Considerations

The establishment of suitable design criteria is of primary importance in the design of any structure, and experienced judgment is most important in this area. Design criteria follow from evaluation of various functional and operational requirements as well as site and environmental conditions, and are eventually translated into design loadings as described in subsequent chapters. Judicious selection of appropriate construction materials is critical to the long-term performance of the structure.

This chapter begins with a review of general design considerations affecting the overall planning, design, and construction of a marine facility. In following sections, site selection and layout criteria are briefly reviewed, and specific facility type requirements are discussed. Environmental conditions and means of describing them for design purposes are reviewed, and a final section on materials gives a general description of marine construction materials and criteria for their selection.

3.1 DESIGN CRITERIA AND GENERAL CONSIDERATIONS

Design criteria for marine structures should be established after careful consideration of various operational, functional, and navigational requirements, environmental and site conditions, and physical and regulatory constraints, as summarized in Table 3.1. Most of the items in the table are self-explanatory, and they are often dictated by the owner/operators and their requirements, existing conditions within the port, or regulatory constraints of government agencies. Structural design criteria in particular should include the following items:

- Overall dimensions and configuration.
- Deck elevation.
- Water depth alongside plus dredge and scour allowance.
- Deck, cargo, and equipment loads.
- Berthing loads, maximum vessel, speed and approach angles, and so on.
- Mooring loads, maximum vessel, wind and current, wave action, and so on.
- Design life and durability requirements.
- Materials and construction methods.

GENERAL DESIGN CONSIDERATIONS 45

Table 3.1. General design considerations for marine facilities.

Site Conditions
- Topography.
- Bathemetry: soundings.
- Subsurface data: geologic history, soil properties, depth to rock, etc.
- Seismicity.

Environmental Conditions
- Meteorology: normal and extreme, wind, rainfall, temperature.
- Oceanography: normal and extreme, wave, tide, current, ice, water chemistry, seiche or harbor surge, etc.
- Frequency and probability of storm conditions.

Operational Considerations
- Vessel data, sizes, types, frequency, berth occupancy time, loading and servicing requirements.
- Vehicle data, sizes, types, capacities, operating dimensions (turning radii, etc.).
- Trackage, cranes, loaders, railroad, capacity, weights, windage, gage, speed, reach and swing, etc.
- Special equipment, mooring hardware, capstans, loading arms, product lines, etc.
- Services and utilities, shore connections, fire protection and safety equipment, lighting and security, electrical power, piping.
- Cargo storage area.

Functional Considerations
- Dredging, scour and siltation, propeller wash.
- Vessel traffic and traffic control systems (VTS).
- Landside access, remoteness, roadways, airports, etc.
- Maintenance practices; cathodic protection, damage repair, etc.

Navigational Considerations
- Channel depths and widths.
- Vessel approach conditions.
- Nav-Aids.
- Availability of tugs.

Constraints
- Harbor and pier head lines.
- Regulatory; water quality standards, oily ballast, dredge disposal, fill, etc.
- Permits and licensing.
- Availability of materials and equipment.
- Existing facility: changed usage or upgrading limitations.

- Allowable stresses and factors of safety.
- Applicable codes and standards.

Note that the first six items in the list are a direct function of the design vessel or vessels and their cargo handling requirements, the first three involving spatial requirements and the next three loads and forces. Load evaluation is an area where marine structures differ most from traditional civil engineering structures;

46 DESIGN OF MARINE FACILITIES

Figure 3.1. Generalized loads and environmental factors affecting pier design.

so individual chapters of this book have been devoted to direct structural loadings and berthing and mooring loads. These loads in turn depend upon the establishment of other design criteria such as design storm conditions, and vessel operational limits such as berthing velocities and berth approachability and occupancy criteria, all of which are probabilistic in nature and are discussed in subsequent chapters.

Figure 3.1 illustrates various sources of loadings and design considerations for a typical pier structure, which will be discussed in turn in following sections. Because of the exposure of marine structures to the natural elements and the probabilistic nature of vessel impacts and accidental overloads, the design of marine structures traditionally has been characterized by a modest degree of conservatism. However, in more recent years—with the application of computers, a better understanding of environmental forces, a probabilistic approach to design, better quality control of materials, and so on—structures have been built to the economic limit of the established design criteria.

Major port facilities usually have economic design lives on the order of 25

years, with practical service lives of up to 50 years or more with periodic maintenance or upgrading. In fact, many U.S. facilities of World War II vintage currently are being rehabilitated, upgraded, or converted to a new use as an economic alternative to new construction or as an environmental alternative to usurping precious wetlands (see Section 11.1).

Probability and risk considerations are important in the determination of design loads and operating conditions consistent with economic realities. Structures for which governing design loads are associated with natural events such as storm winds and tides usually are designed for an event associated with a statistical return period of 50 to 100 years or more. The statistical return period (T_R), the encounter probability (E_P), and the design life or interval (n) over which the likelihood of encounter is being evaluated are related as follows:

$$E_P = 1 - (1 - 1/T_R)^n \qquad (3\text{-}1)$$

The T_R and n normally are measured in years and E_P in percent. Therefore, in any given year, ($n = 1$), there is a 1% probability of occurrence of the 100-year event. A structure designed for the 100-year event will stand a 22% chance of encountering such an event over a 25-year design life and a 64% probability of encountering the 25-year event (see Table 3.2).

The return period of the event causing some predictable level of damage is often traded off in cost optimization studies allowing for the cost of certain maintenance and repairs to be carried out over time. This type of cost analysis commonly is carried out in the design of breakwaters and shore protection structures, but it may not result in very great cost savings for most pier and wharf structures of interest herein. Other risk criteria used to evaluate cost include: frequency distribution of total damage, mean total damage, and probability of zero damage criteria (1). More elaborate techniques often are applied to the reliability and risk analysis of offshore platform structures, as described by Stahl in reference (2).

Virtually all risk analysis depends upon the determination of the return period of some major storm event. Return periods most often are determined by ex-

Table 3.2. Encounter probability vs. design life and return period.

Design Life (years)	Return Period (years)						
	5	10	25	50	100	200	500
1	.200	.100	.040	.020	.010	.005	.002
5	.672	.410	.185	.096	.049	.025	.010
10	.893	.651	.335	.183	.096	.049	.020
25	.996	.928	.640	.397	.222	.118	.049
50	.999+	.955	.870	.636	.395	.222	.095

trapolating data from relatively short periods of record to long-term probability distributions using any of several extreme value probability laws. Alternatively, the prediction of long-term extreme values of random events may be made from statistical models of the actual physical processes involved (3). Design storm selection criteria and a summary of applicable statistical probability laws are presented by Borgman and Dean in reference (2). Reference (4) presents a general methodology of analyzing wind and wave statistical data and their effect on ship and port operations. In terms of exceedance statistics, we are often more interested in the joint probability of more than one environmental parameter, such as the highest wind speed from a given direction combined with extreme tide height, which in turn may result in the most damaging wave conditions. In this case, a worst-case design storm could be developed based upon the coincidence of meteorological conditions and astronomical tide data.

Of the many government and regulatory agencies that may be involved with planning, permitting, operating, and/or design requirements, the following are of importance in most U.S. port projects. The U.S. Army Corps of Engineers (USACE) is responsible for dredging and maintaining navigable waterways and for granting permits for construction thereon, including inland waterways, under the Rivers and Harbors Act of 1899. The USACE has additional authority for control of dredged and fill material and its transport under the Clean Water Act and the Marine Protection, Research, and Sanctuaries Act of 1972. The U.S. Coast Guard (USCG), in addition to enforcing maritime law, maintains and controls aids to navigation and is responsible for ice breaking operations and the reporting of ice conditions. The Environmental Protection Agency (EPA) oversees or controls development with potentially adverse environmental effects. Environmental protection usually is administered at the state level, and the number of agencies involved and modus operandi vary from state to state. The Coastal Zone Management Act (CZM) is a federally mandated program that provides states with funds for management and improvement of coastal areas. On the local level, a city or town conservation commission and a harbor commission and possibly other groups must review and approve all plans for proposed construction.

Another important general consideration is the possibility of future events that may have a profound effect on a proposed structure's function or integrity, such as future dredging, or channel widening, increased vessel traffic, decreased vehicular access, construction on adjacent sites, and so on. The structure's design should also consider the possibility of future expansion and upgrading (5).

3.2 SITE SELECTION AND LAYOUT

The relative exposure of a site to wind and sea has a major impact on its structural design features. In the past, the majority of piers and wharves were situated in relatively well-protected natural harbors where wave action did not have

GENERAL DESIGN CONSIDERATIONS 49

a significant effect on terminal structures' layout and design. In the past decade or so, with the advent of larger and deeper-draft vessels, marine terminals have been constructed in increasingly exposed locations, such as large natural bays or man-made harbors, with perhaps partial protection from prevailing seas afforded by a breakwater or more distant land masses. Offshore terminals consisting of sea islands or mooring buoy arrangements with a pipeline or conveyor system link to shore may be constructed in essentially fully exposed locations. Depending upon the frequency and duration of storms and limiting sea states for operations, such terminals may have limited or seasonal berth occupancy times. General factors affecting site selection, planning, and layout and design of harbor protection structures are given good overall treatment in references (6) through (8). Other sources of particular interest in harbor and site selection include references (9) and (10).

Port facilities siting also depends importantly on land support, in terms of storage area and access to landside transportation systems and the many other siting considerations outlined in Table 3.1. Land space requirements range from 2.5 to 5.0 acres per berth for a traditional general cargo terminal to 20 to 50 acres per berth for a contemporary container terminal, depending upon the maximum vessel capacity. The proximity of adequately sheltered anchorage areas for vessels waiting for an available berth is another important general consideration.

It is helpful at this point to introduce a few general definitions, but one must remember that many nautical terms may be used loosely with imprecise definitions. A *harbor* is, in general, a relatively protected water body where vessels can seek refuge from storms, transfer cargo, or undergo repairs. Harbors may be natural, such as well-sheltered bays, estuaries, or river locations, or artificial, such as areas created by breakwaters or dredged basins. An *impounded basin* is very well protected and can be entered only through locks that maintain a constant water level within the basin. *Coastal basins* are formed by encircling breakwaters with narrow but direct connection to the open sea or by dredging into the coastline with access through a narrow inlet. A *port* is a location where ships transfer cargo, usually for commercial trade. A port may consist of a single marine terminal or loading facility occupying a small portion of a harbor, or it may encompass the entire harbor with several terminal or ship service facilities. Offshore ports may exist at exposed coastal locations and typically are dedicated to the transfer of a single commodity type. A *channel* is an unobstructed water course of suitable depth for navigation. A *restricted channel* typically is relatively narrow and bounded by shoreline banks on both sides. An *unrestricted channel* is a dredged or naturally deep channel bounded by open water of less than the required navigation depth. A *fairway* is a relatively wide expanse of water with a clear unobstructed water depth suitable for vessel transit. A *roadstead* or *road* is essentially an open anchorage area outside of demarcated channels of suitable depth and width for vessels to swing about at anchor. A

turning basin is a widened portion of a channel suitable for turning a vessel around, usually created by dredging. An *anchorage basin* is a typically well-protected area created by dredging, where usually smaller vessels may anchor.

Within established ports there usually are defined limits to which construction can take place in order to avoid encroachment on navigable waters. The *channel line* demarks the limits of a dredged or natural channel, which generally is buoyed and of a *control* depth and width denoted on nautical charts. Inshore of the channel line, many harbors have a designated *pierhead line*, which denotes the limit of open pier construction and beyond which no fixed construction is allowed. The pierhead line may or may not coincide with the *bulkhead line*, which is the limit to which solid fill structures can be built.

A *dock* is the most general term for a place or structure at which a vessel may be moored. A vessel is *moored* when it is secured at one location, whether it be alongside a fixed structure or riding to a single anchor. Piers, wharves, quays, bulkheads, and so forth, may serve as a dock (these terms are further defined in Section 7.1). A *berth* is a location where a vessel can be safely moored, such as alongside a pier or wharf or within a *slip* between piers or fixed berthing structures such as dolphins or ferry racks.

Figure 3.2 illustrates vessel berth exposure conditions such as within protected harbors, in more exposed semiprotected artificial harbors, and at offshore *sea island* type berths, which are fully exposed and must be abandoned under adverse weather conditions. Figure 3.3 illustrates typical berth type arrangements.

If most of the vessel's length must be accessed for cargo handling, the vessel is berthed *alongside* a pier or wharf structure, which typically extends the full length of the vessel or longer with mooring hardware distributed along the face of the pier. If the vessel's cargo can be transferred from a single location on the vessel, such as at a tanker's manifold connections, an open breasting dolphin-type berth arrangement may be used, where mooring lines are secured to discrete mooring dolphins connected only by personnel catwalks. Where rapid turnaround and loading operations are required, such as for ferry and Ro/Ro operations, a slip-type berth with guide-in dolphins or continuous fender racks may be used. Variations of the three basic berth types include *multiple berth* arrangements, such as alongside continuous *marginal* wharves or between adjacent piers and/or piers set at skew angles to the shoreline. NAVFAC (10) provides further discussion of the relative advantages and disadvantages of various alongside berth arrangements.

Figure 3.4 shows a shore-connected multiple tanker berth capable of berthing six tankers simultaneously.

For single-commodity terminals such as oil terminals and those relegated to a particular vessel type and limited size range, the layout of fender systems, mooring hardware, loading equipment, manifolds, cargo deck space, and so

GENERAL DESIGN CONSIDERATIONS 51

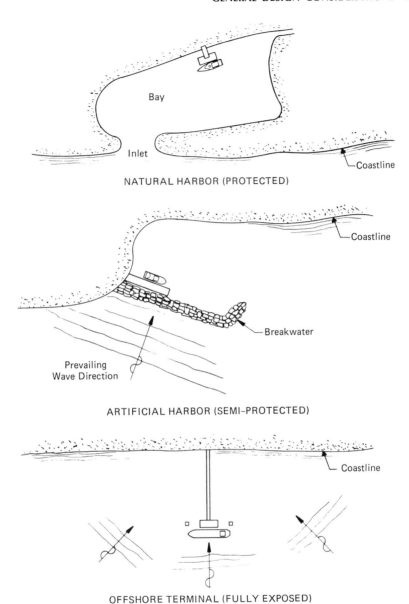

Figure 3.2. Vessel berth exposure conditions.

on, will be relatively straightforward. Where a large range of vessel types and sizes is expected, such as at a general cargo terminal, particular attention must be paid to the size and number of vessels and their relative residence times.

52 DESIGN OF MARINE FACILITIES

ALONGSIDE TYPE

OPEN DOLPHIN TYPE

FERRY (SLIP) TYPE

Figure 3.3. Berth arrangements.

Figure 3.5 illustrates various design considerations in the general layout of a typical vessel berth. Figure 3.6 illustrates general berth dimensions as recommended by the U.S. Navy (11) for multiple berthing.

The minimum slip width between adjacent piers should not be less than about 300 feet for larger vessels or two times the beam of the largest ship plus tug allowance, provided the vessel can be turned outside of the slip. The slip width

GENERAL DESIGN CONSIDERATIONS 53

Figure 3.4. Six-berth products pier for vessels up to 50,000 DWT. (Photo courtesy of Frederic R. Harris, Inc.)

also must consider: access for smaller service vessels such as tugs; exposure to wind, wave, and currents; line-handling methods; and the need to move berthed vessels while they are moored. An allowance of from 10% to 20% of the LOA of the largest vessel, or approximately 50 feet minimum, should be provided between vessels berthed alongside end to end. The term *winding ship* in port refers to the practice of rotating a vessel at a pier so that the opposite side of the vessel will be against the pier. Obviously if this kind of operation is required, then a clear distance equal to the vessel's *LOA* plus operating room for tugs, perhaps 100 to 200 feet minimum, must be provided. The minimum turning diameter for a vessel being warped around a pier or dolphin is 1.2 × *LOA*.

Water depths must have a sufficient *under keel clearance* with respect to the vessel's full load draft and range of tide. An absolute minimum of four feet at low water should be provided to account for vessel trim and tidal variations. Other factors affecting minimum keel clearance include character of the bottom, siltation allowance, exposure to waves, vessels with vulnerable appendages or

54 DESIGN OF MARINE FACILITIES

Figure 3.5. Berth layout design criteria.

water intakes below their keel lines, vessels with extreme trim, extreme tidal variations, and so forth. The consequences of grounding on a rock bottom versus soft sediment should be carefully considered in setting the keel clearance, especially in areas with large tidal ranges that usually have extreme tides which can fall several feet below normal low water. The length of the dredged berth

GENERAL DESIGN CONSIDERATIONS 55

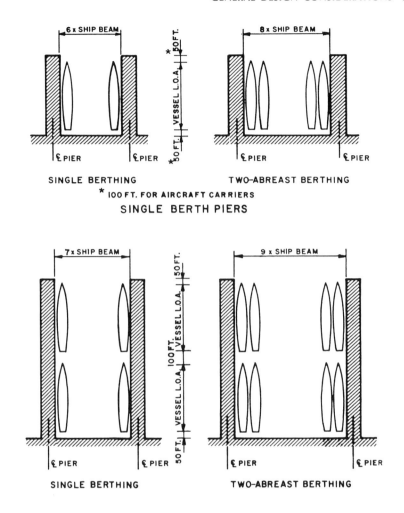

Figure 3.6. Vessel slip widths for multiple-berth piers. (After NAVFAC (11).)

alongside a pier or wharf should be a minimum of $1.25 \times LOA$ where tug assistance is provided, up to $1.5 \times LOA$ where tug assistance is not provided. The dredged berth should be at least 1.25 times the beam of the largest vessel to use the berth (9). For turning basins, a minimum diameter of $4 \times LOA$ should be provided for unassisted vessels, down to $2 \times LOA$ where tugs assist.

Further description of vessel maneuvering area and navigation requirements can be found in references (6) through (10) and (12) through (14). Additional discussion of berth requirements and mooring line layout is provided in Section

6.1, and of pier and wharf general dimensions and features in Chapters 4 and 7.

3.3 FACILITY TYPE REQUIREMENTS

Most marine facilities are dedicated to a specific purpose and have particular requirements dictated by the vessel type, dimensions, and operating characteristics and by the means of handling and conveying the vessel's cargo. The following factors, which are a consequence of vessel type and cargo transfer operations, determine the facility's overall layout and functional design criteria:

- Berth length and number and location of vessel loading points.
- Apron width and need for short-term storage on the pier or wharf.
- Storage area required, including the need for open backland and covered transit shed areas.
- Cargo-handling equipment requirements and traffic flow.
- Exposure to sea conditions as it affects cargo-handling operations.
- Deck elevation as it affects cargo operations throughout the tidal cycle, and the frequency and consequences of storm flooding.
- The need for buildings such as storage sheds, waiting areas, customs and administrative offices, longshoremen's shelters and toilets, and so on.

In addition to the above considerations, some facility types are particularly adaptable to mixed-use, automation, and future expansion possibilities. General trends in the development of modern marine terminals are reviewed in reference (5). Today's ideal new terminal should be adaptable to multipurpose use and able to accommodate the largest loads normally encountered.

The following discussion reviews specific functional requirements of the various facility types introduced in Chapter 1.

General cargo or break-bulk cargo terminals usually receive vessels equipped with their own cargo-handling equipment. The ship's cargo gear is necessarily of limited reach, and cargo is placed on deck near the vessel awaiting conveyance by forklifts, mobile cranes, trailers, or other mobile equipment. General cargo-handling methods are covered in reference (15). Long berth lengths are preferred, as self-unloading may tie up a single berth for several days. Marginal wharves are preferred to allow continuous access for the mobile equipment and space for operations. Apron widths of 60 feet minimum to 120 feet maximum are typical. Cargoes may be packaged in a variety of ways, such as on pallets, bailed, bundled, or individually packaged. Even standardized containers may be hauled by break-bulk ships. Refrigerated cargoes carried by special reefer ships are often containerized in standard 20- or 40-foot containers with their own refrigeration system. Contemporary general cargo facilities typically are multiuse facilities. Minimum single berth lengths of 600 feet or more should be considered.

GENERAL DESIGN CONSIDERATIONS 57

Container terminals are rapid turnaround intermodal facilities that typically require relatively large backland areas for the stacking and temporary storage of containerized cargoes. Typical container dimensions are given in Section 4.1. The containerized cargo system was first employed by the Sea Land Corporation in 1956, and early container ships required a 500- to 600-foot berth and approximately 7 to 10 acres of total yard area. Today, a single berth is on the order of 900 to 1000 feet long and ideally requires 25 acres or more of space. Frequently more than one gantry-type crane will load/unload the same vessel simultaneously. The apron width is largely determined by the crane gage plus backreach and rail setback. Crane gages vary with their vintage and type but usually fall within the range of 50 to 150 feet, resulting in apron widths of from approximately 80 to 160 feet. In addition to rail-mounted cranes, which load and unload the vessels, various types of rubber-tired mobile equipment, such as straddle carriers, forklifts, and tractor-trailers, are used to move and stack containers. Railroad spurs also may be an integral part of container terminals. Optimization of marshaling areas where containers are stored on trailer chassis and the traffic flow are critical elements in efficient container terminal operations.

Roll on/roll off (Ro/Ro) facilities originally were intended as large ocean-going ferries for tractor-trailer service. Vessels usually are equipped with their own ramps, which may be straight stern and/or bow ramps for end-on loading only, or slewing-type quarter ramps that allow the vessel to unload from alongside. A variation of Ro/Ro service is the pure vehicle carrier, which primarily delivers new automobiles to their market. Vehicle carriers usually are equipped with two or more side ramps for self-unloading operation. Today's Ro/Ro vessel often also carries containers on deck, and oceangoing Ro/Ro service requires facilities tailored to this mixed use. Quay and wharf decks must be clear and unobstructed and able to support the localized reactions of the vessel's ramps. The deck elevation relative to the tide range and the vessel's loaded and unloaded threshold height are critical for efficient full-time operation. Quay decks may be sloped to accommodate varying ramp angles and threshold heights through the tide cycle. Many Ro/Ro facilities provide adjustable fixed transfer bridges or floating transfer bridges, referred to as *link spans*, to overcome the problem of large tidal range. International standards have been developed for dimensioning and details of Ro/Ro ship-to-shore connections (16), and additional information on Ro/Ro ramp and system characteristics can be found in references (17) and (18).

Ferry terminals service passengers and highway vehicle traffic. They require a small to moderate backland area for short-term queuing of traffic and for passenger waiting areas. Vehicles usually are loaded end-on via transfer bridges similar to those used in Ro/Ro operations. Passengers are boarded through side doors via boarding platforms and gangways, or sometimes over the transfer bridge in smaller-scale operations. The most important design considerations

for a ferry terminal are its transfer bridge and a high energy absorption/low maintenance fender system, because of the frequent and high-speed nature of the ferry operation. Ample guide-in dolphins should be properly located to facilitate rapid and safe berthing maneuvers. Another very important consideration in ferry terminal design is scour protection from propeller wash, which is an especially acute problem at the toe of bulkheads (see Section 4.6). Additional information on ferry terminal facilities can be found in reference (19).

Passenger terminals have experienced a recent upsurge in construction in concert with the burgeoning cruise ship trade. Alongside berthing space is required, and special attention must be given to passenger boarding structures and waiting areas. Separate means of loading vessel stores and provisions also must be provided. Unlike most cargo-carrying vessels, passenger ships do not change draft significantly between loaded and unloaded conditions. A crude rule of thumb requires approximately 10 square feet of space per passenger in waiting areas. Direct connection with ground transportation links also is necessary.

Barge carrier systems, generally known as LASH for "lighter aboard ship" (other proprietary systems include SEABEE and BARCAT), are designed for barge-handling afloat where the vessel route typically connects with inland waterways or shallow water terminals. LASH vessels also may berth alongside where vessels are equipped to carry containers on deck. Vessels alternatively may moor offshore to load or unload their barges. Typical barge dimensions for some of the more active operators are given in Chapter 2 (see Table 2.8).

Dry bulk cargo facilities often are dedicated to handling a specific commodity, such as ores, minerals, coal, cement, grains, fertilizers, wood chips, and so on. They usually are loaded and unloaded by special equipment with conveyor systems to storage and distribution areas. Because of the nature of the cargo and the fact that it can be readily conveyed on a mechanized system, and because of the generally large size and deep draft of bulk carriers, bulk carrier berths often are built offshore. Offshore berths at exposed locations must be designed to survive the most severe sea conditions expected over their design life, as well as maximum vessel berthing and mooring loads under the worst allowable operating conditions with the vessel in berth. The storage and distribution area may be a considerable distance from the vessel berth. Very high surcharge or deck loads may result from the piling up of bulk cargoes. Bulk carriers often need to be shifted in position along the berth length to allow ship loading equipment to access all of their holds. Bulk ships undergo a very large change in draft and trim between loaded and unloaded conditions. Some bulk ships, such as certain colliers (coal carriers), are equipped with self-unloading booms and conveyors. Self-unloading vessels may discharge their cargo directly to the quay or wharf deck or into hoppers or bins, or to smaller barges for inland distribution. Some dry bulk cargoes may be mixed with water and pumped

through pipelines in slurry form, but this practice is limited because mixing of large volumes of water in a vessel's holds places limitations on its cargo-carrying capacity.

Liquid bulk cargoes include crude oil and petroleum products (by far the most common) and various chemical solutions and some foodstuffs. Liquefied compressed gases such as natural gas (LNG) and propane gas (LPG) and others are a very special type of liquid bulk cargo; they are very hazardous, requiring low temperature and/or high pressure storage vessels, and are subject to strict regulation (20). These vessels are similar in berthing requirements to typical crude oil tankers, in that the transfer of cargo usually takes place at a single fixed-manifold location. A typical tanker berth requires only two breasting dolphins flanking a central loading platform for handling hose connections, and additional mooring points ashore or on dolphins for bow and stern lines. It is important to consider the size range of vessels that may use the berth, as that has a major effect on dolphin location and size. Many tankers are of a very large size, and thus must berth and transfer their cargo from offshore terminals such as sea-islands or from single point moorings (SPMs) of various types. The provision of fire protection systems and the need for booster jumps and possible heating of pipelines are important considerations in tanker terminal design. Tankers normally discharge cargo with their own shipboard pumps, but, depending upon the distance and the elevation of the storage yard, they may require additional pump stations. Typical tanker pumping rates range from 22,000 barrels per hour (bbl/hr.) for a tanker of 25,000 DWT to approximately 75,000 bbl/hr. for a 200,000 DWT tanker. Guidelines for the layout and design of supertanker berths can be found in references (21) and (22), and for SPMs in reference (23).

Shipbuilding and ship repair yards have unique requirements for new construction and repairs, and shipyards may be specialized in either activity or carry out both. Ship construction requires large open layout and storage areas, machine shops and administrative buildings, large traveling cranes, and a means of launching such as building ways, building basins, or transfer onto dry docks. Dry docks are a central feature of ship repair yards and have their own special mooring, access, and operational requirements, as described in Chapter 10. Both shipyard activities require alongside berthing space for fitting out newly launched vessels or conducting repairs afloat. Vessel berths require crane service, usually in the form of revolving portal-type cranes which typically range from 25 tons to 65 tons capacity. For new construction, revolving-type cranes of 300 or more tons have been built. Fitting out berths also requires extensive utility, electrical, and ship services systems, which often have a major impact on the pier or wharf configuration. Fitting-out activities may be carried out from piers with two-side berthing, as immediate continuous access to shore supplies is not constantly required. Access for workers and minimization of deck clutter

are important considerations. High-capacity corner bollards should be provided for alongside propeller testing. Further descriptions of shipyard facilities and operations can be found in references (24) and (25).

Military bases for the berthing of warships are designed to specific naval requirements such as given in the U.S. Navy's NAVFAC design manuals, which are cited throughout this book (10, 11, 15). Military installations are largely self-sufficient and typically require extensive utility and ship services similar to those of shipyards. In addition, potable water and sewage requirements are high, as crews live aboard vessels while in port, even during extensive overhauls. Security and fire protection requirements are generally more stringent than for other facility types. Contemporary warships are typically of lighter hull construction, and fitted camels and/or floating fender systems often are employed, in addition to the pier's own fendering.

Fishing and commercial small craft such as tugs, workboats, small barges, and miscellaneous service craft usually have minimal berthing facility, backland, and service requirements. Water and electrical power often are the only utilities provided. Fishing vessels require a means of discharging their cargo and loading ice and supplies. This is often done from davit hoists or small jib cranes mounted at fixed locations. Larger fishing vessels may be unloaded by mechanized-conveyor-type loading devices. Belt-type conveyors frequently are utilized to convey fish to the processing facility, after they have been offloaded into buckets or containers at the vessel. All commercial small craft generally require relatively protected locations and rugged fender systems, which are subject to continual abuse. Fishing boats and workboats often raft together, thus helping to minimize alongside berth length for large fleets. Vessels typically are boarded via vertical ladders built into the side of the wharf or pier structure. Small passenger vessels such as sightseeing and commuter boats, however, require gangways or direct walk-aboard access as provided by floating dock systems.

Marinas are small craft facilities for yachts or recreational vessels that are generally under 50 to 60 feet in length. They require sites that are well protected from wave action, as well as adequate backland area for parking. As the general design of marinas and small craft harbors is beyond the scope of this text, the reader is referred to references (19) and (26) through (28) for additional information and guidelines.

Additional information on specific facility requirements for commercial merchant ships can be found in references (6), (7), (8), and (29). Reference (30) presents important information for the design of facilities that handle hazardous cargoes.

3.4 ENVIRONMENTAL CONDITIONS

Environmental influences on marine structures design include the effects of wind, waves and harbor motions, tide and storm surges, currents, ice, seismic-

GENERAL DESIGN CONSIDERATIONS 61

ity, water quality factors such as salinity, dissolved oxygen, pH, pollutants, and so forth, and climatic conditions such as temperature range, precipitation, and visibility, as well as fouling and biodeterioration. These factors are all discussed in this section, and a more detailed overview of these factors as they relate to structural design has been provided by this author in reference (31). Additional textbook treatment of waves and oceanographical factors can be found in the general coastal engineering literature cited at the end of Chapter 1.

Environmental loadings are time-dependent, covering a spectrum ranging from fractions of a second to several hours. Table 3.3 summarizes the time domains of various environmental influences that are periodic in nature. A dynamic analysis is required if the unit response of the structure to any disturbing force is significant, and if any of the structure's natural periods nearly coincides with the period of the disturbing force. Mathematical models for computer application have been developed to describe the various complex interactions of the natural elements—for example, to predict the wave climate within a harbor. General descriptions of the applications of mathematical models and computer simulations in harbor and coastal engineering applications can be found in references (32) through (35).

The following discussion describes the various environmental factors and means of defining them for design purposes.

Wind

Prevailing and extreme wind speeds and directions and their frequency of occurrence at a given site are of utmost importance. Certain sites may be subject to intense tropical storms such as hurricanes or typhoons relatively infrequently, whereas others may experience less severe but more frequent extratropical cyclones, such as the winter storms common to the mid latitudes. The wind climate at a given site may exhibit regular and dramatic seasonal variations. The

Table 3.3. Approximate range of periods of environmental forces

Environmental Force	Range (sec)
Wind turbulence	0.05–20
Currents	
Unsteady velocities in tidal flows	0.1–1.0
Vortex shedding (slender objects)	0.3–2.0
Wave forces	
Wind waves and swell	1.0–20+
Drift force (wave grouping)	20–100
Long waves, seiche, and tsunamis	20–4000+
Tidal flows	50,000+

local topography may serve to greatly modify wind conditions, resulting in localized funneling or *jets*. At some sites in proximity to high hillsides or mountains local anabatic (upslope) or katabatic (downslope) winds may be experienced.

In addition to imposing direct velocity forces against fixed structures, equipment, and moored vessels, winds generate waves and may have a profound influence on berthing operations. Of all the factors affecting the berthing maneuvers of large vessels, wind usually is considered the most significant by vessel masters. This is especially true for vessels riding high in ballast condition. Wind data usually are expressed in terms of percent frequency of occurrence and mean speed by direction, usually over a typical one-year period. This information is conveniently summarized on a diagram called a wind rose. Knowledge of mean annual and seasonal winds is important in berthing and loading operations, especially when there is sufficient fetch for waves to develop, and in other aspects of port operations. For structural design purposes, however, extreme wind speeds associated with longer return periods, such as the 50- and 100-year storms, are of primary importance.

Figure 3.7, reproduced from reference (36), shows an example of long-term probability distributions of wind speeds over open ocean for several areas of offshore oil drilling interest. In these curves, the maximum wind is defined as the average of the maximum measured wind speeds over a one-minute interval. The maximum gust will be approximately 1.4 times the sustained maximum wind speeds given. The data from which these distributions were plotted were generalized for large areas and are not site-specific.

Within harbors and nearer to the coast, extreme wind speeds generally are less than at offshore locations. Extreme wind speed and direction data for the United States and its territories can be found in references (37) and (38) and in airport meteorological records maintained by the National Weather Service. Recommendations for collecting and interpreting wind data at remote sites where record data are scarce can be found in reference (4).

Table 3.4 illustrates approximate ranges of threshold wind velocities as they affect marine terminal operations. The wind scale employed is the Beaufort Scale, commonly used by seamen to describe wind and sea conditions. The operational limits are based on wind action alone, and where waves are present threshold velocities may be reduced. Berthing of large vessels is generally affected when winds exceed approximately 20 knots, but high-speed ferries may continue to operate in winds up to approximately 40 knots. Container and gantry-type cranes generally are affected when sustained wind speeds exceed approximately 25 knots. It often is assumed for design purposes that larger vessels will vacate their berth when winds exceed around 50 to 60 knots. This should not be taken as a rigid assumption, and the local conditions should be studied carefully to assess the possibility of a vessel's being "trapped" at berth under extreme storm conditions.

GENERAL DESIGN CONSIDERATIONS 63

Figure 3.7. Long-term wind speed vs. return period for selected offshore regions. (From reference (36).)

Winds exhibit both vertical and horizontal (temporal) variations in speed, so both a reference height and a time duration must be specified to define the design wind speed.

Wind speed data and gust factors for design are discussed further in Section 6.3 in connection with the evaluation of mooring loads. The wind climate in general and its effects on structures are treated in detail in references (39) and (40).

Waves and Harbor Oscillations

As with winds, information on both the annual and the extreme wave climate for a given site must be obtained. Most often wave climate data are compiled from forecasting methods using wind record data and knowledge of the local topography and bathymetry. Short-term data usually are presented in terms of percent frequency of occurrence via histograms and/or cumulative distributions expressing yearly averages for each month. Long-term wave statistics usually

Table 3.4. Wind speed vs. operational criteria.

Beaufort Scale/ Seaman's Descriptions	Wind Speed (knots)	Effect on Operations*	
		Vessel	Facilities
0 Calm	0–1		
1 Light air	1–3		
2 Light breeze	4–6		
3 Gentle breeze	7–10		
4 Moderate breeze	11–16		
5 Fresh breeze	17–21	↑	↑
6 Strong breeze	22–27	Berthing limit	
		↓	Crane operations cease ↓
7 Near gale	28–33	Tugboat limit	↑
8 Fresh gale	34–40		Loading arms disconnected
		Ferry operations cease	↓
9 Strong gale	41–47	Emergency mooring lines	
10 Whole gale	48–55	Larger vessels put to sea	Facilities secured, cranes lashed, etc.
11 Storm	56–63		
12 Hurricane	64–71		

*Because of wind alone, wave action at exposed locations may result in greater limitations.

are given in terms of maximum wave heights versus statistical return period in a manner similar to Figure 3.7 for extreme winds. Where the effects of cumulative damage can be related to the number of wave cycles of a given wave height, as is required in fatigue analysis, then plots of wave height versus cycles over a specified period may be prepared, as illustrated in Figure 3.8, reproduced from reference (41). Such information must include both the normal wave climate and extreme event components. Normally this information is required only for offshore terminal structures or where the dynamic response to wave action is significant. The wave heights must be applied over an appropriate range of wave periods in order to evaluate the dynamic response in terms of stress level and cycles to failure, as discussed in Section 7.3. Another way of defining the wave climate for design purposes is in the form of sea spectra or the wave height frequency spectrum (discussed below). Sources of measured wave data for coastal North America have been compiled by Edge et al. (42), and means of analyzing long-term wave data statistics can be found in reference (4).

The three most important parameters controlling wave development at sea are: the sustained wind speed; the fetch length, or distance over which the unrestricted wind blows; and the duration of time for which the wind blows. For a given wind speed, the maximum wave height may be limited by the fetch or the duration. If the fetch and the duration are sufficient, for a given wind

GENERAL DESIGN CONSIDERATIONS 65

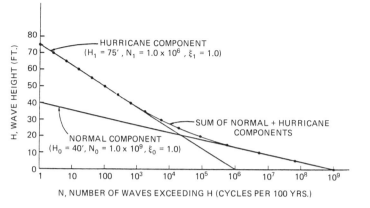

Where: H_0 is the maximum normal wave height over period T.
H_1 is the maximum hurricane wave height over period T.
N_0 is the number of wave cycles from normal distribution over period T.
N_1 is the number of wave cycles from hurricane distribution over period T.
T is the duration of the long-term wave height distribution.
ξ_0 is the parameter defining the shape of the Weibull normal distribution. Value of 1.0 corresponding to the exponential distribution results in a straight line.
ξ_1 is the parameter defining the shape of the Weibull hurricane distribution.

EXAMPLE WAVE HEIGHT DISTRIBUTION OVER TIME T

Figure 3.8. Example of wave height distribution over time. (Reproduced with the permission of the American Petroleum Institute (41).)

speed, a more or less steady-state condition known as a fully developed sea (FDS) will be attained, whereby average wave heights do not increase any farther. It has been well established that within a fully developed sea (FDS) condition in deep water, the probability distribution of wave heights very nearly follows the extreme-value Rayleigh distribution of statistics. The significant wave height (H_s) is defined as the average of the highest one-third of the waves present in an FDS. Table 3.5 shows the relationship between H_s and other statistical heights.

Near the coast and in inland waters, wave growth is further constrained by

Table 3.5. Relation of wave height parameters to significant wave height based upon Rayleigh distribution.

Average wave height	0.64
Root mean square (RMS) value	0.707
Significant wave height	1.00
Average of highest 10%	1.28
Average of highest 1.0%	1.67
Highest	1.87

water depth, bottom friction, restricted fetch lengths and widths, intervening land masses, and so forth. In the open ocean, fetch lengths and durations required for FDS conditions are seldom attained for sustained wind speeds exceeding 40 to 50 knots, and the developing sea state tends to be steeper and more confused than under FDS conditions. The USACE *Shore Protection Manual* (43) provides instructions and graphs for predicting wave heights under specific conditions. Such calculations can be carried out by an experienced harbor engineer for wind waves within bays or harbors of simple geometry and relatively uniform depths. However, for major facilities and those with greater exposure and complicated bathymetry, wave climate predictions should be carried out by qualified oceanographers.

Figure 3.9 illustrates the basic nomenclature for a simple sinusoidal wave form. The height (H) is the vertical distance from trough to crest, and the length (L) is the horizontal distance between crests. The period (T) is the length of time elapsed between the passage of successive crests. The wave crest elevation above the still water level (SWL) is denoted by η_c. There are various wave theories used to describe wave kinematic properties, and selection among theories depends upon the relative water depth, the degree of accuracy required, and the application. The most basic wave theory is the small-amplitude sinusoidal wave or, alternatively, the linear or Airy wave theory, named for its originator. According to basic sinusoidal wave theory, the sea surface elevation (η) relative to the SWL at any time (t) is given by:

$$\eta(t) = \frac{H}{2} \cos(kx - \omega t) \qquad (3\text{-}2)$$

where $k = 2\pi/L$ is known as the wave number, and $\omega = 2\pi/T$ is the circular frequency.

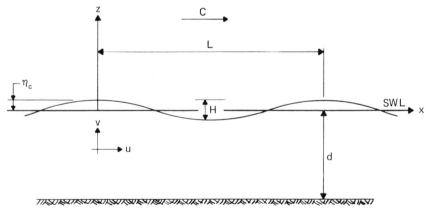

Figure 3.9. Wave definition sketch.

The wave phase velocity, or celerity (C), is the speed at which the wave form propagates, given by:

$$C = L/T = \sqrt{\frac{g}{k} \tanh kd} \qquad (3\text{-}3)$$

In deep water the group velocity (C_g), which is the speed at which a train of waves advances, corresponding to the rate of transmission of energy, is equal to one-half the celerity. In deep water, defined as a depth (d) greater than $L/2$, equation (3-3) reduces to:

$$C = 5.12\ T \qquad (3\text{-}4)$$

for C in feet per second (fps) and T in seconds. Hence:

$$L = 5.12\ T^2 \qquad (3\text{-}5)$$

for L in feet.

In shallow water where d is less than approximately $L/20$, the celerity becomes independent of the wavelength and depends only on the water depth, and is found from:

$$C = \sqrt{gd} \qquad (3\text{-}6)$$

where g = acceleration of gravity.

In transitional water depths, $L/2 > d > L/20$, equation (3-3) applies, and in general all of the wave properties undergo characteristic changes within the given relative water depth regimes. As the wave form passes, a hypothetical water particle follows an essentially circular orbit in deep water, which becomes more elliptical in shallower water. The maximum vertical and horizontal partical velocities (v and u) at the surface in deep water are given by:

$$u_{\text{max.}} = v_{\text{max.}} = \frac{\pi H}{T} \qquad (3\text{-}7)$$

where $u_{\text{max.}}$ occurs at the peak wave crest position, positive forward, and the trough position, negative backward, (see Figure 3.9), and $v_{\text{max.}}$ occurs at approximately the quarter wavelength positions upward, positive, with an approaching crest and downward, negative, after the crest passage. Similarly the maximum particle accelerations are given by:

$$\dot{u}_{\text{max.}} = \dot{v}_{\text{max.}} = 2H\left(\frac{\pi}{T}\right)^2 \qquad (3\text{-}8)$$

v_{max} occurs at approximately the quarter-wavelength position ahead of the crest, positive, and u_{max} occurs at the trough position, positive. The maximum sea surface slope is given by $\pi H/L$. A more detailed treatment of wave theories and mechanics is beyond the scope of this text. However, the above relations have been introduced here because of their importance to an understanding of subsequent discussions in this and other sections of this book.

Wave periods typically lie in the range of 2 to 20 seconds, but waves with periods greater than approximately 15 seconds usually are associated only with severe storms or with swell resulting from distant storms, whose periods have lengthened with distance traveled. At sea, wave steepness (H/L) usually ranges from 1/12 to 1/33, whereas near the coast and in inland waters waves are generally steeper, around 1/9 to 1/17. When waves approach a limiting steepness of approximately 1/7, they begin to break, spilling their energy in turbulence. When waves enter progressively more shallow water approaching a limiting depth, the height begins to increase, and water particle orbits become more elliptical than the circular orbits of deep water waves, prior to breaking at around the limiting depth of $d_b = 1.28H$. The relative depth at breaking d_b/H is decreased by steep bottom slopes, depending upon the wave's initial deep water characteristics H_0 and L_0. Wave crest elevations (η_c) above the still water level (SWL) are generally between 55% and 75% of their total height (H).

The waves so far discussed have been assumed to be of uniform dimensions and periods (i.e., "regular" waves); but, with the possible exception of some long-period ocean swells, the situation is far different from this for wind waves. A common way of representing the irregular nature of the sea surface is to use a wave spectrum, representing the superposition of a large number of regular waves of varying heights and periods. Wave spectra normally are presented as plots of the spectral density, $S(\omega)$, versus frequency. The spectral density may be given in terms of the square of wave heights or amplitudes, or of the double height or half amplitudes, depending upon the user's preference. The spectral density is proportional to the sea surface energy. The abscissa of the plot is the wave frequency ($f = 1/T$), commonly given in terms of circular frequency, ω. The area under the spectrum corresponds to some statistical wave height parameter such as H_s when the double-height spectrum is used. For the wave height spectrum, the area under the curve equals the root mean square (RMS) value of the wave heights present.

Several spectral formats commonly are used, depending upon the application. The Bretschneider height spectrum (44), for example, is especially useful in many civil engineering applications, and was developed to correspond with the Rayleigh distribution of wave heights. Values of H_s and T_s must be determined from wave forecasting curves or other techniques, whereas other spectral formats such as the Pierson-Moskowitz (45) and the JONSWAP spectrum (46)

GENERAL DESIGN CONSIDERATIONS 69

have wave forecasting parameters such as wind speed and fetch length built into them. The TMA spectrum (47) is especially useful for shallow-water near-shore applications. Michel (48) has presented a simplified review of sea spectra and their applications. Figure 3.10 shows a family of Bretschneider spectra for a given site under specified wind conditions. The site is somewhat representative of local wind wave conditions found inside the larger protected harbors of the United States.

Actual measured spectra may be compiled from record data of selected events or time intervals using spectral analysis techniques such as the fast Fourier transform or autocorrelation method. The measured spectra can be compared with any of the standard spectral formats to see which gives the best representation of conditions at the site. It is not unusual for measured spectra to exhibit more than one spike, especially for distant ocean swell (of very narrow bandwidth) superposed upon local wind-generated seas. Studies of actual measured

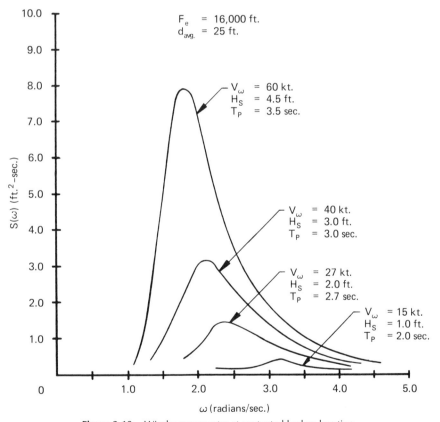

Figure 3.10. Wind wave spectra at protected harbor location.

spectra at a site taken over sufficient periods of time may reveal high levels of energy at unexpected frequencies that could result in harbor or vessel resonant motions.

The wave spectra thus far discussed are essentially unidirectional, describing only the distribution of heights and periods and nothing about the direction of travel. Although three-dimensional directional spectra (49) have been developed and applied in certain instances, such as hydraulic model tests, their current use in harbor design is limited. The directionality of component waves in a given sea state can be accounted for by the application of a directional spreading function. Textbook treatment of random sea processes can be found in references (50) and (51).

Wave grouping is another phenomenon that is not described by spectra. There is a distinct tendency for higher waves to occur in groups with some degree of periodicity, which may result in a periodic setup of the water level inside the surf zone, known as surf-beats, and a set-down of the water level immediately offshore. The period of wave groups is on the order of five to ten times that of the regular wave period, which puts it within the usual range of harbor surging and moored vessel resonance.

As ocean waves propagate into increasingly shallow water, their properties are modified by the effects of refraction, shoaling, and random breaking. Refraction refers to the bending and consequent change in direction of wave crests as they begin to "feel bottom" and align themselves with the bottom contours. As waves progress shoreward, the wave motion is retarded by bottom friction in shallow water. Wave heights may be initially reduced but then increased, and their lengths decreased by the effects of shoaling. Random breaking of the higher and steeper waves offshore modifies the spectrum of waves reaching the coastline.

The change in height of a wave due to refraction and shoaling effects is commonly expressed in terms of the ratio of the modified wave height (H) to the unaltered offshore wave height (H_0), as refraction and shoaling coefficients (K_r) and (K_s), respectively. The shoaling and refraction coefficients are a function of the water depth and deep water wavelength ratio (d/L_0), and the wave height at any given location must be found by applying K_r and K_s over a series of incremental distances. In general, although wave heights, lengths, and directions may be altered, their periods remain essentially unchanged.

Waves traveling around a fixed object will be altered in height and direction by diffraction effects, resulting in waves of diminished height spreading into the area behind the object. Diffraction of large waves around a breakwater can result in notable currents being established along the lee of the breakwater. The diffracted wave height is given by the diffraction coefficient ($K_d = H/H_0$). A useful rule of thumb for waves propagating around the end of a narrow but long breakwater is that $K_d \cong 0.5$ along an imaginary line drawn from the end of the

breakwater in the direction of travel of the incident wave train. Wave heights inside the lee of the breakwater will be rapidly diminished, whereas heights actually will be increased approximately 10% in a narrow zone just off the tip of the breakwater outside of the imaginary line.

Waves impinging on a fixed object such as the natural shoreline or a structure will be partially reflected and partially dissipated, depending upon the nature of the reflecting surface. Reflection coefficients (K_{rf}) range from nearly 1.0 for smooth vertical walls extending above the water surface to around 0.5 for riprap stone slopes of 1 vertical to 2 to 3 horizontal, to approximately 0.1 for natural beaches with relatively flat slopes. When the angle (α) between a line normal to the structure face and the wave crest is greater than 45°, the angle of reflected wave travel will equal the angle of wave incidence. When α is between 45° and 20°, there will be a Mach-type reflection if the angle between the structure normal and the wave crest is greater than α; and when α is less than 20°, there will be no reflection. Reflected waves will interact with other incoming waves resulting in waves of greater height, which can be determined from:

$$H = \sqrt{H_i^2 + H_j^2 + \cdots H_n^2} \qquad (3\text{-}9)$$

where H is the resultant height and H_i, H_j through H_n are the heights of the incident waves from various sources. When waves are reflected normal to a vertical wall, they form a standing wave pattern in front of the wall, with a height equal to twice the incident wave height. Waves may pass over submerged barriers or over top breakwaters, resulting in transmitted waves of reduced height as given by a transmitted wave coefficient (K_t). Figure 3.11 schematically illustrates the transformation of waves approaching and within a semiprotected artificial harbor.

Wave properties also are modified by currents, being generally steepened by opposing currents and flattened by following currents. As a rule of thumb, waves cannot propagate against a current with a speed in knots greater than three-fourths the wave period in seconds. Modest changes in tidal levels also may result in dramatic changes in wave conditions at a given site. Clearly, the wave conditions possible within a harbor under a range of environmental conditions are very complex.

Wave height criteria must be compared to limiting vessel motion criteria, which vary with vessel size, type, and orientation to the seas. Limits on vessel motions also may be related to limits imposed by safety considerations such as risk of damage to vessel and berth and personnel injury, and to limits imposed by operational considerations such as the ability to efficiently transfer cargo. Allowable vessel motions also may be stipulated by local experience through regulation or local practice. Vessel response is very complex, depending upon the period of the motion as well as its amplitude, as discussed in Section 6.6.

72 DESIGN OF MARINE FACILITIES

Figure 3.11. Transformation of wave properties from offshore to harbor location.

Allowable wave heights in berthing areas may range from one foot for small craft to four or five feet for larger vessels. At offshore moorings where vessels are always oriented head to sea, cargo transfer operations may continue with waves up to six to eight feet high.

Goda (51) notes that any harbor where H_s is less than approximately three feet, even under design storm conditions, can be considered a calm harbor, and offers the following recommendations for harbor tranquillity:

1. The harbor should have a broad interior.
2. Any portion of the harbor that can be viewed through the entrance should have a natural beach or wave-absorbing structure.
3. Small craft should not be viewable from the open ocean at any angle.
4. A portion of the waterfront should be wave-dissipating in character.
5. Wave reflection from the back side of vertical breakwaters should be avoided, and the use of energy-absorbing quay walls such as perforated or slit caissons should be considered.

Problems of wave agitation in harbors have been studied by several investigators, notably LeMéhauté (52, 53). In addition to locally wind-generated

waves, waves caused by vessel wakes, long-period waves from ocean swell, and harbor resonant oscillations all may contribute to the confused water motions observed in most harbors. Fully enclosed and semienclosed harbor basins, even large bodies of water such as Lake Erie, are subject to resonant oscillations—known variously as seiche, surge, or ranging action—as a result of the water surface being set in motion by some disturbing force, such as wind stress, moving atmospheric pressure systems, incident wave trains at the harbor mouth, and even the movement of large vessels, and continuing to slosh back and forth at its natural period. The sloshing motion may take place about a single nodal point (the point where there is no vertical motion) or have several nodal points, resulting in a very complicated sloshing pattern. Figure 3.12 illustrates typical harbor modes of oscillation.

For an essentially closed rectangular basin of constant depth, the natural period of the first mode (single node) is given by:

$$T_n = \frac{2L}{\sqrt{gd}} \qquad (3\text{-}10)$$

where L is the basin length in the direction of the motion. For a similar open-ended basin with the nodal point at the opening, the natural period is twice that given by equation (3-10). Most harbors and bays that exhibit such motions have periods in the range of a few minutes to one hour or more. Although vertical motion is nil at a nodal point, the horizontal water movement is greatest, and even for low-amplitude seiches the horizontal displacements of vessels located near nodal points can be very large. Such action has been responsible for the breaking of mooring lines and other damage in larger vessels.

In summary, the wave climate within a harbor is a result of the interactions of waves from the following sources, as transformed by the effects of refraction, shoaling, diffraction, reflection, and so on:

1. Locally generated wind waves, limited by harbor fetch lengths and water depths.
2. Vessel wakes, limited by vessel size and speed restrictions.
3. Penetration of ocean swell and longer-period waves through the harbor entrance, limited by the offshore wave climate and coastal and harbor bathymetry.
4. Seiches and harbor resonance phenomena, governed by harbor geometry, bathymetry, and meteorological/oceanographical disturbing forces.

The general problem of defining the wave climate for a harbor has been addressed by Seymour (54).

Marine facilities at certain locations—the southern Alaska coast and the Ha-

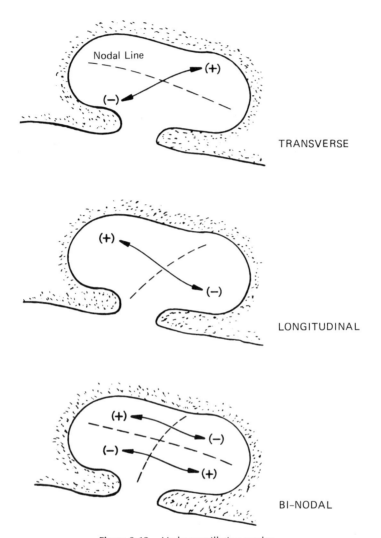

Figure 3.12. Harbor oscillation modes.

waiian Islands, for example—may be subject to inundation by seismically generated sea waves, known as tsunamis. These waves are of very long length and travel at very high speeds at sea. Their heights at sea are virtually undetectable, but when such a wave strikes a coastline, the run-up of water can attain great elevations, and the uprush or velocity forces from the bore that is formed can be devastating; so the potential for tsunami damage at a given site must be addressed (55, 56).

Tides, Storm Surge, and Water-Level Variations

Tide is the periodic rise and fall of water levels along the coast in response to the gravitational attraction of the moon and sun. The range of tide is a very important factor in the planning and design of marine facilities. Tide ranges around the world vary from one foot or less at offshore islands to 40 feet or more inside certain constricted bays or basins. Figure 3.13 shows typical tide curves for several harbors, selected to illustrate a variety of tide range and type.

Semidiurnal tides manifest themselves in twice-daily highs and lows of approximately equal height; a diurnal tide has one well-defined high and low per day, usually with a much smaller high and low in between; whereas a mixed tide falls somewhere in between these extremes. Approximately twice a month higher-than-average high water combined with lower-than-average low water (i.e., a larger range) will occur, a phenomenon known as spring tides, due primarily to the alignment of the sun and moon and consequent enhancement of tide-generating potential. Conversely, twice a month smaller ranges of tide, called neap tides, alternate with periods of spring tides. Springs and neaps are usually on the order of 10% to 20% greater or smaller, respectively, in range than the mean tide range. The sun and moon go through a characteristic pattern of positions relative to the earth over a period of approximately 18.6 years, known as the metonic cycle. Therefore, tide data must be averaged over such a length of time, known as a tidal epoch, in order to be accurate. Daily tidal predictions for most world ports are readily available (57).

The establishment of a fixed reference plane is of obvious importance in port engineering. Nautical charts usually refer to mean low water (MLW), which is the average of all low waters over a tidal epoch for a semidiurnal tide, as on the U.S. East Coast, and mean lower low water (MLLW), which is the average of the lowest lows in areas subject to diurnal or mixed tides, such as the U.S. Gulf and West Coasts. Admiralty charts often refer to the lowest astronomical tide (LAT) for soundings, which is the lowest predicted tide over a long period of time such as the metonic period. The highest astronomical tide is designated the HAT. In determining design water levels, the average lowest and highest spring tide levels averaged over a long time may be cited, designated the MLWS and the MHWS, respectively. Other tidal reference planes also are in use, and a thorough discussion of tidal datums can be found in references (58) and (59).

Primarily because of long-term changes in the overall sea level, the mean tide level at a given location will vary over time. In the United States the National Geodedic Vertical Datum (NGVD) was established as a fixed plane of reference in 1929, and corresponded to mean sea level (MSL) at that time. The NGVD of 1929 should not be confused with the North American Datum of 1927, which is cited on nautical charts and refers to horizontal controls. Because the NGVD is a fixed plane of reference relative to the land, some engi-

76 DESIGN OF MARINE FACILITIES

Figure 3.13. Typical tide curves for selected ports. (Source: NOAA, NOS (57).)

neers prefer to cite it on their drawings. The MLW or the MLLW is still used by most port engineers, and preferably the current relation between the MLW and the NGVD is given on drawings.

Measurements taken worldwide indicate an overall rise in sea level relative to the land, on the order of four to five inches per century over the past 100 years, and there are indications that the rate of rise is increasing. A recent

National Academy of Sciences study (60) indicated that a minimum sea level rise of 0.4 foot and a maximum of 0.7 foot above 1986 levels are likely by the year 2020. After the year 2000 the rate of rise is likely to increase dramatically because of global warming effects, with a maximum of nearly a two-foot increase possible by 2050. Port and harbor engineers should keep themselves informed of current knowledge of the sea level rise and consider its long-term effects on their designs.

Extreme water levels, denoted as extreme high water (EHW) and extreme low water (ELW), usually are not associated with extreme astronomical tides alone, but rather with a combination of large astronomical tides and storm surge effects, especially for coastal areas bordered by a wide continental shelf. Storm surges are associated with periods of intense meteorological activity (low atmospheric pressure), which cause a pile-up of water against the coast due to an inverted barometer effect, wind stress, deflection of wind-generated currents by the Coriolis force, and sometimes resonant effects when the movement of the low pressure center coincides with the shallow-water wave speed locally. Also, along open coastlines (e.g., beaches) there is an additional setup due to wave action. A detailed explanation of all these factors is beyond the scope of this text (see references 61–63); suffice it to say that the increase in water level due to storm surge alone can be on the order of several feet, as occurs on the U.S. Atlantic (64) and Gulf Coasts. When this surge coincides with times of high astronomical tide (65), exceptionally high water levels result, allowing larger waves to penetrate farther inland than they normally would. In most civilized areas, historical records of extreme high water are available, and they often are presented on a return-period, frequency-of-occurrence basis, as for extreme winds and wave conditions. In remote areas, the potential for such action must be assessed.

Figure 3.14 illustrates the relationship between various tidal datums, storm surge components, and the design water level (DWL) for extreme events. In general, extreme water levels may result in overtopping and flooding, increased hydrostatic pressures and buoyancy effects, increased soil pressure and drainage problems, and an increase in the level of action of berthing, mooring, and wave loads. The entire range of water levels must be considered in determining deck elevations and fender system requirements and in the calculation of soil pressures as well as environmental and mooring loads.

Currents

Current refers to the horizontal movement of water, which is generally associated with the vertical rise and fall of the tide (i.e., tidal currents) or with water level differences such as river or hydraulic currents. Currents impose velocity (drag) forces on fixed structures and moored vessels, and they affect navigation, transport ice and debris, cause scouring or deposition of bottom material, and

Figure 3.14. Design water levels.

affect corrosion rates. Like tides, tidal current velocities vary periodically, being similar to the sinusoidal-like tide curves shown in Figure 3.13. Currents associated with rising tides are called flood currents, and those associated with falling tides are ebb currents. Offshore tidal currents are rotary in nature, changing direction through all compass directions throughout the tide cycle. Near the shore and inside bays and estuaries, tidal currents are more typically reversing in nature, flowing in opposite but relatively fixed directions with the ebb and flood. The direction in which a current flows is referred to as the set, and its average speed as the drift. The time of maximum reversing-type currents is generally near midtide; but currents may lag or precede water levels, so that water movement may be significant even when the tide itself is changing. Like tides, daily tidal current predictions are published for many ports (66), but unfortunately current velocities usually are reported only in ship channels or at selected reference stations; so in areas where strong currents are noted, site-specific measurements must be taken.

Current velocities usually exhibit a vertical profile with the highest velocities near the surface, although this is not necessarily the case. In some stratified estuaries where fresh water and seawater mix and in river mouths, a current shear may exist, where the current at the bottom may be in the opposite direction to that at the surface, depending upon the tide stage. Wind stress currents, usually noticed only under extreme wind conditions, are primarily surface currents, and for design purposes may be approximated as varying linearly from a maximum of 1% to 3% of the sustained wind speed at the surface to zero at the sea bottom. Wind stress currents may exceed one knot under certain circumstances and must be added vectorially to the prevailing tidal currents. Tidal current velocities increase approximately in proportion to the increase in water level associated with extreme tides and storm surges. The effect of peak river discharge rates due to storm runoff or spring thaws must be considered at estuaries and river sites, where ebb currents typically are stronger than flood currents. Figure 3.15 illustrates some typical vertical velocity profiles.

It is generally considered that a current of 0.5 foot per second (fps) or more, as measured at one foot above a flat bottom, will erode fine to medium sands. The threshold velocity is greater for silts and gravels. Additional information on sea floor scour can be found in reference (67) and on the general effects of currents on structures in reference (31). Current loads and effects on structures are discussed in Sections 4.5 and 6.4.

Ice

Ice can be a significant factor in the design of facilities located in temperate climates, and generally governs the design of arctic structures. The most important parameters determining ice effects on structures are the ice's thickness,

80 DESIGN OF MARINE FACILITIES

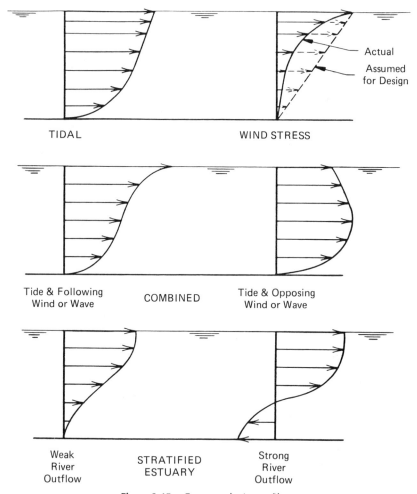

Figure 3.15. Current velocity profiles.

concentration, persistence, and variable mechanical properties. Ice may exert horizontal thrust, impact, and vertical uplift forces on structures, and it contributes to abrasion and freeze–thaw damage to structures. The U.S. Coast Guard is responsible for monitoring ice conditions in navigable waterways.

Information on ice and its distribution can be found in the numerous publications of the USACE Cold Regions Research and Engineering Laboratory (CRREL) and the International Association for Hydraulic Research (IAHR). Arctic ice engineering is beyond the scope of this text; interested readers are referred to the proceedings of the International Conferences on Port and Ocean Engineering under Arctic Conditions (POAC). See also Appendix 2, and references (68) through (73), which cover arctic and ice engineering problems in

general. Site-specific data on desired ice properties often are difficult to find. Because the mechanical properties of ice are highly variable, and ice is an anisotropic material, many assumptions often must be made in estimating ice forces. The design of structures to resist ice forces is discussed in Section 4.5.

Seismicity

Marine structures typically are designed to accommodate relatively high lateral loads, so earthquake loads often are not of major significance except in severe earthquake hazard zones or for more sensitive structures such as offshore platforms. Immersion in water provides some increased damping over land structures. Lateral stability under seismic loads should be verified in any case, and particular attention should be given to the stability of cranes, transit sheds, or other sensitive structures that may be mounted on the pier. Seismic design criteria usually are given in local building codes. At remote sites, specialist advice must be sought as to the level of the earthquake hazard and maximum probable ground motions and accelerations. Section 4.5 provides additional discussion of seismic design for marine structures.

Water Quality and Climatic Conditions

Water properties such as temperature, salinity, pH, dissolved oxygen (O_2) content, turbidity, the presence of pollutants, and seasonal variations of all of the above may have a profound influence on the corrosion and biofouling history of a structure. Salinity, which is basically the dissolved salt content, customarily is given in parts per thousand (ppt), designated ‰. In the open ocean, salinities range from 32‰ to 38‰, the mean being around 35‰. In coastal waters the range usually is somewhat less, say 28‰ to 32‰. For a concise overview of the importance of water quality factors, the reader should see reference (31). Instructions for obtaining oceanographic data can be found in reference (74).

Seasonal temperatures, precipitation, and visibility all can have important impacts on structure design. Temperature extremes affect ice formation, corrosion rates, and the thermal expansion/contraction of structures. An allowance must be made for runoff and drainage of rainfall. For facilities located along rivers, heavy rains can bring increased water levels and currents. Visibility is an important factor in berthing vessels and should be weighed in the selection of design berthing velocities.

Fouling and Biodeterioration

Biological environmental factors that must be considered include the growth and accumulation of certain sessile organisms that affix themselves to a structure

and directly attack materials such as wood, for use as a food source or substrate. Within the immersed and tidal zones, organisms that are generally referred to as marine borers may actively destroy unprotected timber; whereas above the tidal zone, wood may be attacked by fungi and other forms of decay, as discussed in Section 11.2.

Various marine bacteria can cause corrosive conditions, such as hydrogen sulfide production in bottom muds under anaerobic conditions. Although fouling generally does not lead to the deterioration of structural materials, it can add considerably to the effective mass and projected areas of structural elements, thus increasing wave and current loads, and it also greatly hampers inspection. Fouling generally can be classified as soft fouling, which includes algae (grasses) and soft-bodied organisms such as tunicates, hydroids, sponges, and anemones, and hard fouling, consisting of barnacles, mussels, and other hard-shelled animals. Soft fouling has a specific gravity of around 1.0 whereas hard fouling, which can grow to a foot or more in thickness, usually has a specific gravity of 1.3 to 1.4. Fouling growth typically diminishes rapidly with depth below the surface and often follows a temporal (seasonal) pattern. General treatment of fouling and its prevention can be found in references (75) through (77), and biodeterioration is discussed in reference (78).

3.5 MATERIALS SELECTION

Proper selection and application of materials with respect to environmental conditions are of great importance in waterfront/ocean construction. The primary materials used are common to all civil engineering construction: timber, steel, concrete, and stone. Other materials of interest to port engineers include: rubber and elastomers employed as resilient fendering; synthetic materials used for pile protection and filter fabrics; metals used for special hardware and fastenings, such as irons, steel alloys, stainless steels, and aluminum; and occasionally copper–nickel alloys and other exotic metals for special purposes. Table 3.6 summarizes applications of various materials commonly used in harbor construction.

Important considerations in materials selection include:

- Structural properties, such as density, strength, ductility, fatigue, and impact resistance, and their changes under temperature extremes.
- Durability: the natural resistance to the marine environment and other deteriorating agents, as well as protective measures needed and long-term maintenance requirements.
- Compatibility: physical and chemical interaction with other materials and the ability to be integrated structurally with other materials.
- Cost and availability, including transportation and handling and long-term

GENERAL DESIGN CONSIDERATIONS 83

Table 3.6. Usual applications of materials in harbor construction.

Steel: Piles, sheet pile (bulkheads), miscellaneous hardware and fittings, fastenings, grating, pipe, crane rails (use in decking and superstructure generally is limited to deep water terminal platform structures).
Concrete: Piles, sheet piles (bulkheads), deck systems, walls, caissons.
Timber: Piles, sheet piles (bulkheads), decking and deck framing, fendering, mats, cribbing.
Stone: Breakwaters, slope protection (use in wall construction generally is limited to existing older structures).
Miscellaneous metals:
　Aluminum: Gangways, ramps, boarding platforms.
　Cast iron: Mooring hardware and miscellaneous castings.
　Stainless steel: Fastenings (bolts and pins).
Elastomers and plastics: Fendering, fender facing, bearing pads, wraps and cladding, geotextiles (filter fabrics).
Bituminous materials: Paving, revetments.
Masonry and block work materials: Paving.
Specialty items: Chain, wire rope, synthetic fiber lines, protective coatings, grouts.

maintenance costs and the availability of required member sizes, desired shapes, and so on.

An in-depth treatment of marine materials for construction of harbor and coastal structures can be found in references (77) and (79) through (81). The following discussion is intended to highlight some important aspects of the selection and specification of the most common structural materials.

Specifications for structural steel should consider strength, ductility, fatigue and fracture requirements, weldability and connection methods, and corrosion protection measures. Table 3.7 summarizes various grades of structural steel used in marine construction in accordance with the standard specifications of the American Society for Testing Materials (ASTM). As with land-based structures, the most common grade of steel is mild carbon steel such as ASTM, A-36. Although many of the higher-strength grades such as A-242, A-441, and A-588 exhibit varying degrees of improved atmospheric corrosion resistance, their added cost, their availability, and their lower ductility—plus the fact that most designers prefer to stay with certain minimum metal thicknesses (i.e., higher strength is not required)—contribute to their less frequent use, compared to the commonly used steel. ASTM, A-690 is available in H-pile and sheet pile sections and is alloyed with 0.52% copper and 0.54% nickel, which give it two to three times greater resistance to corrosion within the splash zone as compared with ordinary structural steels.

All structural steels and steel hardware used in the marine environment must be protected against corrosion by protective coatings, claddings, galvanizing, or cathodic protection measures, as described in Chapter 11. Even where proper

Table 3.7. Selected structural steels for marine structures.

General classification	ASTM description	Description	F_y^1 (ksi)	$F_T^{1,2}$ (ksi)	Chemical composition (%)³	Remarks
Ordinary strength (mild steel)	A-36 A-131 (ABS)	Structural steel. Structural steel for ships.	36 34*	58–80 58–71	.20 to .29 C max. .20 Cu when specified .16 to .23 C max. Al, Co, V, Ni, Mo, Cu for higher strength and toughness	General purpose, standard structural (carbon) steel, good for formability, weldability. *Available to 51 ksi, F_y, many grades for strength and toughness, good formability and weldability.
High strength–low alloy (HSLA)	A-242	High strength–low alloy structural steel.	50	70	.12 C max. Cr, Ni, Cu, Ti	Atmospheric corrosion resistance 4 to 6 times carbon steel w/o Cu, good coating adherence, good weldability.
	A-588	HSLA structural steel with 50,000 psi min. yield pt. up to 4" thick.	50	70	.20 C max. Cr, Ni, Cu, V	Similar to A-242
	A-441	HSLA structural manganese–vanadium steel.	42–50	63–70	.22 C max. Cu, V, Ni, Cu when specified	Atmospheric corrosion resistance 2 times carbon steel w/o Cu, good weldability.
	A-572	HSLA columbium–vanadium steels of structural quality.	42–65*	65–80*	.21 to .26 C max. Co, V, Ni, Cu when specified	High strength, good formability. *Higher strength grades not recommended for welding.
	A-633	Normalized HSLA structural steel.	42–60	63–100	.21 C max., Cr, Mo, Ni, Cu V, Ni, Cb for high strength/toughness	High strength with good impact resistance and toughness, low temperature service.

High yield strength (HYS)	A-514	High yield strength quenched and tempered alloy steel plate suitable for welding.	90–100	110–130	.12 to .20 C max., Cr, Mo, Cu, Ti, B	High strength, superior corrosion resistance. Available in plates only. Heat treated. Low H_2 welding techniques required.
Pipe and tubes	A-53 (Gr. B)	Welded and seamless steel pipe.	35	60	Similar to A-36	—
	A-500	Cold formed, welded and seamless carbon steel, structural tubing in rounds and shapes.	33–46	45–62	Similar to A-36	—
	A-501	Hot formed, welded and seamless structural tubing.	36	58	Similar to A-36	—
Piles and sheet piles	A-252	Welded and seamless pipe piles.	30–45	50–66	Similar to A-36	—
	A-328	Steel sheet piling.	38.5	70	Similar to A-36	
	A-690	HSLA steel H-Piles and sheet piling for use in marine environments.	50	70	.22 C max., Cu, Ni	Superior corrosion resistance within splash zone (A-690).
Special Applications	A-709	Structural steel for bridges.	36–100	58–130	Various grades equal or exceed: A-36, A-572, A-588, A-514	Unified specification with good corrosion resistance; can specify: impact, toughness, and lowest service temperature.

Notes:
1. Strengths, mechanical properties, and chemistry vary with specified grade and material thickness.
2. All steels above, except HYS, have minimum elongations of > 20% in 2 inches.
3. All steels above have varying amounts of: C, Mn, P, S, and Si; other significant alloying elements are listed. Maximum carbon (C) content (%) is also given.

corrosion protection is provided, a nominal corrosion allowance, usually on the order of $\frac{1}{16}$ to $\frac{1}{8}$ inch of extra metal thickness, is provided consistent with anticipated corrosion rates, coating performance, and structure life. Aluminum alloys for gangway and boarding structures usually are selected on the basis of a suitable balance of strength, weldability, and corrosion resistance properties.

Concrete is an excellent material for marine construction when properly made and applied. Concrete specifications should consider strength and permeability requirements and mix proportions, and the use of admixtures as they affect durability requirements such as resistance to sulfate, freeze–thaw attack, and so on. Quality control in mixing and placing concrete is essential to achieving the desired properties. Reinforcing steel properties, placement, and details also are critical. Concrete mixes are adapted to special applications such as cast-in-place structural concrete, precast and precast/prestressed, massive (cyclopean) concrete, tremie concrete, pumped and prepacked concrete, and gunite or shotcrete applications. Table 3.8 lists representative mix requirements for some typical applications. Marine concrete and its applications have been covered in articles by Gerwick in references (77) and (82) and in the numerous technical papers collected in references (83) and (84). Design guidance for mix proportions, cover requirements and details, and so on, can be found in references (85) and (86).

The single most important factor contributing to the durability of marine concrete is impermeability. Concrete is most susceptible to freeze–thaw damage, especially in temperate to cold climates with large tide ranges, and to abrasion damage; so a dense, durable, high-quality mix is desired. Structural concrete preferably contains Portland Type II cement for is sulfate-resisting qualities, or Type V cement where high sulfate resistance is required. Type I and Type III, which often is preferred for precast work, also may be used, provided that the tricalcium aluminate (C_3A) content is within the range of 4% to 10%, and other mix proportions ensure a durable concrete with adequate protection against corrosion of the reinforcing steel. Mix proportions must be adjusted to achieve the specific design strength and to meet permeability requirements. In general, the minimum 28-day compressive strength (f'_c) should be at least 4000 psi for cast-in-place (CIP) structural elements and 5000 to 6000 or more psi for prestressed structural elements. Lower strengths may be specified for massive concrete and pile fill applications. Higher strengths should be specified where severe scouring action or tide zone exposure exist.

In order to achieve the above strength and durability requirements for normal structural concrete, the following mix proportions generally will apply: the cement content will be within the range of 600 to 700 pounds per cubic yard, the maximum water/cement ratio (W/C) will be from 0.4 to 0.45, the maximum aggregate size will be around 0.75 inch, and air-entraining admixtures should

GENERAL DESIGN CONSIDERATIONS 87

Table 3.8. Typical marine concrete applications and mixes.

Application	Exposure zone*	Minimum compressive strength @ 28 days (f_c) (psi)	Minimum cement content[+] (lb./cy.)	Maximum W/C ratio	Maximum aggregate size (inches)	Admixtures/ Remarks
Cast-in-place structural elements	Atmospheric, splash, and tidal	4000 to 5000	611-658	0.45-0.40	1-1 1/2	Air entraining
Precast/pre-stressed structural elements	All zones	5000 to 7000	705	0.40	3/4	Air entraining Water reducing
Massive structures	All zones	4000	611-658	0.45	1 1/2-3	Air entraining Plasticizing Pozzolans
Tremie	Submerged	4000 to 6000	705	0.50-0.45	3/4	Air entraining Plasticizing
Shotcrete	Splash and tidal	5000 to 6000	705	0.55-0.50	N.A.	Set Accelerator

*In general, the more severe the exposure conditions, the greater the impermeability that is required—hence mix proportions that favor higher strength and density.
[+]Cement may be of Portland Type I, II, or III, provided that C_3A content is within 4% to 10%. Type II is generally preferred for its sulfate resistance.

be used to obtain air contents of from 4% to 8% by volume, depending upon aggregate size and exposure. Air entrainment is essential in freeze–thaw climates. Aggregates should be sound and hard, and other mix proportions should be as required by normal civil engineering practice to obtain the needed strengths, and per reference (87) to meet durability requirements.

In addition to strength and durability requirements, concrete mixes may need to be proportioned to meet certain workability and consistency requirements affecting concrete placement, and density requirements for special lightweight or heavyweight applications. Concrete properties can be greatly modified by the use of various admixtures. Polymer modified concretes (PMC) are used in applications where high strength and high bond are required and the mix is to be placed in relatively thin layers. Such applications include toppings, overlays, and various types of patching and repairs. Further discussion of concrete durability, protection, and repair applications is provided in Chapter 11. The reader is referred to the literature cited for additional discussion of special-purpose concrete mixes.

Corrosion of the reinforcing steel is a common cause of the spalling and degradation of concrete. An adequate concrete cover, typically from 2 to 3 inches, over the reinforcement, combined with a dense durable mix that maintains a high alkalinity at the steel surface, normally suffices for most marine exposures. Where deck surfaces are exposed to road salts and/or to the storage of salts or other corrosive bulk materials, fusion bonded epoxy coated reinforcing steel may be specified. There are still questions regarding the bond strength of concrete with the epoxy coated steel, and special consideration should be given to bond when epoxy coated steel is specified.

Timber is the traditional material of waterfront construction. It is durable, is convenient to work with, and possesses good impact resistance and the ability to distribute loads effectively. Its chief enemies are rot and attack by marine organisms, and to a lesser degree abrasion and wear such as that caused by moving ice. Proper design and installation of fastenings are critical to the construction of sound timber structures. Table 3.9 summarizes the properties of selected timber frequently used in waterfront construction. Southern yellow pine and Douglas fir are the woods most commonly specified for marine construction in the United States. Tropical hardwoods such as greenheart and ekki (azobe) are desirable because of their natural resistance to borer attack and high strength. Further discussion of timber applications and biodeterioration is presented in Chapter 11.

Timber is specified to meet minimum strength, density, and grading requirements. Moisture content and impact resistance are important considerations. Lumber is graded for quality according to such properties as maximum number of knots, slope of grain, and other characteristics that may affect its strength, in accordance with grading rules established by various agencies depending

GENERAL DESIGN CONSIDERATIONS 89

Table 3.9. Properties of timber commonly used in marine construction*

Timber type	Moisture content (air-dry) (%)	Elastic modulus in bending (psi)	Proportional limit in compression parallel to grain (psi)	Proportional limit in compression perpendicular to grain (psi)	Shearing strength parallel to grain (psi)	Unit weight[‡] (pcf)
Douglas fir (coast type)[+]	12	1,950,000	5,850	870	1,160	34±
Southern yellow pine (long leaf)[+]	12	1,990,000	6,150	1,190	1,500	36±
Greenheart (*Ocotea radiaei*)	14±	3,700,000	10,140	2,090	1,830	66±
Azobe (Ekki) (*Lophira procera*)	12.2	3,000,000	10,260	2,870	2,650	65.5

*Data compiled from U.S. Forest Products Laboratory data and from Tropical Woods Importers data. Timber properties vary greatly with moisture content, and in general strength properties will be reduced and unit weights increased for higher moisture contents of undried woods.
[+]Pressure treatment may reduce strength properties by as much as 30% to 50% of values given. Refer to NAVFAC, DM-25.6 (91).
[‡]Pressure treatment increases unit weight from approximately 2% to 25%, depending upon treatment type and retention.

upon the species. The Southern Pine Inspection Bureau (SPIB) and the West Coast Lumber Inspection Bureau (WCLIB), for example, have grading rules for southern yellow pine and Douglas fir, respectively. Allowable stresses are determined based upon the structural grade, in accordance with the *National Design Specification for Wood Construction* (88). Preservative treatment of softwoods is essential, and means of joining and fastening timbers must be specified. Steel bolts and hardware must be hot-dipped galvanized and meet minimum size requirements.

Timber treatment is either by oil-type or waterborne preservatives, which usually are applied by pressure methods. Oil-type preservatives typically are either coal-tar creosote and creosote oil solutions or preservative chemicals such as pentachlorophenol dissolved in a nonaqueous carrier. In the United States today, the use of oil-type preservative treatments in waterfront construction is very limited because of environmental concerns. Waterborne preservatives should be of a leach-resistant type such as chromated copper arsenate (CCA), of which there are three types depending upon the relative percentages of chromium and arsenic present, or ammoniacal copper arsenate (ACA). The treatment is applied under pressure, usually via the full-cell process, wherein the wood is subjected to a vacuum while the treatment chamber is filled so as to maximize retention. Preservative treatment usually is specified in accordance with the American Wood Preservers Association standards (89). Retentions, which are measured in pounds per cubic foot of preservative retained, are described further in references (89) and (90). Common retentions for marine construction are summarized in Table 3.10.

Stone such as granite block, the most durable of all waterfront materials, was widely used around many harbors worldwide until the early part of the twentieth

Table 3.10. Commonly specified timber preservative treatments for southern yellow pine and coastal Douglas fir used in marine construction.

	Preservative		
Member use	CCA	Creosote and creosote coal tar	Pentachlorophenol
Round piles			
Moderate exposure to borers	1.5–2.0	20.0	—
Severe exposure to borers	2.5	20.0	—
Immersed and tidal zones	2.5	25.0	—
Splash zone	0.6	12.0	0.60
Above splash zone	0.4	10.0	0.50

*All retentions in pounds per cubic foot.

century. Cost and availability greatly restrict its use today in massive structures, but smaller-size irregular stone still is employed as riprap slope protection. Stone is specified to meet minimum density requirements (usually on the order of 160 pounds per cubic foot), as well as for soundness under chemical attack and abrasion and for water absorption characteristics. Stone (riprap) slope protection also must meet minimum size, weight, gradation, and placement requirements.

Environmental deterioration usually exhibits both temporal and spatial distribution. The temporal distribution manifests itself in seasonal and long-term variations in marine borer activity, fouling, and corrosion rates, as well as infrequent storm or overload and impact damage and general wear and fatigue effects with time. The spatial distribution varies with structure type, orientation, and exposure, but generally on almost all structures a characteristic pattern emerges with elevation relative to water level planes. This general vertical zonation is illustrated in Figure 3.16, which also shows some typical forms of deterioration found within the defined zones.

Figure 3.16. Vertical zonation of deterioration modes.

The *splash zone* usually extends from the plane of the mean high water (MHW) to the upper limit normally attained by salt spray from spring tides and normal wave action. This zone usually is characterized by heavy corrosion of steel structures and spalling of concrete in freezing climates. Above the splash zone, in the *atmospheric zone* where salt spray normally does not penetrate, timber structures are subject to rot and general biodeterioration. Immediately below the splash zone, the *tide zone* is defined as the area between MHW and the average or lowest low tide plane, MLW or MLWS. The tide zone is characterized by the heaviest attack on concrete structures, especially freeze–thaw damage in northern climates, and by moderate to heavy corrosion of steel. Below mean low water in the constantly *submerged zone*, corrosion rates usually are moderate, and concrete usually fares well, but timber is subject to attack by marine organisms. Below the mudline in the *buried zone*, most materials are well protected from biodeterioration or physicochemical attack.

In general, the distribution and severity of corrosion and biodeterioration are affected by water properties such as temperature, salinity, dissolved oxygen content, pH, and pollutants, as well as the tide range and current velocities. Further discussion of modes of deterioration is provided in Section 11.2. General criteria for waterfront construction with particular regard to material durability are provided by NAVFAC (91).

REFERENCES

1. Borgman, L. E. (Aug. 1963). "Risk Criteria," *Proc. ASCE*, J.W.H.D., Vol. 89, No. WW-3.
2. McClelland, B. and Reifel, N. D., eds. (1986). *Planning and Design of Fixed Offshore Platforms*. Van Nostrand Reinhold, New York.
3. Borgman, L. E. (1975). "Extremal Statistics in Ocean Engineering," *Proc. ASCE*, C.E.O. III, Spec. Conf.
4. PIANC (1979). "Report of Working Group I; Methods for Analysing Wind, Wave and Swell Data . . . " (ICORELS), Annex to Bull. No. 32, Vol. I, Brussels, Belgium.
5. PIANC (1987). "Development of Modern Marine Terminals," Report of a Working Group of PTC-II, Supp. to Bull. No. 56, Brussels, Belgium.
6. Agerschou, H., Lundgren, H., and Sorensen, T. (1983). *Planning and Design of Ports and Marine Terminals*. Wiley, Chichester, U.K.
7. Bruun, P. (1981). *Port Engineering*, 3rd ed. Gulf Publishing, Houston, Tex.
8. Quinn, A. D. F. (1972). *Design and Construction of Ports and Marine Structures*. McGraw-Hill, New York.
9. Thoresen, C. A. (1988). *Port Design—Guidelines and Recommendations*. Tapir Publishers, Trondheim, Norway.
10. NAVFAC (1981). *Harbors*, Design Manual 26.1. Dept. of the Navy, Naval Facilities Engineering Command, NAVFAC DM-26.1, Alexandria, Va.
11. NAVFAC (1980). *Piers and Wharves*, Design Manual 25.1. Dept. of the Navy, Naval Facilities Engineering Command, NAVFAC DM-25.1, Alexandria, Va.

12. PIANC (1980). "Final Report of the International Commission for the Reception of Large Ships" (ICORELS), Report of Working Group IV, Supp. to Bull. No. 35, Brussels, Belgium.
13. ——— (1975). *Port Approach Design—A Survey of Ship Behaviour Studies*. National Ports Council, London.
14. Kray, C. J. (Feb. 1973). "Design of Ship Channels and Maneuvering Areas," *Proc. ASCE*, Vol. 99, No. WW-1.
15. NAVFAC (1981). *Cargo Handling Facilities*, DM-25.3. Dept. of the Navy, Alexandria, Va.
16. PIANC (1978). "Report of the International Commission for the Standardization of Roll-on/Roll-off Ships and Their Berths."
17. ISO (1983). *Roll-on/Roll-off Ship to Shore Connection*. International Standards Organization, ISO 6812-1983, Geneva.
18. ICHCA (1978). *Ship and Shore Ramp Characteristics*."
19. NAVFAC (1981). *Ferry Terminals and Small Craft Berthing Facilities*," DM-25.5. Dept. of the Navy, Alexandria, Va.
20. ——— (1976). "Source Book on Environmental and Safety Considerations for Planning and Design of LNG Marine Terminals," Task Comm. Report, WPCOE Div., New York.
21. PIANC (1973). *Big Tankers and Their Reception*. Int. Oil Tankers Commission, Brussels, Belgium.
22. OCIMF (1978). *Guidelines and Recommendations for the Safe Mooring of Large Ships at Piers and Sea Islands*. Witherby and Co., London.
23. USACE (Dec. 1976). "Report on Design of Single-Buoy Mooring Terminals," Dept. of the Army, O.C.E., Washington, D.C.
24. MarAd (1983). "Report on Survey of U.S. Shipbuilding and Repair Facilities," U.S. Dept. of Trans., Maritime Administration.
25. Acker, H. G. and Bartlett, F. G. (1980). "Ship Construction," in *Ship Design and Construction*. SNAME, New York.
26. Dunham, J. W. and Finn, A. A. (Dec. 1974). "Small Craft Harbors: Design, Construction and Operation," SR-2, U.S. Army, CERC, USACE, Fort Belvoir, Va.
27. ——— (1969). *Small Craft Harbors*, Manual and Report on Engineering Practice #50. ASCE, New York.
28. Chauncy, C. A. (1961). *Marinas—Recommendations for Design Construction and Maintenance*. NAEBM, Greenwich, Conn.
29. AAPA (1973). *Port Planning, Design and Construction*. American Association of Port Authorities, Washington, D.C.
30. PIANC (1985). "Dangerous Goods in Ports—Recommendations for Port Designers and Operators," Report of a Working Group of PTC-II, Supp. to Bull. No. 49, Brussels, Belgium.
31. Gaythwaite, J. W. (1981). *The Marine Environment and Structural Design*. Van Nostrand Reinhold, New York.
32. Delft Hydraulics Laboratory (1985). "Dynamic Behavior of Moored Ships in Harbours," Working Group on Moored Ships in Harbours of DHL, *The Dock and Harbour Authority*.
33. Jensen, O. J. and Warren, I. R. (Oct. 1986). "Modelling of Waves in Harbours: Review of Physical and Numerical Methods," *The Dock and Harbour Authority*, London.
34. Fleming, C. A. (May 1986). "Mathematical Modelling Marine Engineering Design," *World Dredging and Marine Construction*, Vol. 22, No. 5.
35. Koufitas, C. G. (1988). *Mathematical Models in Coastal Engineering*. Pentech Press, London.
36. ——— (Apr. 1974). *OCS Oil and Gas—An Environmental Assessment*, A Report to the President by the Council on Environmental Quality, Executive Office of the President, Vol. I. Washington, D.C.
37. ANSI (1982). *American National Standard Minimum Design Loads for Buildings and Other Structures*, A58.1. American National Standards Institute, New York.

38. Batts, M. E. et. al. (May 1980). *Hurricane Wind Speeds in the U.S.*, NBS Building Science Series No. 124. U.S. Dept. of Comm., Washington, D.C.
39. ASCE (1987). "Wind Loading and Wind-Induced Structural Response," State of the Art Report by Comm. on Wind Effects of the Structural Div., ASCE, New York.
40. Simiu, E. and Scanlan, R. H. (1978). *Wind Effects on Structures: An Introduction to Wind Engineering*, Wiley, New York.
41. API (1987). "Recommended Practice for Planning, Designing and Constructing Fixed Offshore Platforms," RP-2A, American Petroleum Institute, Washington, D.C.
42. Edge, B. L. et al. (1977). "Compendium of Coastal Wave Data in North America," *Proc. ASCE*, Ports '77, Spec. Conf., New York.
43. USACE (1984). *Shore Protection Manual*, Vols. I and II. Dept. of the Army, CERC, Vicksburg, Miss.
44. Bretschneider, C. L. (1959). "Wave Variability and Wave Spectra for Wind Generated Gravity Waves," U.S. Army Beach Erosion Board, TM-118.
45. Pierson, W. J. and Moskowitz, L. A. (1963). "A Proposed Spectral Form for Fully Developed Wind Seas Based on the Similarity Theory of S.A. Kitaigorodskii," U.S. Navy Oceanographic Office, T.R. Contract No. N62306-1042.
46. Hasselmann, K. et al. (1973). "Measurements of Wind Wave Growth and Swell Decay During the Joint Sea Wave Project (JONSWAP)," Report No. 12, Deuthshes Hydrographisches Institut, Hamburg.
47. Hughes, S. A. (Dec. 1984). "The TMA Shallow Water Spectrum Description and Applications," USACE, CERC, TR-CERC-84-7, Vicksburg, Miss.
48. Michel, W. H. (Jan. 1968). "Sea Spectra Simplified," *Marine Technology*, SNAME.
49. Wiegel, R. L. ed. (1981). "Directional Wave Spectra Applications," *Proc.* of Conf. at U.C.L.A., Berkeley, ASCE, New York.
50. Sarpkaya, T. and Isaacson, M. (1981). *Mechanics of Wave Forces on Offshore Structures*. Van Nostrand Reinhold, New York.
51. Goda, Y. (1985). *Random Seas and Design of Maritime Structures*. University of Tokyo Press, Tokyo.
52. LeMéhauté, B. (1962). "Theory of Wave Agitation in a Harbour," *Trans. ASCE*, Part 1, No. 127, Paper 3313.
53. LeMéhauté, B. (1977). "Wave Agitation Criteria for Harbors," *Proc. ASCE*, Ports '77, Spec. Conf., New York.
54. Seymour, R. J. (1977). "Defining the Wave Climate for a Harbor," *Proc. ASCE*, Ports '77, Spec. Conf., New York.
55. Camfield, F. E. (Feb. 1980). "Tsunami Engineering," USACE, CERC, SR-6, Fort Belvoir, Va.
56. Wiegel, R. (1964). *Oceanographical Engineering*, Prentice-Hall, Englewood Cliffs, N.J.
57. NOAA (Annual). *Tide Tables* (4 vols. cover most of the world). U.S. Dept. of Comm., NOAA, NOS.
58. Marmer, H. A. (1951). "Tidal Datum Planes," USCGS, S.P. No. 135.
59. Harris, D. L. (Feb. 1981). "Tides and Tidal Datum in the U.S.," CERC, SR-7, Ft. Belvoir, Va.
60. NRC (1987). *Responding to Changes in Sea Level—Engineering Implications*. National Research Council, National Academy Press, Washington, D.C.
61. Bodine, B. R. (May 1971). "Storm Surge on the Open Coast: Fundamentals and Simplified Prediction," USACE, CERC, TM-35.
62. Bretschneider, C. L. (1966). "Engineering Aspects of Hurricane Surge," in *Estuary and Coastline Hydrodynamics*, Ippen, ed. McGraw-Hill, New York.

63. Hansen, J. (May 1978). "Wave Set-up and Design Water Level," *Proc. ASCE*, WW-2, Vol. 104.
64. Ho, F. R. (1976). "Hurricane Tide Frequencies along the Atlantic Coast," *Proc. ASCE*, 15th Coastal Engineering Conf., Honolulu.
65. DeYoung, R. K. and Pfafflin, J. R. (1975). "Recurrence Intervals of Abnormally High Tides by Superposition of Storm Surges over Astronomical Tides," *Proc. ASCE*, C.E.O. III, Spec. Conf.
66. NOAA (Annual). *Tidal Current Tables*, (2 vols. cover Atlantic and Pacific Coasts of North America). U.S. Dept. of Commerce, NOAA, NOS.
67. Herbich, J. B., et al. (1983). *Seafloor Scour—Design Guide for Ocean Founded Structures*, Marcel Dekker, New York.
68. Chen, A. T. and Leidersdorf, C. B., eds. (1988). "Artic Coastal Processes and Slope Protection Design," State of the Art Report by The Technical Council on Cold Regions Engineering of the ASCE, New York.
69. Cammaert, A. B. and Muggeridge, D. B. (1988). *Ice Interaction with Offshore Structures*. Van Nostrand Reinhold, New York.
70. Carstens, T., ed. (June 1980). "Working Group on Ice Forces on Structures," SR-80-26, prepared by Int. Assoc. for Hyd. Res., U.S. Army, CRREL.
71. Caldwell, S. R. and Crissman, R. D., eds. (1983). "Design for Ice Forces," Tech. Council on Cold Regions Engineering, ASCE, New York.
72. Wortley, C. A. (May 1978). "Ice Engineering Guide for Design and Construction of Small Craft Harbors," University of Wisconsin, Sea Grant, AR417.
73. Eranti, E. and Lee, G. C. (1986). *Cold Region Structural Engineering*. McGraw-Hill, New York.
74. U.S. Navy (1968). *Instruction Manual for Obtaining Oceanographic Data*, H.O. #601. U.S. Navy Oceanographic Office.
75. ——— (1956). "Marine Fouling and Its Prevention," Report to U.S. Navy prepared by Woods Hole Oceanographic Inst.
76. Benson, P. H. et al. (1973). "Marine Fouling and Its Prevention," *Marine Technology*, SNAME, New York.
77. Myers, J. J., Holm, C. H., and McAllister, R. F., eds. (1969). *Handbook of Ocean and Underwater Engineering*. McGraw-Hill, New York.
78. U.S. Navy (1965). *Marine Biology Operational Handbook*, NAVDOCKS, MO-311, Washington, D.C.
79. Whiteneck, L. L. and Hockney, L. A. (1989). *Structural Materials for Harbor and Coastal Construction*. McGraw-Hill, New York.
80. USACE (Feb. 1983). "Construction Materials for Coastal Structures," CERCL SR No. 10, Ft. Belvoir, Va.
81. BSI (1984). *British Standard Code of Practice for Maritime Structures*, Part I, General Criteria. British Standards Institution, BS6349, London.
82. Fintel, M., ed. (1985). *Handbook of Concrete Engineering*. Van Nostrand Reinhold, New York.
83. ACI (1988). "Concrete in the Marine Environment," *Proc.* of 2nd Int. Conf. ACI, SP-109, Detroit, Mich.
84. ACI (1980). "Performance of Concrete in the Marine Environment," ACI, SP-65, Detroit, Mich.
85. ACI (1978). "Guide for the Design and Construction of Fixed Offshore Concrete Structures," ACI 357R, Detroit, Mich.
86. FIP (1985). *Design and Construction of Concrete Sea Structures*, Federation Internationale de la Precontrainte. Thomas Telford, London.

87. ACI (1978). "Guide to Durable Concrete," ACI 201.2R-78, Detroit, Mich.
88. NFPA (1986). *National Design Specification for Wood Construction*, incl. supp. *Design Values for Wood Construction*. National Forest Products Association, Washington, D.C.
89. AWPA (1979). *All Timber Products—Preservation Treatment by Pressure Processes*. AWPA Std. D1-74. Washington, D.C.
90. ——— *Wood Preservations: Treating Practices*, Federal Specification TT-W-571. U.S. Dept. of Agriculture, Forest Products Laboratory.
91. NAVFAC (1981). *General Criteria for Waterfront Construction*, DM-25.6 Dept. of the Navy, Alexandria, Va.

4
Operational and Environmental Loads

Piers, wharves, and dry docks typically are designed to support relatively heavy transient loads as compared with other types of civil engineering structures, and a relatively large lateral load capacity is a common feature of waterfront construction. This chapter considers the various sources of direct structural loadings imposed upon marine structures. Because of the importance and complexity of berthing and mooring loads, which usually govern the design lateral load capacity of waterfront structures, separate chapters (5 and 6) have been devoted to those loads, and soil pressures and geotechnical considerations are covered in Chapter 8.

This chapter reviews the usual range of deck and cargo live loads, as well as loads due to vehicles and mobile equipment frequently used on piers and wharves. Loads due to rail-mounted, fixed, and other specialized material handling equipment also are discussed. Port buildings and various types of elevated superstructures may be constructed on piers and wharves, and environmental forces resulting from wind, waves, currents, ice, and earthquakes may act directly on a structure and its appurtenances. Separate sections cover those topics. Finally, miscellaneous load sources that must be considered in the design of marine structures are identified.

4.1 CARGO LOADS

Determination of appropriate uniform live loads and vehicular wheel loads is very important to the economic design of waterfront structures. Many existing facilities need to be upgraded in this area because of contemporary transportation practices. In the past, general cargo facilities were designed for uniform live loads on the order of 500 to 600 psf, and for standard highway trucks and for forklift trucks of perhaps 5 to 10 tons capacity. Today, with containerized cargoes predominant, deck loads of 800 to 1200 psf, truck cranes, and forklift vehicles of up to 30 tons capacity are common. Table 4.1 summarizes representative ranges of deck and equipment loads for various types of facilities. Often uniform live loads govern foundation design, and wheel loads govern slab thicknesses. Vehicular and equipment loads are covered in subsequent sections. Live load reduction factors of 25% to 33% of the total uniformly distributed load sometimes are taken for pile foundation design (1). Impact factors do not normally apply to uniform live loads.

Table 4.1. Typical deck loads for piers and wharves.

Facility type	Uniform live load (psf)	Vehicular load class	Rail-mounted equipment	Special equipment, concentrated loads/remarks
Fishing and light commercial	250–400	H-10 to HS-20 Small F.L.T.	—	Fixed davit/jib boom hoist. Light conveyer.
General cargo	500–1000	HS-20 10–20 ton F.L.T. 70–140 ton T.C.	AREA E-70 25–50 ton Portal crane	May handle occasional containers, bulk, and palletized cargoes.
Container	800–1200	HS-20, 30 ton F.L.T. Straddle carrier Transtainer	Container crane	T.C. up to 300 tons may be used at some facilities.
Dry bulk cargo	1000–3000	HS-20 Loaders/scrapers	AREA E-80 Specialized loaders/bulk handling cranes	Wide variety of specialized equipment and conveyer systems may be used.
Liquid bulk cargo	60–100 on catwalks and access ramps	—	—	Cargo transfer usually at fixed manifold locations. Design for hose towers, product lines, and other fixed equipment requirements.
Ship repair and outfitting	600–1000	HS-20 70–140 ton T.C.	Portal crane full revolving 25–300 ton	50 ton concentrated load. Extensive utility systems.

Note: F.L.T. = forklift truck; T.C. = truck crane.

Table 4.2. Typical cargo weights vs. stacking heights.*

Cargo/Packaging	Packaged density (pcf)	Stack height (ft.)	Live load (psf)
General cargo, average/on pallets	32	8 to 16	256 to 512
Timber/bundled:			
Softwoods	to 40	10 to 20	400 to 800
Hardwoods	to 72	10 to 20	720 to 1440
Paper/baled	50	to 10	500
Dry goods: cotton, wool/baled	20 to 50	to 10	200 to 500
Produce: fruits, vegetables, grains/bags, cartons, cases	31 to 58	to 10	310 to 580
Fertilizers/bags	52 to 59	to 6	312 to 354
Cement and Lime/bags, barrels	50 to 73	to 6	300 to 438
Oils and paints/barrels/boxes	35 to 70	to 6	210 to 420
Rope (Manila)/coils	42	to 6	252
Metal products/loose bars and coils	75 to 225	to 10	750 to 2250
Automobiles	7.5 to 15	—	15

*Representative weights and stack heights are given. Specific applications should be investigated independently where storage loads are critical. Values given were averaged from various sources.

Table 4.2 presents typical cargo weights for various materials and stacking heights. General cargo occupies, on the average, around 70 c.f. per long ton (l.t.), corresponding to a mean density of 32 pcf, and normally ranges from around 20 c.f./l.t. for tightly packed/high density items to 140 c.f./l.t. for loose/low density items. (Seawater occupies 35 c.f./l.t.). Uniform live load criteria and design loads for miscellaneous materials can be found in references (2) through (5). Quinn (1) suggests designing the floors of general cargo storage areas for around 500 psf for most purposes, with facilities for specialized goods designed for more or less—such as wool and cotton for 300 to 400 psf and metal products for 600 to 800 psf. General cargo can be shipped in boxes, crates, bales, barrels, and so on, or with no packaging at all.

One common way to bundle general cargo is to use pallets. A pallet is essentially a wooden or metal base frame upon which the cargo sits, and it usually is secured with metal banding. The pallet facilitates lifting and transport by forklift vehicles, and usually has provisions for being lifted by slings. Pallets generally are loaded to a maximum height of around 8 to 10 feet, but may be stacked up to approximately 16 feet under certain circumstances. Pallets can be

of any size. Some typical standard sizes per the U.S. Standards Institute (USSI) are as follows:

Rectangular			Square
24 × 32	32 × 48	48 × 60	36 × 36
32 × 40	36 × 48	48 × 72	42 × 42
36 × 40	40 × 48	88 × 108	48 × 48

Dimensions are in inches. Palletized cargoes normally require deck load capacities around 400 to 600 psf for stowage of unstacked pallets.

General cargo commonly is shipped in containers, which usually are handled by special shoreside cranes and loaded onto the beds of trailer trucks. Containers may be of various sizes, usually 8 feet wide × 8 feet high and from 10 feet to 40 feet long. They may be constructed of either aluminum or steel, and empty tare weights range from approximately 6200 lb. to 7900 lb. and from 4200 lb. to 4800 lb., for standard 40-foot and 20-foot units, respectively. Various standards organizations, including USSI and the International Standards Institute (ISO), as well as shipping companies, have tried to standardize container dimensions (see Table 4.3).

Loaded containers can weigh from 5 l.t. up to 30 l.t., corresponding to approximately 210 psf for a single container and 420 psf, 630 psf, and 840 psf for containers loaded to capacity and stacked two, three, and four high, respectively. Container handling facilities usually are designed for a minimum of 800 to 1200 psf, and the deck system design may be governed by the very high wheel loads of container handling straddle carriers and forklift trucks (to be discussed). Even general cargo vessels frequently carry some containers on

Table 4.3. Dimensions and capacities of typical intermodal containers.

Type/ description	Exterior dimensions (ft.) (width × height × length)	Volume capacity (c.f.)	Load capacity (lb.)
ISO/USSI	8 × 8 × 40	2,090	50,000 to 55,000
standards	8 × 8 × 30	1,560	45,000
	8 × 8 × 20	1,040	38,000 to 44,800
	8 × 8 × 10	490	23,000
European	8 × 8 × 6.67	329	16,000 to 18,000
standards	8 × 8 × 5	248	12,000
SeaLand	8 × 8.5 × 35	2,088	45,000
Matson	8 × 8.5 × 24	1,415	42,000 to 46,000
Conex	6.25 × 6.88 × 4.25	135	9,000
(Military standard)	6.25 × 6.88 × 8.5	295	9,000

deck; so such facilities should be designed with limited container handling and stowage capacity in mind. Accordingly, most contemporary general cargo facilities should be designed for a minimum uniform live load of 600 to 1000 psf and multipurpose container Ro/Ro facilities for a minimum of 1000 psf (5).

For fishing and light commercial use, including passenger/ferry use, a uniform live load capacity of from 250 psf to perhaps 400 psf usually is sufficient. Piers limited to pedestrian use, such as marina access, commuter boat piers, and so forth, should be designed for a minimum of 100 psf. Where automobiles or light trucks are allowed, a uniform live load of 200 and 250 psf should apply. Gangways up to around four feet in width should be designed to safely accommodate a minimum of 40 psf, and wider gangways for 60 to 100 psf. In general, catwalks and access ramps should be designed for from 60 psf to 100 psf, depending upon the application, the designer's judgment, and local building code requirements. AASHTO (6) requires 85 psf over walkway surfaces of pedestrian bridges. NAVFAC (3) requires that brows and boarding platforms be designed for a minimum of 75 psf, and further limits the deflection of brows (gangways) to span/240 under full live load plus dead load conditions. Such structures should also be checked for local concentrated loads, depending upon their intended use. Concentrated loads, other than wheel loads, may range from 1000 lb. for a personnel access ramp to 50 tons or more for a ship repair pier, and usually are applied to a single beam. Similarly, uniform live loads may be applied to deck systems parallel to and midway between beams.

Bulk cargo densities range from 37 to 49 c.f./l.t. for crude oil, from 45 to 50 c.f./l.t. for dry bulk commodities such as grain, and to 25 c.f./l.t. or less for ores such as coal and to 10.5 c.f./l.t. for iron ore. Liquid bulk cargoes are handled by hoses and pipelines and exert local concentrated loads at pipe supports, bends, hose towers, and manifold connections. Grains, salt, fertilizer, sand, coal, iron ore, and so on, may be stored temporarily in piles on the wharf deck. Unless stored in bins or hoppers, the total deck load will be limited by the material's natural angle of repose and the width of the pier or wharf apron. Table 4.4, reproduced from reference (4), gives typical dry bulk densities and angles of repose for common bulk materials.

For pier widths of 50 to 100 feet, the peak load can reach 2000 to 3000 psf for materials such as sand and coal piled 20 to 30 feet high. Areas where certain ores are stockpiled may be subject to loads of 4000 to 6000 psf. The use of low retaining walls can further increase stockpile heights. Saturated unit weights should be considered as appropriate for uncovered materials. Stockpiling of such materials on wharf decks is not normally done, but the surcharge load on the earth behind the wharf and hence the lateral earth pressure on the wharf bulkhead or quay wall can be extremely high, and the overall slope stability as well as bulkhead pressures must be checked carefully (see Chapter 8). Deck loads for military installations are covered in reference (3) and for offshore structures in reference (7).

Table 4.4. Typical dry bulk densities and angles of repose.

Material	Dry bulk density (t/m^3)	Angle of repose (degrees)
Ores		
Iron (Limonite)	2.24 to 3.00	35 to 40
Copper (Copper pyrites)	2.56	38 to 45
Lead (Galena)	2.56 to 2.76	35 to 40
Zinc (Zincblende)	1.50 to 1.79	38
Aluminium (Bauxite)	1.33	28 (when dry) 49 (in 8% moisture)
Tin (Cassiterite)	1.63 to 1.99	35 to 38
Chromium (Chromic iron)	2.39 to 2.56	33 to 40
Magnesium (Magnesite)	1.44	35
Manganese (Manganite)	1.79 to 2.39	35 to 45
Basic chemicals		
Sulphur	1.12 to 1.20	35 to 40
Phosphate rock	1.03	30 to 34
Kaolin	0.90 to 0.94	30 to 35
Solid fuels		
Coal	0.72 to 0.90	30 to 45
Coke	0.36 to 0.51	37
Building materials		
Natural aggregates	1.28 to 1.60	30 to 40
Granite (chippings)	1.20 to 1.24	35
Sand	1.79 to 1.89	30 to 40
Limestone	1.63	34
Waste products		
Domestic refuse	0.56	
Scrap iron	1.0 to 1.6	35
Foodstuffs (normally stored in sheds or silos)		
Cereal	0.51 to 0.76	40
Sugar	0.78	40
Salt	0.90	45
Soya bean	0.82	35 to 60
Copra	0.51	35

Note: 1 t/m^3 = 62.47 pcf. From reference (4); reproduced with permission of the British Standards Institution.

4.2 VEHICULAR LOADS

Contemporary tire-mounted and crawler track-mounted equipment often governs the design of the deck system. The designer should investigate any special equipment that will be used at an early stage with the owner's operation personnel, and should be sure to verify specific equipment characteristics with the manufacturer. Turning radii and operating limits must be verified, in addition to wheel loads. Manufacturers' general brochures can be dangerously mislead-

ing in regard to deck loads. For example, the average tread pressures usually listed for crawler-type cranes are wholly inadequate for slab design. Even without a load most crawler cranes have trapezoidal, nearly triangular, load profiles beneath their tracks, with load peaks up to two times or more the average bearing pressure. Wheel loads under the lift axles of forklift trucks designed to lift loaded containers are quite severe.

The contact area of rubber-tired equipment can be estimated from the axle load, tire size, and inflation pressure. Maximum inflation pressures for most heavy equipment are in the range of 80 to 125 psi. The American Association of State Highway and Transportation Officials (AASHTO) (6) provides information on estimating contact areas and wheel load distribution to deck elements. Truck cranes lifting "on rubber" may only increase their tire inflation pressures on the order of 10%, but loads under outrigger floats may be considerably higher.

Virtually all cargo handling facilities are designed to accommodate a standard highway truck of 20 tons gross weight as designated by the HS-20-44 class loading of AASHTO (6), illustrated in Figure 4.1. Other vehicle load classes

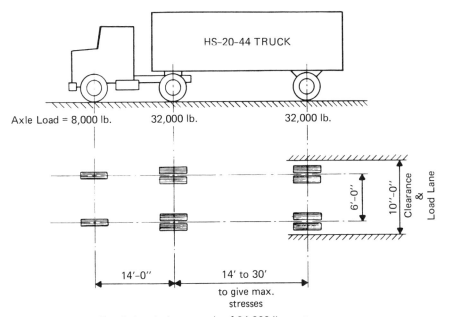

For timber decks, one axle of 24,000 lbs. or two axles of 16,000 lbs. each, spaced at 4 feet c.c. may be used instead of the 32,000 lb. axle shown, whichever produces the greatest stress. Refer to AASHTO (6) for detailed description of load application.

Figure 4.1. Standard highway tractor-trailer truck wheel loads per AASHTO (6).

also are designated by AASHTO (6). As an alternate load case, pier decks also should be checked for a military truck loading consisting of two 24,000-lb. axles spaced four feet apart. Most states issue special oversize vehicle permits, usually with a gross weight upper limit of around 80,000 lb. for multiple-axle vehicles. Single and steering axles usually are limited to 22,400 lb. and tandem axles to 36,000 lb. Such vehicles, along with other special cargo handling equipment, should be considered in the design of major cargo facilities. Tractor-trailer trucks typically have a turning radius on the order of 40 to 45 feet. Additional information on turning room for various vehicle types can be found in reference (8). Impact factors applied to vehicle wheel loads on piers and wharves are usually less than for highway bridges, typically for a 10% to a 20% increase (3), because of slower speeds and restricted travel. In addition to impact, longitudinal forces due to traction and braking must be taken into consideration. AASHTO specifications should be used as a design guide for general vehicular loads and their application and distribution. Additional information on vehicular and moving equipment loads can be found in references (9) and (10).

Table 4.5, from Sembler (11), presents some generalized data for various types of rubber-tired equipment that may be found in a contemporary container handling port. Individual wheel loads of up to 100,000 lb. are possible. Forklift trucks commonly are found in ports, and generally have lifting capacities in the range of 5 to 30 l.t., up to 40 tons, with typical wheel bases of 8 to 10 feet and spacings of 6 to 8 feet, and a corresponding turning radius of from 13 to 16 feet. A pier designed for an HS-20-44 truck loading usually will check out safely for up to an 8- to 12-ton capacity forklift truck. NAVFAC (3) recommends that general cargo piers be designed for a 20-ton forklift truck. Container terminals should be designed to accommodate the largest forklift equipment, of around 30 l.t. capacity, such as that shown in Figure 4.2. In addition, container terminals usually are equipped with straddle carriers and transtainers, which have wheel loads on the order of 30,000 lb. and up to 100,000 lb., respectively. A typical container yard straddle carrier is shown in Figure 4.3.

In general, a horizontal force of 10% of the wheel loads of all rubber-tired equipment should be applied longitudinally at deck level to account for traction and braking forces. An additional lateral force of 10% should be considered for truck cranes to account for slewing forces and windage. A vertical impact factor of 20% should be applied to the wheel loads of all lifting equipment. Impact factors apply to all deck system structural elements but not to the piles themselves.

Truck cranes are a versatile and common type of lifting equipment, ranging from 10 to 15 tons up to 300 tons rated lifting capacity. The larger sizes can be used to lift containers, and at 300 tons capacity can lift a 30-ton container at up to a 100-foot radius. During such a lift, the loads under the outrigger floats are

Table 4.5. Material handling equipment wheel loads.

Identification Code	Capacity (a)	Drive Axle Load (lbs.)	No. of Wheels and Wheel Loads (lbs.)	Design Wheel (lbs.)
Highway truck forklifts	H-20-44	—	16,000	16,000
A	30 L.T.	185,000	4-46,150	92,500 (b)
B	40 S.T.	180,000	4-45,000	90,000 (b)
C	30 L.T.	159,000	4-39,800	79,600 (b)
A	20 L.T.	152,000	4-38,000	76,000 (b)
C	20 L.T.	127,000	4-31,750	63,500 (b)
D	20 S.T. @ 30"	84,000	4-21,000	42,000 (b)
E	15 S.T. @ 48"	82,700	4-20,700	41,450 (b)
F	15 S.T. @ 48"	79,600	4-20,000	40,000 (b)
G	15 S.T. @ 48"	78,500	4-19,700	39,300 (b)
H	7.5 S.T. @ 48"	48,700	4-12,100	24,400 (b)
J	7.5 S.T. @ 48"	43,200	4-10,800	21,600 (b)
Straddle carriers				
K	30 L.T.	—	4-27,000	27,000
L	40 S.T.	—	4-26,000	26,000
M	20 L.T.	—	4-18,000	18,000
N	20 L.T.	—	6-13,000	13,000
Transtainers				
Max. load	40 L.T.	—	4-100,000	100,000
Max. load	30 L.T.	—	4-90,800	90,800
Ave. working load	25 L.T.	—	4-84,500	84,500
Empty	0	—	5-60,000	—

Note: (a) L.T. = Long Tons S.T. = Short Tons
(b) With dual wheels closer together, effect on pavement would probably be sum of two wheels.
From Sembler (11).

extremely high and must be properly distributed to the deck structure and foundation via mats or special floats. Such equipment preferably is used in solidly filled wharves or quays; otherwise, for economy of deck design, it must be relegated to designated heavy lift areas. Individual wheel loads while traveling are similar to trailer truck loadings, but the total vehicle weight will be much greater for the larger-size truck cranes. NAVFAC (3) requires that all general berthing piers be designed to accommodate a minimum 70-ton capacity truck crane, and outfitting and repair piers a 140-ton truck crane. Maximum outrigger float loads for over-the-corner capacity lifts are on the order of 150,000 lb. for

106 DESIGN OF MARINE FACILITIES

Figure 4.2. Large capacity forklift truck in use at containership terminal.

a 70-ton crane and 230,000 lb. for a 140-ton crane. Truck cranes in the 50-ton to 140-ton capacity range have wheel bases between 18 and 20 feet, with about 8 to 9 feet of track width and around 20 feet center to center for floats in the lifting position. Figure 4.4 gives wheel and float loads for a representative truck crane. Additional data on truck crane loadings and dimensions can be found in references (4) and (5) and the manufacturer's product literature.

Another type of vehicle capable of transporting very heavy loads, used pri-

Figure 4.3. Container handling straddle carrier.

OPERATIONAL AND ENVIRONMENTAL LOADS 107

CRANE TRAVELING — BOOM OVER FRONT

140-TON TRUCK CRANE

FLOAT SIZE
2'-2" x 2'-2"
FLOAT AREA
4.69 SQ. FT.

PLAN — OUTRIGGERS EXTENDED

1. DATA SHOWN FOR OUTRIGGER FLOAT LOADS ARE FOR OVER SIDE AND OVER REAR LIFTS. FOR OVER FRONT LIFTS, A FRONT BUMPER FLOAT IS REQUIRED.
2. BOOM IS OVER THE CORNER FOR WHICH FLOAT LOAD IS GIVEN.
3. FOR EQUAL RADII, RATED LOADS VARY ACCORDING TO BOOM LENGTH.

CRANE TRAVELING — RUBBER TIRE WHEEL LOADS (LBS)							
TIRES		BOOM LENGTH (FT.)	TOTAL WEIGHT (LBS.)	BOOM OVER FRONT		BOOM OVER REAR	
NO.	SIZE			EACH FRONT SINGLE TIRE	EACH REAR DUAL TIRE	EACH FRONT SINGLE TIRE	EACH REAR DUAL TIRE
12	14.00-24	50	168,821	6,711	35,494	13,759	28,447

OUTRIGGER FLOAT LOADS (LBS)						
RATED LOAD (LBS.)	RADIUS (FT.)	BOOM LENGTH (FT.)	FLOAT NUMBER			
			(1)	(2)	(3)	(4)
280,000	12	50	172,500	225,500	233,500	164,000
138,360	25	↑	172,500	225,500	233,500	164,000
114,400	30		163,900	214,200	221,800	155,800
74,900	40	↓	146,600	191,700	198,500	139,400
55,100	50	50	133,700	174,800	181,000	127,100
43,400	60	60	131,100	171,400	177,500	124,600
35,100	70	70			↑	↑
29,200	80	80				
25,200	90	90				
21,400	100	100				
18,500	110	110				
15,900	120	120				
14,000	130	130				
12,200	140	140				
10,500	150	150				
9,050	160	160				
7,700	170	170				
6,650	180	180				
5,550	190	190				
4,500	200	200				
4,300	200	210				
3,500	200	240	↓	↓	↓	↓
2,650	200	270	131,100	171,500	177,500	124,600

Figure 4.4. Wheel and float loads for typical 140-ton truck crane. (Reproduced from NAVFAC (3).)

marily in ship construction and repair facilities, is the load transporter (see Figure 4.5). Such vehicles are used to transport very heavy and large loads such as ship modules or power plant units, and can be custom-made to carry loads of several hundred tons. Individual wheel loads usually are similar to those of

108 DESIGN OF MARINE FACILITIES

Figure 4.5. Heavy load transporter in use at shipbuilding yard. (Photo Courtesy of KAMAG Corporation of America.)

a tractor-trailer truck. The use of such equipment must be investigated on an individual-case basis.

Other types of rubber-tired equipment include: front end loaders and scrapers, commonly employed in bulk handling facilities; rough terrain cranes, which typically can lift up to 40 tons; and man-lift "cherry pickers," which have relatively light wheel loads. Again, such equipment should be investigated on a per-case basis with the specific manufacturer's data. Maximum wheel loads on a single tire are on the order of 50,000 lb. for the large balloon tires of scrapers and heavy duty dump trucks. Wherever heavy equipment is used, the designer should check its total loaded weight as a distributed load, in addition to designing for maximum wheel loads.

OPERATIONAL AND ENVIRONMENTAL LOADS 109

Automobiles present relatively light loads. The floors of parking garages, for example, often are designed for an equivalent uniform live load of only 50 to 75 psf. However, the design of Ro/Ro and vehicle handling piers should consider much higher uniform loads, as well as the localized deck loads imposed by a vessel's ramps.

Crawler cranes mounted on steel treads or tracks occasionally are used on piers and wharves. Such equipment should not be allowed to travel directly over concrete or timber decks; instead, timber mats usually are laid down in the path of the tracks, to protect the deck surface and aid in load distribution. Crawler cranes are usually in the same capacity range as truck cranes—20 to 300 tons. As previously stated, the load distribution under their tracks is nearly triangular, resulting in high localized loads. In addition, an impact factor of at least 20% should be applied to the total vertical load to account for possible lift off the crane track. Peak tread pressures can exceed 100 psi. Tread widths are usually on the order of 36 to 54 inches, with a track base of 12 to 16 feet.

For virtually all traveling cranes or other boomed lifting equipment, the actual lifting capacity drops off rapidly with the boom angle or the distance from the center of rotation. Therefore, the rated capacity of such equipment generally is much larger than the nominal loads to be handled. The peak ground pressure under a track, outrigger float, wheel, and so on, usually occurs for an over-the-corner lift. The manufacturer's load/capacity charts are based upon rated safe lift capacities versus reach and usually do not include allowance for wind or impact. Therefore, when the proposed use of such equipment is critical to the design, the resulting loads must be examined carefully.

4.3 RAIL-MOUNTED AND MATERIAL HANDLING EQUIPMENT

Rail-mounted equipment commonly seen on piers and wharves includes: railroad trains, portal-type revolving cranes, gantry types, container cranes of various configuration, and a wide variety of specialized loaders used in bulk handling. Cranes and loaders less commonly may be fixed in position, such as the classic "hammerhead" revolving-type crane.

Railroad loadings are given by The American Railway Engineering Association (AREA) in its *Manual for Railway Engineering* (12). The standard load case is a "Coopers train," consisting of two locomotives in tandem followed by an infinite number of loaded freight cars. In E-80 loading, the string of freight cars is represented by an equivalent linear load (on two rails) of 8000 plf. Other classes of loading are in direct proportion: class E-70 has a load of 7000 plf, and so on. For most piers, however, which have relatively short spans, the wheel loads of the heaviest locomotive usually govern. Locomotives gen-

erally range from 45 to 145 tons in weight, usually distributed evenly over a minimum of eight wheels mounted on two trucks. NAVFAC (3) uses a standard locomotive weight of 120 tons with a corresponding single wheel load of 15 tons. AREA (12) provides additional railroad load data. A locomotive crane with a 40-ton capacity at a 12-foot reach will have a maximum corner wheel load on the order of 35 tons. Although impact factors vary with span length according to AREA specifications, an impact factor of 20% as used by NAVFAC should be sufficient for pier and wharf design purposes. The standard rail gage is $4'8\frac{1}{2}''$ center to center of rails.

Portal-type revolving cranes, such as the one shown in Figure 1.5, are essential features of shipbuilding and repair facilities and also are seen in some general cargo facilities. Such cranes may be equipped with a jib boom to increase their reach, and generally are level-luffing, allowing the boom angle to be changed without changing the height of the load. Revolving portal cranes have typical gages from 20 to 50 feet and rated capacities (boom up) on the order of 25 to 100 tons or more, although shipbuilding cranes of 300 tons, full revolving, capacity are in use. Figure 4.6 illustrates some typical portal crane wheel loads for general reference. Large—especially high—cranes should be checked for peak wheel loads under storm conditions, as wind heeling moments may produce higher wheel loads than those experienced during rated lifting. The manufacturer's wheel load data should also include an increase for normal allowable operating wind speeds, which should be verified by the designer. Longitudinal forces due to traction, braking and slewing, and lurch loads also must be considered. Crane trackage design conditions and loads are discussed in greater detail in Section 7.6.

Fixed-base pier cranes may range in capacity from a few tons for small boat docks to 50 tons or more for shipyard applications. In addition to their normal dead, live, and wind loadings, an impact factor of 100% of the capacity load in the vertical direction and 20% of the capacity load in the horizontal direction should be taken into account for slewing loads of revolving-type cranes.

Traveling gantry cranes or overhead bridge cranes that travel on fixed elevated tracks commonly are employed in shipyards or stockpiling-type facilities. Such cranes usually have gages (spans) of 100 feet or more, and spans of 555 feet have been built. Lifting capacities for large shipyard cranes range from 300 to 1500 tons. Mazurkiewicz presents an in-depth discussion of shipyard (dry dock) cranes (13).

Container cranes are conspicuous features of virtually all major seaports. Container cranes may be of various configurations, of which the standard A-frame (see Figure 1.2) and low-profile (see Figure 4.7) types are the most basic. These basic types can be modified for longer spans, back reaches, or lift heights. Most container cranes have a lifting capacity of 40 l.t. and gages of 50 to 150

OPERATIONAL AND ENVIRONMENTAL LOADS 111

NOTE: ONE TON = 2240 LB.

Figure 4.6. Portal crane wheel loads. (Reproduced from NAVFAC (3).)

112 DESIGN OF MARINE FACILITIES

Figure 4.7. Low-profile-type container crane in use at Port Elizabeth, New Jersey. (Photo Courtesy of PACECO, Inc.)

feet. Some cranes have additional capacity, say up to 65 tons, for handling more than one container at a time. Other relevant features of container cranes are their outreach, backreach, and number and spacings of wheel trucks and wheels (see Figure 4.8). Wheel loads are often on the order of 90 to 120 tons at approximately five-foot spacings with a total of 10 to 20 wheels per rail.

A wide variety of specialized rail-mounted equipment is employed in bulk cargo handling terminals, such as the catenary continuous loader shown in Figure 1.3. There are also traveling ship loaders and bucket elevators that utilize various conveyors, grab devices, and bucket and screw mechanisms. Their large and complicated superstructures and mode of operation often require that special attention be given to lateral wind and operational loads. In general, the designer must work closely with the equipment manufacturer during the design stage in order to verify loadings and other relevant operating criteria. Most often maximum wheel loads will be limited to around 120 to 150 tons because of the practical design capacity of the wheels themselves and local contact stresses in standard rail sections. Section 7.6 proves further discussion of wheel loads and crane rail track design.

Other types of bulk handling equipment include conveyor systems of the belt or bucket type that are supported by fixed towers or frameworks and may have

OPERATIONAL AND ENVIRONMENTAL LOADS 113

ITEM	CONTAINER CRANE CAPACITY (Long Tons)		
	40	40	40
1. GAUGE - G	50'-0"	50'-0"	60'-0"
2. LONGITUDINAL CLEARANCE - C	50'-0"	50'-0"	50'-0"
3. WHEELBASE - W	54'-0"	54'-0"	54'-0"
4. NUMBER OF WHEELS PER LEG	8	6	6
5. WHEEL SPACING	5'-0"	5-0"	5'-0"
6. OUTREACH - O	115'-0"	115'-0"	115'-0"
7. BACKREACH - R	50'-0"	40'-0'	115'-0"
8. WHEEL LOADS (Kips)			
a. MAXIMUM WITH 100% LIVE LOAD (OPERATING WIND)			
(1) WHEEL A (WATERSIDE)	75●	90●	76●
(2) WHEEL B (LANDSIDE)	67●	81●	96●
b. MAXIMUM STOWED WITH BOOM UP (STORM WIND)			
(1) WHEEL A (WATERSIDE)	48♦	68♦	79▼
(2) WHEEL B (LANDSIDE)	82♦	100♦	110▼

● 10 PSF WIND EQUIVALENT TO WIND VELOCITY OF 55 MPH (48 KNOTS)
♦ 30 PSF WIND EQUIVALENT TO WIND VELOCITY OF 95 MPH (83 KNOTS)
▼ 50 PSF WIND EQUIVALENT TO WIND VELOCITY OF 123 MPH (107 KNOTS)

Figure 4.8. Container crane wheel loads. (Reproduced from NAVFAC (3).)

pivoted or hinged movable arms, such as linear and radial loaders, elevator systems, aerial cableways, tram car systems, and so on. A description of the numerous types of bulk handling equipment is beyond the scope of this book. Additional general descriptions of marine material handling equipment can be found in reference (1) and in an article by Boylston in reference (14). Descriptions of specific installations can be found among the numerous papers pub-

114 DESIGN OF MARINE FACILITIES

lished in the several ASCE ports specialty conferences, and the PIANC navigation congresses (see Appendix 2). In addition to static wind and live loads, certain loading towers and elevated conveyor structures may be subject to dynamic response under wind loadings, as described by Phang (15).

Liquid bulk cargoes such as petroleum products and slurries are transported by pipelines that terminate at manifolds and/or hose handling towers, where the connection to the vessel is made. A typical hose tower installation is shown in Figure 4.9. In addition to the gravity loads of pipes running full, pipelines must

Figure 4.9. Hose loading tower with 8- and 12-inch flow boooms at Finnart Refinery, Scotland. (A Ralston photo courtesy of Woodfield Systems, Ltd.)

be designed for horizontal thrust at bends and elbows. Expansion loops and/or joints must be provided for materials that flow at controlled and variable temperatures. Kemper (16) provides a useful discussion on the design of loading arms for bulk liquid handling.

4.4 PORT BUILDINGS AND SUPERSTRUCTURES

The most common type of building constructed on a pier or wharf, usually on filled-type wharves, is the transit shed. These sheds typically are large single- or sometimes double-story structures used to temporarily store noncontainerized cargoes between shipments. Such structures often have long clear spans and inside clear heights of 16 to 22 feet (NAVFAC (3) requires 20 feet minimum) for the stocking of palletized cargoes. Prefabricated metal buildings may be used. Single floor levels are preferred, but in passenger and multiuse terminals two or more floor levels may be used. Building column lines should be made to line up with pier bent spacing as much as possible. Columns should be protected against collision by forklifts and other cargo-moving equipment, for example, by using concrete jackets around their lower levels. Sometimes the upper stories or roofs of these buildings are fitted with booms or davits for assisting in cargo handling. The design of these structures is in accordance with usual building codes and standards, with the pier or wharf serving as the building's foundation. Quinn (1) addresses the general design of port buildings, and additional load criteria for buildings can be found in references (2) and (3). Floor loads are suited to the cargoes being stored (see Section 4.1). In areas where local building codes do not exist, national building codes as found in references (17) and (20) may be utilized.

Other fixed structures commonly found on piers include: light towers and standards; storage tanks or towers for sandblast grit, and so on; hoppers and bins for bulk commodities; and support structures for various material handling equipment previously discussed. Handrails, ladders, gangways, and boarding platforms are discussed in Chapter 7. For wind loads on miscellaneous fixed structures, references (2), (21), and (22) are useful. Wind loads on buildings or elevated superstructures may be critical to the design of marine structures, and design windspeeds and appropriate drag coefficients should be selected carefully. Wind loads acting directly on pier or wharf structures themselves usually are not of consequence compared to the usual design loads. However, for lightly framed walkway piers or trestles or single long-span catwalks not subject to other lateral design forces, some provision must be made for lateral resistance. Reference (10) recommends that a minimum transverse wind load of 300 plf be applied to the design of reinforced concrete bridges including pedestrian bridges, in lieu of a more detailed analysis. In addition, an uplift/overturning force of 20 psf over the deck area should be applied at the windward quarter point of the structure width.

4.5 ENVIRONMENTAL LOADS

Environmental loads that may act directly upon marine structures include: wind, wave, current, ice, and ground motions such as earthquake or at some locations bottom "mud slides." As a general rule, horizontal design loads on vessel berthing structures are governed by vessel berthing and mooring loads, and wind, wave, and current forces acting directly on the structure usually can be neglected. In some instances, however, they cannot be, such as: wind loads on buildings, cranes, or elevated superstructures (see preceding sections); wave loads on structures at exposed/offshore locations; and structures situated in strong currents. Evaluation of hydrodynamic loads on fixed structures is a complex subject whose detailed treatment is beyond the scope of this text. However, the subject is well covered elsewhere, and the following paragraphs are intended to review basic principles and direct the reader to other sources for a comprehensive treatment.

Wave and Current Loads

Wave loads are dynamic in nature, but for the range of water depths of interest herein they generally can be represented as equivalent static loads. The nature of wave loadings and the approach to the calculation depend upon the structure or member dimension relative to the wave length. For example, when the structure width or pile diameter (D) is small with respect to the wavelength (L), the wave motion is relatively unaffected by the presence of the pile, and the resultant force on the pile is due to water particle velocities (drag forces) and accelerations (inertial forces). When the structure width relative to the wavelength is large enough that it affects the wave form, then the diffraction and scattering of the incident wave must be considered. The resulting forces are primarily inertial in nature in that they primarily depend upon the water particle accelerations. If the structure is very wide with respect to the wavelength, such as a continuous wall, then the incident wave is reflected, and forces are treated as a rise in hydrostatic pressure head.

The generally accepted ranges of application of wave force calculations, in terms of the structure or member width to wavelength ratio (D/L), are as follows:

- For $D/L \leq 0.2$, drag and inertia forces dominate; use the Morison equation.
- For $D/L > 0.2$, diffraction effects become increasingly important; use diffraction theory.
- For $D/L > 1.0$, pure reflection conditions exist; treat the structure as a seawall.

For pile-supported structures with $D/L \leq 0.2$, loads on individual piles are computed by the Morison equation (23), where the total force (F) per unit length of pile is the sum of the drag (F_D) and inertial (F_I) components, as given by:

$$F = F_D + F_I = C_D \frac{\gamma}{2g} DU|U| + C_M \frac{\gamma \pi D^2}{4g} \frac{\partial U}{\partial t} \quad (4\text{-}1)$$

where γ = unit weight of water, g = acceleration of gravity, U = horizontal water particle velocity component, D = pile diameter, and C_D and C_M are the drag and inertia force coefficients, respectively. For cylindrical piles C_D is dependent upon the Reynolds number (N_R) of flow and usually ranges between 0.6 and 1.2, and C_M is usually within the range of 1.3 to 2.0. For sharp-edged square piles and H-piles, C_D and C_M are on the order of 2.0 and 2.5, respectively. For other piles shapes, the reader is referred to references (4) and (24). Due consideration should be given to the effects of fouling in increasing projected area and surface roughness. In order to obtain the total instantaneous force, the water particle velocities and accelerations must be integrated over the water depth (d), and to find the total maximum force, F_D and F_I must be added vectorially. Refer to the wave force definition sketch, Figure 4.10. The maxi-

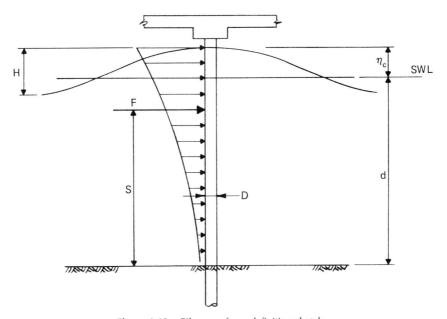

Figure 4.10. Pile wave force definition sketch.

mum drag force occurs at the wave crest position, and the maximum inertial force occurs at about $L/4$ ahead of the crest. In general for shallow water and steep waves F_D predominates, and in deep water and for long, low waves F_I dominates.

The Morison equation can be rewritten in terms of the incident wave height (H) to find the total force (F_T) on the pile, as:

$$F_T = \frac{\gamma}{2g} C_D K_D D H^2 \vec{+} \frac{\gamma}{2g} C_M K_M D^2 H \qquad (4\text{-}2)$$

where K_D and K_M are drag and inertial force factors representing the integration of water particle velocities and accelerations from the bottom to the free surface, as given by:

$$K_D = \frac{1}{H^2} \int_0^{\eta_C} U|U| \, dz \qquad (4\text{-}3)$$

where η_C is the wave crest elevation; and:

$$K_M = \frac{\pi}{2H} \int_0^{\eta_S} \frac{\partial U}{\partial t} \, dz \qquad (4\text{-}4)$$

where η_S is the free surface elevation such that F_I is a maximum. Both K_D and K_M depend upon the period parameters: d/T^2 and H/T^2 where T = wave period. Nondimensional graphs and tables are available to aid in solving the above equations, such as those in references (25) through (27).

The wave kinematic properties are dependent upon the selection of an appropriate wave theory, which in turn is largely dependent upon the relative water depth at the site. General treatment of wave theories and more in-depth treatment of forces on pile structures can be found in references (28) and (29). To find the total overall force on the structure, a succession of wave positions must be plotted. Ball and Hall (30) studied current drag forces on pile groups at varying orientations to the flow. They found that C_D is a function of yaw angle and may vary by up to 40% in yawing the group through 45°. They further concluded that the total drag force is significantly reduced by reducing the number of piles and increasing their diameter. They note that such a reduction in the number of piles should also result in a reduction in siltation and the flow-field divergence that influences vessel mooring forces. Worked examples of wave forces on pile structures can be found in references (1), (25), (26), and (31) through (33). Guideline design criteria for evaluating wave loads on fixed offshore structures are given in references (24) and (34).

In the special case of shallow water where the wave breaks on the pile struc-

ture, the velocity forces govern, and the breaking wave force (F_B) can be found from:

$$F_B = \frac{\gamma}{2g} C_B D U_C^2 \qquad (4\text{-}5)$$

where C_B is the breaking wave drag force coefficient, which can be taken as 1.75 for a cylindrical pile (25), and U_C is the wave crest particle velocity. Note that for shallow water the wave crest velocity approaches the wave celerity at breaking, as given by:

$$U_C = \sqrt{gd_b} \qquad (4\text{-}6)$$

where d_b is the water depth at breaking, which is approximately equal to the breaker height (H_B). Steady current velocities should be added vectorially to the wave particle velocities. Note that under storm conditions surface currents of 1% to 3% of the sustained wind speed are possible, and when combined with increased tidal currents due to storm surge effects (see Section 3.4) may result in a significant increase in loads.

Transverse lift forces are generated by the periodic shedding of vortices from alternate sides of the pile. The magnitude of the lift force (F_L) is given by:

$$F_L = \frac{\gamma}{2g} C_L D U^2 \qquad (4\text{-}7)$$

where C_L is the lift force coefficient, which can be taken as approximately $= C_D/3$ for circular cylinders, and U is the wave and/or steady current water velocity. The lift force is oscillatory in nature and is seldom of consequence to the overall structure's stability or integrity. However, in strong currents where the vortex shedding frequency is near the natural frequency of pile motion, resonant vibrations may occur. Individual piles are especially susceptible during construction (35). Flow-induced in-line oscillations may also occur when vortex shedding and pile response become locked-in. The frequency of vortex shedding (f_v) is given by the Strouhal number (N_S) as:

$$N_S = \frac{f_v D}{U} \qquad (4\text{-}8)$$

For cylinders in steady currents and long waves, N_S has an approximate value of 0.2.

The Kuelegan-Carpenter number (N_{KC}) also may be used to predict vortex shedding regimes in unsteady flows, and is given as:

$$N_{KC} = \frac{U_{max} T}{D} \qquad (4\text{-}9)$$

where U_{max} is the average maximum water velocity over the depth, and T is the wave period. For $N_{KC} < 3.0$, no eddies are formed. Detailed treatment of flow-induced oscillations of marine structures can be found in reference (36).

In addition to drag, inertia, and lift forces, horizontal members near the surface may be subject to impulsive, vertical wave slam loads caused by the sudden rise in sea surface elevation. Slam loads are of significance only to local member design and can be estimated from:

$$F_S = \frac{\gamma}{2g} C_S D U_V^2 \qquad (4\text{-}10)$$

where F_S is the vertical slam force per unit length of member; C_S is the slam force coefficient, which can be taken as $C_S = \pi$ for cylindrical members; and U_V is the vertical water particle velocity, which, according to small amplitude wave theory, has a maximum value of $\pi H/T$ in deep water.

For large-diameter structures such as isolated cells or caisson units where $D/L > 0.2$, diffraction theory must be employed to calculate wave loads. MacCamy and Fuchs (37) presented a general solution for a bottom-supported, surface-piercing cylinder based on diffraction theory of light waves. The theory has been extended to structures of arbitrary geometry by Garrison et al. (38) and Hogben and Standing (39). According to reference (28), the total force is given by an equation of the form:

$$F = \frac{\pi}{8} \gamma H D^2 \tanh\left(\frac{2\pi d}{L}\right) C_M \cos(\omega t - \delta) \qquad (4\text{-}11)$$

where all of the terms are as previously defined for pile structures, and ω is the wave circular frequency $= 2\pi/T$, and δ is the wave phase angle. Mogridge and Jamieson have presented graphical solutions for forces and moments on both square (40) and circular (41) cylinders. The results of physical model studies usually are presented in terms of dimensional analysis in the general form:

$$\frac{2 F_{max}}{a^2 H} = f\left(\frac{2\pi a}{L}, \frac{d}{a}\right) \qquad (4\text{-}12)$$

where a is some characteristic structure dimension. Apelt and MacKnight (42) discuss the results of a particular investigation of caisson structures for a large offshore bulk coal loading berth. For a theoretical treatment of diffraction theory, the reader is directed to references (28) and (29).

When the structure width is longer than the incident wavelength, $D/L > 1.0$, it can be treated as a continuous seawall where the wall is designed to resist the maximum wave force per unit length along the wall. Wave forces on walls are treated in virtually all general textbooks on coastal engineering, and references (25) and (26) provide step-by-step instructions and graphical aides to expedite the calculations. Figure 4.11 illustrates wave pressure profiles at walls. Wave forces due to reflected, nonbreaking waves generally are calculated after the methods of Sainflou and Miche-Rundgren, which account for partial reflection, as described in reference (25). In general, it is assumed that a standing wave, sometimes referred to as a clapotis, equal to twice the height of the incident wave is formed in front of the structure. The standing wave profile oscillates about an elevated still water level (SWL), resulting in an increase and a decrease of pressure at the toe of the wall, corresponding to the wave crest and trough positions, as given by:

$$P = \gamma d \pm \frac{\gamma H}{\cosh\left(\frac{2\pi d}{L}\right)} \qquad (4\text{-}13)$$

The total force per unit length of wall is obtained from the area of the net pressure prism. The peak dynamic pressure is at the SWL. The increase, "setup," in the water level above the initial SWL is given by:

$$h_0 = \frac{\pi H^2}{L} \coth\left(\frac{2\pi d}{L}\right) \qquad (4\text{-}14)$$

Reflected wave conditions can be assumed when $H < 1.7\,d$. In shallow water where waves may break against the wall, the Minikin method (43) generally is employed. The peak dynamic pressure at the SWL is given by the semiempirical relation (in foot and pound units):

$$P_M = 101\gamma \frac{H_B}{L_0} \frac{d_S}{d_0} (d_S + d_0) \qquad (4\text{-}15)$$

where d_S is the water depth at the structure, d_0 is the water depth at one wavelength from the wall, and L_0 is the wavelength at the location of d_0. For a sloping bottom, the wavelength at the wall is reduced in accordance with linear wave theory.

The total dynamic force is centered about the SWL and is given by:

$$F_M = \frac{P_M H_B}{3} \qquad (4\text{-}16)$$

122 DESIGN OF MARINE FACILITIES

Figure 4.11. Wave force pressure profiles on vertical walls.

This force is added to the hydrostatic pressure to obtain the total pressure profile shown in Figure 4.11. Peak dynamic pressures of over 2000 psf are not unusual in seawall design. Complete instructions and graphs for applying the Minikin method can be found in references (25) and (26). The method of Goda (44) can be applied to the general case of breaking or nonbreaking wave forces against walls and considers the random nature of wave loading. The Minikin method generally yields more conservative results except for the case of long waves and shallow water depth. Chu (45) recently compared these methods and made recommendations for their application.

In general, pier deck elevations should be located above the highest expected wave crest elevation. Where this is not feasible, wave uplift pressures should be considered in the design. Notable investigations of wave uplift pressures on pierlike structures include the work of ElGhamry (46), Wang (47), and French (48). All of these investigators found uplift pressures characterized by a high localized initial peak pressure of short duration followed by a slowly varying uplift pressure (positive upward then negative downward) of lower magnitude but of longer duration. The slowly varying pressure is of greatest significance to the overall global uplift force on a structure and its pile foundation, whereas the peak pressures may be critical to local member design such as the planks of a timber pier deck (see Figure 4.12). Wave uplift data often are normalized in terms of the relative pressure head, $P/\gamma d$, which is dependent upon the wave celerity (C), wave height (H), under-deck clearance above the SWL (S), and relative location along the wave profile. (See definition sketch, Figure 4.13.) French noted maximum values of the slow varying pressure head nearly equal to the static pressure head at the deck or $\gamma(H - S)$. ElGhamry placed an upper empirical bound on the peak pressure of five times the incident wave height above the SWL.

Recent work by Lai and Lee (49) has included the effect of bottom slope and end reflecting boundary, as would be the case for a marginal wharf-type structure. They developed a numerical model and normalized their data in terms of the total uplift force: Fu = uplift pressure × contact area, and the hydrostatic force: Fs = total weight of water above the platform (the shaded area in Figure 4.13). They reported maximum values of Fu/Fs in the range of 1.0 to 2.0 for flat bottoms and from 3.0 to 8.0 for sloping bottoms and end walls. It should be noted that all of the cited studies were based on "solitary" waves whose celerity is controlled by the water depth, with their entire crest height above the SWL. Clearly there is potential for very high uplift pressures below marginal wharves with steep bottom slopes that are subject to overtopping waves. Leitass (50) has reported on studies of wave-induced air pressure cycling damage to stone slope protection below marginal wharves. Experimental studies have shown a 50% decrease in peak pressures when only a 0.5% area of pressure relief openings relative to deck area were provided.

124 DESIGN OF MARINE FACILITIES

Figure 4.12. Storm wave action on timber pier. Note jet of spray through deck at wave crest location.

Ice

Ice imposes the following kinds of loads and effects on structures: uplift and pile jacking loads, lateral thrust, impact from drift ice, increased weight and windage due to ice accumulation, abrasion of timber and concrete in the tidal zone, and expansion forces of ice in captive pockets. Even structures in temperate climates subject to occasional ice covers of, say, 6 to 12 inches thickness or more should be designed to resist nominal ice uplift forces. This is especially true for light-timber pier construction, such as that shown in Figure 4.14.

Wortley (51) has presented minimum suggested ice uplift loads on piles of

Figure 4.13. Wave uplift definition sketch.

various dimension and for various ice thickness. Wortley's work is based upon a "first crack" analysis whereby radial cracks extend outward from the center of the pile to a circumferential crack at a distance (a) called the radius of load distribution. Wortley notes that for marina piles in the Great Lakes this crack typically occurs approximately 6 inches out from the face of the pile, so that for estimating purposes it can be taken as the pile radius plus 6 inches. The uplift loads are directly proportional to ice flexural strength. Using Wortley's method and assuming an ice flexural strength of 200 psi, ice uplift loads on a 12-inch-diameter timber pile would range from approximately 8000 lb. to 30,000 lb. for ice thicknesses of 12 and 24 inches, respectively. For a 12-inch-diameter steel pile, the uplift forces would be approximately 9000 and 33,000 lb. respectively. Increasing the pile diameter, however, does not dramatically increase the uplift force. For example, a 24-inch-diameter steel pile in a 24-inch-thick ice sheet would be subject to 37,000 lb. uplift, and a 36-inch-diameter pile to 42,000 lb. uplift, as compared with 30,000 lb. for the 12-inch-diameter pile. Effective pile adhesion values are generally less than 50 psi for timber and steel and generally less than 70 psi for concrete piles.

Figure 4.14. Light-timber pier encumbered with ice. Note coverage of cross-bracing and ice collars on piles.

126 DESIGN OF MARINE FACILITIES

Lateral thrust due to moving ice can be extremely high. Dams and structures in confined areas are subject to lateral thrust due to thermal expansion of ice that can be on the order of 20 to 30 tons per lineal foot. The calculation of lateral ice thrust usually is based upon an equation of the form:

$$F_i = C_i f_{ic} A_{ic} \qquad (4\text{-}17)$$

where C_i is a coefficient depending upon the structure shape and configuration, the rate of load application, and so on, which normally ranges between 0.3 and 0.7; f_{ic} is the ice compressive strength, usually assumed to be between 100 psi and 400 psi; and A_{ic} is the contact area, which is equal to the pile diameter or structure width times the ice thickness. As can be seen, equation (4-17) leaves room for a lot of assumptions! In reality, however, the lateral force of the ice sheet often may be limited by the driving force rather than the compressive strength of the ice, such as wind and current acting on an ice sheet along the coast or within an open bay. In this case, the wind or current velocity and shear stress are the governing parameters. Here the driving force can be calculated using the well-known drag force equation (see Section 6.3), which for wind reduces to:

$$F_{iw} = .0034 \, C_{iw} V^2 A_{is} \qquad (4\text{-}18)$$

where F_{iw} is the force in pounds; C_{iw} is the wind shear stress drag coefficient, which normally ranges from .002 to .01; V is the wind velocity in knots; and A_{is} is the surface area of the ice sheet. For current-driven ice, the equation becomes:

$$F_{ic} = C_{ic} U^2 A_{is} \qquad (4\text{-}19)$$

where C_{ic} ranges from around .01 to .1, and U is in feet per second units. Figure 4.15 shows the force per unit length of structure for an effective ice sheet length of one nautical mile for wind and current action based upon the above relationships.

Impact loads due to drift ice can be even more damaging when borne by strong currents. In order to estimate ice impact loads, it is necessary first to assume a weight of ice floe and an impact velocity. Wind-driven ice may travel at 1% to 7% of the mean wind speed, but usually it is less than 3%. Ice borne by currents can be assumed to travel at the water surface velocity. Cammaert and Tsinker (52) studied ice impact forces on marine structures and postulated that approximately one-half of the kinetic energy of an ice floe striking a rigid

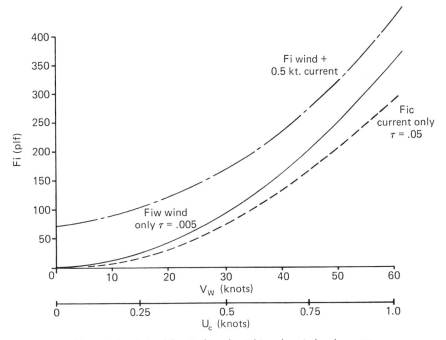

Figure 4.15. Lateral thrust of ice sheet driven by wind and current.

pier will be dissipated in progressively crushing the ice, and then the piece will rotate and pass by.

There is a growing body of literature on designing for ice forces, especially under arctic conditions, of which references (53) through (58) are recommended for their general treatment and practical application.

Seismic Loads and Ground Motions

Because most piers and wharves are designed to resist relatively large lateral forces, they are relatively rigid, with natural periods on the order of 0.5 second or less, and thus offer good resistance to earthquake ground motions. Exceptions may be flexible structures such as those consisting of vertical cantilever piles and certain deep water platform structures and any structures exposed to high earthquake hazard. Near-shore pier and wharf structures should be checked for seismic forces in accordance with standard highway bridge codes such as AASHTO (6). The calculations should include structure dead load only, and a range of live load conditions usually between 10% and 50% of the full design

live load. In general, the maximum ground acceleration should be applied along each principal structure axis nonconcurrently. The API recommended practice for offshore platforms includes a generalized response spectrum for preliminary evaluation of earthquake response. API requires that an acceleration spectrum equal to two-thirds of the maximum applied along the principal axis be applied simultaneously in the orthogonal direction, and one-half of the maximum be applied simultaneously in the vertical direction. This should be done with each horizontal direction serving as the principal axis. Zone factor multipliers are given for U.S. offshore locations. Crane stability and superstructures and/or other susceptible appurtenances should be checked for possible resonant effects and separation forces using standard methods of calculation.

Where preliminary quasistatic methods indicate potentially high forces, a response spectrum or full dynamic analysis may be required. In this case, the loads are a function of the structure response. The reader is referred to references (7) and (33) for treatment of dynamic analysis of offshore structures subject to seismic loads.

At certain river delta locations, subject to the rapid accumulation of soft sediments, structures may be affected by mud slides or longer-term soil flows. Such situations require detailed geotechnical studies. The reader is referred to the discussion on stability of submarine slopes by Kraft and Ploessel in reference (7) for a detailed overview of this subject.

4.6 OTHER LOAD SOURCES

Other load sources also may act on marine structures. Soil pressures and excess pore water pressures due to hydrostatic pressure heads behind walls and bulkheads are addressed in Chapter 8. Temperature, shrinkage, creep, and support settlements are treated in the same manner as for typical land-based structures. Section 7.3 presents some discussion of thermal expansion effects peculiar to marine structures. Dynamic amplification of cyclic, random, impulsive, and long-term cyclic loads results in a progressive increase of deflections and a reduction of material strength such as via fatigue and vibration effects. In general, dynamic effects will not be significant where some natural frequency, $f_n = 1/T_n$, of the structure or its elements is less than one-half or greater than three times the forcing frequency. Where dynamic response does occur, the loads are a function of the structure's mass and stiffness properties, and a more detailed analysis is required. This subject is introduced in Section 7.3. Dynamic analysis is not usually required of most near-shore/in-harbor structures, however, because of their great rigidity and ability to resist lateral loads. Impact factors are applied to impulsive and moving loads, which are then treated as static loads.

In contrast to most land-based structures, marine structures must be designed

to resist relatively large transient live loads and lateral forces. Impact and overloading often result in progressive cumulative damage. Therefore, impact resistance and resiliency also should be considered in determining allowable stresses and material requirements. Redundancy and ductility should be provided wherever possible so that the structure cannot fail in a single catastrophic mode. Equivalent lateral load capacities for typical piers and wharves generally range from approximately 500 to 1000 plf for vessels generally under 2000 DT, and from 2000 to 4000 plf or more for the largest oceangoing vessels. These loads are applied horizontally at deck level and apply only to structures that are continuous over the entire vessel berth length. The lateral capacity also will vary, in particular with vessel type, site exposure, and other factors. In addition to carrying out specific load calculations for a given structure, it is useful for the designer to calculate the equivalent lateral load capacity for comparison with other structures of similar type and function, to ensure that an adequate minimum lateral capacity is provided.

Another source of loading not previously discussed is the strong discharge current produced by a vessel's propeller, generally referred to as prop-wash. This is an important consideration at ferry and Ro/Ro terminals or other quick-turnaround-type operations, as a source of bottom scour and erosion. At certain locations, it also can be a source of local member loading, such as a current drag force producing bending moments in piles. Figure 4.16 illustrates the general geometry of the propeller discharge jet or screw-race. There is a turbulent zone of flow establishment immediately behind the propeller where the maxi-

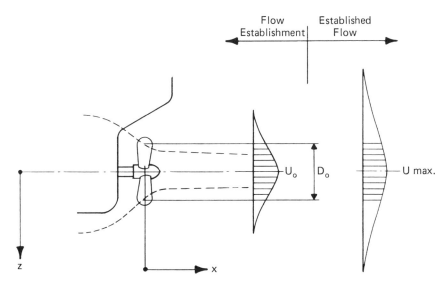

Figure 4.16. Propeller wash definition sketch.

mum initial discharge velocity U_O is confined to a small area slightly less than the propeller diameter (D_P). Further beyond, the flow spreads out and becomes more uniform, with a velocity Umax along the central axis.

There are several empirical formulas for predicting wash velocities at a given distance behind and below the propeller. The simplest formula requiring the least information about the vessel propulsion system, used by Berg and Cederwall (59) for the maximum velocity in the turbulent zone, is:

$$U_O = .95\, nD_P \qquad (4\text{-}20)$$

where U_O is in meters/second, n is propeller revolutions per second, and D_P is the propeller diameter in meters. Note that for single unducted propellers as used on most oceangoing ships, an effective propeller diameter of $.707\, D_P$ should be used in equations (4-21) and (4-22) below. The maximum flow velocity at a distance (x) behind the propeller, according to Blaauw and van de KAA (60), can be found from:

$$U\text{max} = U_O \left(\frac{2cx}{.707 D_P}\right)^{-b} \qquad (4\text{-}21)$$

where b and c are coefficients having average values of approximately 1.0 and 0.18, respectively, for meter and second units. Prop-wash currents in excess of 10 knots are possible at piles or bulkheads in ferry terminals. The drag force (F_C) on the pile can be found from the drag force equation (4-1), which reduces to:

$$F_C = C_D U^2 A_P \qquad (4\text{-}22)$$

where the current velocity is in feet per second, F_C is in pounds, and the pile projected area (A_P) is in square feet. C_D ranges from 0.6 to 1.2 for cylindrical piles to 2.0 or more for square and H-piles. For an otherwise heavily loaded or deteriorated pile, the added bending stresses could result in serious overstressing. Reference (61) contains several valuable papers on prop-wash problems as related to bottom scour and harbor erosion.

REFERENCES

1. Quinn, A. D. F. (1972). *Design and Construction of Ports and Marine Structures*. McGraw-Hill, New York.
2. ANSI (1982). *American National Standard Minimum Design Loads for Buildings and Other Structures*, A58.1-1982. NBS, ANSI, New York.
3. NAVFAC (1980). *Piers and Wharves*, DM-25.1. Alexandria, Va.

4. BSI (1984). *British Standard Code of Practice for Maritime Structures*, Part 1 General Criteria, BS6349. London.
 5. PIANC (1987). "Development of Modern Marine Terminals," Report of a Working Group of PTC-II, Supp. to Bull. No. 56, Brussels, Belgium.
 6. AASHTO (1983). *Standard Specifications for Highway Bridges*, 13th ed. plus annual supplements to date. Washington, D.C.
 7. McClelland, B. and Reifel, M. D., eds. (1986). *Planning and Design of Fixed Offshore Platforms*. Van Nostrand Rienhold, New York.
 8. AASHTO (1973). "A Policy on Design of Urban Highways and Arterial Streets," Washington, D.C.
 9. ——— (Dec. 1981). "Recommended Design Loads for Bridges," by the Comm. on Bridges of the Struct. Div., *Proc. ASCE*, Vol. 107, ST-7.
10. ACI (1988). "Analysis and Design of Reinforced Concrete Bridge Structures," ACI 343R-88, Detroit, Mich.
11. Sembler, E. L. (Oct. 1974). "Pavement Costs in the Big Load Era," *World Ports*.
12. AREA (1985). *Manual for Railway Engineering*, Vols. I and II. American Railway Engineering Association, Washington, D.C.
13. Mazurkiewicz, B. K. (1980). *Design and Construction of Dry Docks*. Trans Tech Publications, Clausthal-Zellerfeld, FRG.
14. Taggart, R., ed., (1980). *Ship Design and Construction*. SNAME, New York.
15. Phang, M. K. (1977). "Dynamic Response of Elevated Conveyor Structures," *Proc. ASCE*, Ports '77, Spec. Conf., Long Beach, Calif.
16. Kemper, W. (1980). "All-Metal Loading Arms for Bulk Liquid Handling," *Proc. ASCE*, Ports '80, Spec. Conf., Norfolk, Va.
17. BOCA (1987). *The BOCA Basic National Building Code*. Building Officials and Code Administrators International, Country Club Hills, Ill. (issued every three years).
18. UBC (198-). *Uniform Building Code*. International Conference of Building Officials, Whittier, Calif.
19. NBC (198-). *National Building Code*. National Conference of States on Building Codes and Standards, Herndon, Va.
20. SBC (198-). *Standard Building Code*, Southern Building Code Congress International, Birmingham, Ala.
21. ASCE, (1987). "Wind Loading and Wind-Induced Structural Response," Comm. on Wind Effects of the Structural Div., New York.
22. Simiu, E. and Scanlan, R. H. (1986). *Wind Effects on Structures*. Wiley, New York.
23. Morison, J. R. et al. (1950). "The Force Exerted by Surface Waves on Piles," *Petroleum Trans.*, No. 189.
24. DNV (1977). *Rules for the Design, Construction and Inspection of Offshore Structures*," Det Norske Veritas, Oslo, Norway.
25. USACE (1984). *Shore Protection Manual*, Vol. II. CERC.
26. NAVFAC (1982). *Coastal Protection*, DM-26.2. Alexandria, Va.
27. Dean, R. G. (Nov. 1974). *Evaluation and Development of Water Wave Theories for Engineering Application*. Vols. I and II, USACE, CERC, SR No. 1.
28. Sarpkaya, T. and Isaacson, M. (1981). *Mechanics of Wave Forces on Offshore Structures*. Van Nostrand Reinhold, New York.
29. Wiegel, R. L. (1964). *Oceanographical Engineering*, Prentice-Hall, Englewood-Cliffs, N.J.
30. Ball, D. J. and Hall, C. D. (May 1980). "Drag of Yawed Pile Groups at Low Reynolds Numbers," *Proc. ASCE*, Vol. 106, No. WW-2.
31. Gaythwaite, J. W. (1981). *The Marine Environment and Structural Design*. Van Nostrand Reinhold, New York.

32. Myers, J. J. et al., eds. (1969). *Handbook of Ocean and Underwater Engineering*. McGraw-Hill, New York.
33. Dawson, T. H. (1983). *Offshore Structural Engineering*. Prentice-Hall, Englewood Cliffs, N.J.
34. API (1987). "Recommended Practice for Planning, Designing and Constructing Fixed Offshore Platforms," RP-2A, Washington, D.C.
35. Khanna, J. and Wood, J. S. (Mar. 1979). "Tripod Structures in Fast Currents," *The Dock and Harbour Authority*, Vol. LIX, No. 700.
36. Hallam, M. G. Heaf, N. J., and Wootton, L. R. (Oct. 1978). "Dynamics of Marine Structures: Methods of Calculating the Dynamic Response of Fixed Structures Subject to Wave and Current Action," CIRIA Underwater Engineering Group, Report UR-8.
37. MacCamy, R. C. and Fuchs, R. A. (1954). "Wave Forces on Piles: A Diffraction Theory," USACE, B.E.B. Tech Memo 69.
38. Garrison, C. J. et al. (1974). "Wave Forces on Large Volume Structures—A Comparison Between Theory and Model Tests," OTC paper No. 2137, Houston, Tex.
39. Hogben, N. and Standing, R. G. (May 1975). "Experience in Computing Wave Loads on Large Bodies," *Proc.* OTC, paper No. 2189, Houston, Tex.
40. Mogridge, G. R. and Jamieson, W. W., (1976a). "Wave Forces on Square Caissons," *Proc.* 15th Coastal Engineering Conf., Honolulu.
41. Ibid. (1976b). "Wave Loads on Large Circular Cylinders: Design Method," Hydraulics Lab, Nat. Research Council Canada, Report No. MH-111, Ottawa.
42. Apelt, C. J. and MacKnight, A. (1976). "Wave Action on Large Offshore Structures," *Proc.* 15th Coastal Engineering Conf., Vol. III, Honolulu.
43. Minikin, R. R. (1963). *Winds, Waves and Maritime Structures: Studies in Harbor Making and in the Protection of Coasts*. Griffen, London.
44. Goda, Y. (1985). *Random Seas and Design of Maritime Structures*. University of Tokyo Press, Tokyo.
45. Chu, Ven-hsi (Jan. 1989). "Breaking Wave Forces on Vertical Walls," *Proc. ASCE*, WPCOE Div., Vol. 115, No. 1.
46. El Ghamry, O. (1963). "Wave Forces on a Dock," Report No. HEL.-9-1, University of California, Berkeley.
47. Wang, H. (1967). "Estimating Wave Pressure on a Horizontal Pier," T.R. #R546, USNCEL, Port Hueneme, Calif.
48. French, J. A. (1979). "Wave Uplift Pressures on Horizontal Platforms," *Proc. ASCE*, Civil Engineering in the Oceans IV, Spec. Conf., New York.
49. Lai, C. P. and Lee, J. J. (Jan. 1989). "Interaction of Finite Amplitude Waves with Platforms or Docks," *Proc. ASCE*, WPCOE Div., Vol. 115, No. 1.
50. Leitass, R. (1979). "Wave Damages to Stone Slope under Marginal Wharves," *Proc. ASCE*, Coastal Structures '79, Spec. Conf., Vol. II, New York.
51. Wortley, C. A. (1984). "Ice Engineering Manual for Design of Small Craft Harbors and Structures," University of Wisconsin Sea Grant Inst., SG-84-417, Madison, Wis.
52. Cammaert, A. B. and Tsinker, G. P. (1981). "Impact of Large Ice Floes and Icebergs on Marine Structures," *Proc.* of 6th Int. Conf. on Port and Ocean Engineering under Arctic Conditions, POAC, Quebec.
53. Caldwell, S. R. and Crissman, R. D., eds. (1983). "Design for Ice Forces," A State of Practice Report by Tech. Council on Cold Regions Engineering, ASCE, New York.
54. Cammaert, A. B. and Muggeridge, D. B. (1988). *Ice Interaction with Offshore Structures*. Van Nostrand Reinhold, New York.
55. USACE (1982). "Ice Engineering," EM 1110-2-1612, Dept. of the Army, Office of the Chief of Engineers, Washington, D.C.

56. Eranti, E. and Lee, G. C. (1986). *Cold Regions Structural Engineering*, McGraw-Hill, New York.
57. API (1982). "Bulletin on Planning, Designing and Constructing Fixed Offshore Structures in Ice Environments," Bull. 2N, Washington, D.C.
58. ACI (1985). "State of the Art Report on Offshore Concrete Structures for the Arctic," ACI 357.1R-85, Detroit, Mich.
59. Berg, H. and Cederwall, K. (1981). "Propeller Erosion in Harbors," Bull, No. TRITA-VBI-107, Hyd. Lab., Royal Inst. of Tech., Stockholm, Sweden.
60. Blaauw, H. G. and van de KAA, E. J. (1978). "Erosion of Bottom and Sloping Banks Caused by the Screw Race of Maneuvering Ships," Pub. No. 202, Delft Hyd. Lab., Delft, The Netherlands.
61. PIANC (1987). "Bulletin 1987, No. 58," Brussels, Belgium.

5
Berthing Loads and Fender System Design

The evaluation of vessel impact loads during berthing and the design of protective fender systems are of central importance to the overall structural design of a facility. This area of design still leaves much to the designer's judgment; so it is most important that the basic mechanics and hydrodynamics of the berthing vessel and the fender and pier system response be well understood. This chapter begins with an overview of the berthing ship problems and of factors affecting the magnitude of vessel impacts and the application of appropriate energy coefficients. The effect of the added mass of water entrained by a vessel maneuvering in shallow water is reviewed. Accepted methods of determining berthing energy for design purposes are presented, and types of fender systems and their selection and application are reviewed. Finally, important aspects of fender system design, arrangement, and specifications are considered.

5.1 INTRODUCTION

The problem of evaluating the forces imposed upon fixed structures by an approaching vessel centers around how the mass of the vessel finally is brought to rest alongside. Given the large range in vessel sizes and the speeds at which they are capable of traveling, there is an almost infinite number of possibilities. The designer's problem then is to evaluate the most probable loads that will occur during routine berthing maneuvers, while providing an adequate margin of overload capacity to minimize damage under occasional high-energy accidental berthings. Factors affecting berthing impact forces include: the vessel's mass and velocity; the angle of approach; the relative water depth; vessel controllability and tug assistance; wind, current, and wave action; the structure's configuration; and the load/deflection properties of the structure and its fender system. The usual approach to the problem is to calculate the percentage of the vessel's kinetic energy that must be absorbed by the berthing structure and fender system and evaluate the terminal force, based upon the load/deflection characteristics of the structure/fender system.

The work done by the deflecting dock/fender system is equal to the area under the load deflection curve as given by the integral:

$$E = \int_0^s R(s)\,ds \tag{5-1}$$

where E is the energy absorbed by the system, R the reaction force, and s the distance traveled.

In the design of contemporary fender systems, the deflection of the pier itself usually is neglected, as the pier's movement is generally very small. Certain dolphins that have no fender system attached to them act themselves as fenders, and here the structure's deformation must expend all the energy. The design of berthing dolphins and timber pile cluster dolphins are discussed in Section 7.4.

The energy to be absorbed by the fender system (E_F) usually is calculated by multiplying the vessel's total kinetic energy by a series of berthing coefficients, as given by:

$$E_F = \frac{\Delta}{2g} V_N^2 \times C_e \times C_m \times C_c \times C_s \tag{5-2}$$

where:

Δ = Vessel displacement.
V_N = Component of vessel's velocity normal to pier.
g = Acceleration of gravity.
C_e = Eccentricity coefficient, which accounts for the vessel's rotation, depending upon the angle of approach and point of contact with the pier; it usually ranges from 0.5 to 0.8, and a value of 0.5 often is assumed.
C_m = Virtual mass coefficient, which accounts for added mass due to entrained water; it usually ranges from 1.3 to 1.8, although it can be much higher under certain conditions.
C_c = Configuration coefficient, which accounts for pier type and geometry; it usually ranges from 0.8 to 1.0.
C_S = Softness coefficient, which accounts for the relative stiffness of the vessel hull and fender system; it usually ranges from 0.9 to 1.0

In the most extreme cases, equivalent to a vessel striking an open pier directly head on or side on, C_e and C_c would go to unity, and the total kinetic energy of the vessel plus entrained water would have to be expended in deforming the fenders, pier, and vessel's hull. For convenience herein, we can define an overall berthing coefficient C_b, which is equal to the product of the four coefficients in equation (5-2). The total energy (E_F) to be absorbed by the fender system usually is in the range of 40% to 70% of the vessel's total kinetic energy, or C_b = 0.4 to 0.7. Some designers take C_b = 0.5 for preliminary design estimates. Exceptions include the case of direct end-on berthing where C_b = 1.0,

136 DESIGN OF MARINE FACILITIES

and situations where the vessel under-keel clearance is very small and C_b exceeds 1.0. Figure 5.1 illustrates the range of berthing energy versus vessel displacement for various normal velocities and a range of C_b. This figure can be used for preliminary estimates and to compare the effects of design velocity selection.

Selection of an appropriate normal velocity. (V_N) is of utmost importance because the energy is proportional to the velocity squared, and this is one of the most subjective judgments a designer must make. In addition to being properly designed for the expected usual berthing impacts, fender systems must possess some degree of overload capacity to protect the ship and pier from damage under occasional accident level impacts and to remain functional under a variety of wind, tide, sea, and operational conditions while the vessel is moored. The fender system load/deflection properties may, therefore, be im-

Figure 5.1. Vessel berthing energy vs. displacement curves.

BERTHING LOADS AND FENDER SYSTEM DESIGN 137

portant in determining mooring loads under the influence of environmental forces while the vessel resides in its berth. Guideline recommendations for the design of fender systems can be found in the PIANC "Report of the International Commission for Improving the Design of Fender Systems" (1). General guidance for the design of fender systems also can be found in references (2) through (4).

We next examine the various factors affecting the choice of design parameters and then return to a discussion of their general application, as well as alternative methods of determining design berthing energy and fender system forces.

5.2 THE BERTHING SHIP

The general geometry of a berthing vessel is shown in Figure 5.2. In most cases, the angle of approach (θ) between the vessel's longitudinal centerline and pier face is taken as between 5° to 15° for design purposes, with the exception of certain ferry and Ro/Ro berths and for large vessels where θ is generally less than 3° to 5°. Note that for the case of a vessel proceeding ahead

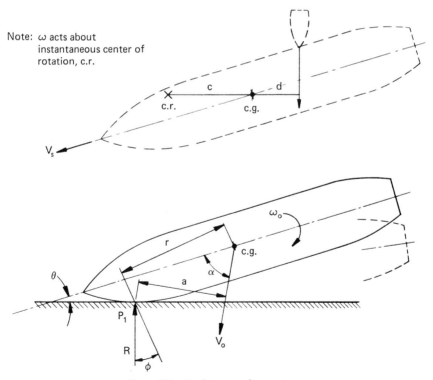

Figure 5.2. Berthing vessel geometry.

138 DESIGN OF MARINE FACILITIES

under its own power at velocity V, the normal velocity $V_N = V \sin \theta$. For large oceangoing ships, the vessel's final approach nearly always is assisted or totally controlled by tugs, and the vessel's resultant sideways translational velocity, V_0, acts nearly normal to the pier face. Under controlled conditions, the angle of approach is usually less than 3°. In addition to being translated sideways toward the pier, the vessel will possess some angular motion (ω_0) about a hypothetical "instantaneous" center of rotation (c.r.). The distance to the instantaneous center of rotation (C) from the vessel's center of gravity (c.g.) is related to the distance (d) from the c.g. to the applied force causing the rotation (i.e., tug thrust, wind or current, etc.) by the following relation: $C = k^2/d$, where k is the radius of gyration of the vessel's mass about its c.g. For most vessels, k usually lies in the range of 0.20 to 0.29 times the vessel's length (L) and varies with the vessel's block coefficient (C_B), being about 0.2 for $C_B = 0.5$ and 0.29 for $C_B = 1.0$.

Referring again to Figure 5.2, the angle between the vessel's resultant velocity V_0 and the vessel's centerline is denoted α, the distance from a line drawn normal to the line of action of V_0 through the c.g. and the point of first impact P is denoted a, and the distance from c.g. to P is denoted r. It can be demonstrated (5) that in relation to the above-defined geometric parameters, the amount of the vessel's kinetic energy to be absorbed by the fender system (neglecting the other berthing coefficients for the time being) is given by:

$$E_F = \frac{\Delta}{2g} V_0^2 \left(1 - \frac{a^2}{k^2 + r^2}\right) + \frac{\Delta}{g} V_0 \omega_0 \left(\frac{k^2 a}{k^2 + r^2}\right)$$
$$+ \frac{\Delta}{2g} \omega_0^2 \left(\frac{k^2 r^2}{k^2 + r^2}\right) \tag{5-3}$$

Neglecting rotation, equation (5-3) reduces to:

$$E_F = \frac{\Delta}{2g} V_0^2 \left(1 - \frac{a^2}{k^2 + r^2}\right) \tag{5-4}$$

and for the case of pure rotation:

$$E_F = \frac{\Delta}{2g} \omega_0^2 \left(\frac{k^2 r^2}{k^2 + r^2}\right) \tag{5-5}$$

Noting that the velocity of the point of the vessel's hull in contact with the fender $V_0 = \omega_0 r$ and substituting into equation (5-5) gives:

$$E_F = \frac{\Delta}{2g} V_0^2 \left(1 - \frac{r^2}{k^2 + r^2}\right) \tag{5-6}$$

By comparing equations (5-3) and (5-6), it can be noted that for equal velocities at the point of contact the amount of energy (E_F) to be absorbed by the fender system will be less for pure rotational motion than for translational motion. Therefore, in general, berthings made through rotating motion are to be preferred to those made through simple translation. It is also interesting to note that forces applied beyond the center of percussion (c.p.) of a rotating body will not change the reaction force at the point about which the body is pivoting. The distance dp from the c.g. to the c.p. is given by $dp = k^2/r$; therefore, tugs pushing on a vessel's stern beyond this distance from its c.g. will not increase the fender reaction. The foregoing equation also assumes that the angle of the reaction force (R) with the normal to the face of fender contact, denoted by ϕ, is larger than the angle of friction, such that sliding at the point of contact does not take place. Sliding expends additional energy and would help to reduce the reaction force. Depending upon the fender stiffness and V_0 and ω_0 at the instant of first impact, the vessel may bounce back and undergo a second impact nearer the vessel's stern. Second impacts not only are undesirable from a berthing operations point of view, but may be of equal or greater magnitude than the first impact and thus must be considered in the evaluation of berthing impacts and in the layout and spacing of fender units. For ferries or Ro/Ro's backing into a slip stern on or head on, there is little or no rotation, $\alpha = 0$, and all of the energy must be absorbed by the fenders. This is also the case for a vessel being translated normal to a pier and striking it perfectly broadside.

The discussion so far may seem somewhat academic from a design point of view, in that exact knowledge of all parameters is not possible. It is important, however, that the designer understand the mechanics of a berthing ship and be able to make reasonable assumptions about the above parameters under given circumstances. For a more rigorous analytical treatment of vessel berthing, refer to Vasco Costa (5, 6).

As a practical matter of design, the rotation of a vessel during berthing is accounted for by the eccentricity coefficient (C_e), already introduced, simplified to the following (7):

$$C_e = \frac{k^2}{k^2 + r^2} \tag{5-7}$$

In this formula, it is assumed that $a = r$ for simplification of the analysis. The error associated with this assumption increases with increase in the distance of the point of impact from the vessel's c.g. A more accurate formula, accounting for the effect of the angle α, is:

$$C_e = \frac{k^2 + r^2 \cos^2 \alpha}{k^2 + r^2} \tag{5-8}$$

Often it is assumed that the point of first contact is at the quarter point, approximately 25% of the vessel's length, from the bow or stern. This distance nearly coincides with the length of parallel midbody (see Section 5.6) for most large vessels. Studies conducted on general cargo ships in Japan (2) indicated that vessels normally have first impacts at points between .05 L and 0.20 L from their ends. For the majority of berthings, the value of C_e was within the range of 0.5 to 0.8. Most vessels' c.g.'s are located very near the midpoint of their LWL or at their midships section. Vasco Costa (5) notes that for large vessels in loaded condition the c.g. is slightly forward of this point, and for ships in ballast it is usually aft of midships.

Fender reactions applied above or below the vessel's vertical center of gravity (v.c.g.) will cause the vessel to heel, thus expending some additional energy. The heeling moment is given by the product Rh, where h is the vertical distance between the v.c.g. and the fender reaction (R). The resisting moment is given by $\Delta GM \sin \psi$, where GM is the transverse metacentric height (see Section 9.2) and ψ is the angle of heel (refer to Figure 5.3). The energy expended in heeling the vessel can be shown in general to be very small compared with the other factors discussed, and it is usual to neglect this effect in design.

The size and geometry of the structure to which the vessel is berthing also affect the energy E_F. For closed piers such as bulkheads and quay walls, water trapped between the wall and the approaching vessel that cannot escape quickly enough from around and under the vessel will act somewhat as a cushion between the wall and the vessel. This is especially true when the quay is much longer than the vessel, and the water depth (d) to vessel draft (D) ratio (d/D) is small. Conversely, for open pile-supported piers and especially for isolated berthing dolphins, the water moves freely past the pier and no cushioning oc-

Figure 5.3. Heeling geometry of berthing vessel.

curs. Piers and wharves may be considered semisolid for these cases: pile-supported wharves below which the bottom slopes rapidly upward, relieving platforms, piers shorter than the vessel's length, and so on. The following values for the so-called configuration coefficient (C_c) are recommended for design:

Pier type	C_c
Closed (solid)	0.8
Semisolid	0.9
Open	1.0

Values as low as 0.6 for closed construction longer than the vessel's length have been proposed (8). Recent research efforts demonstrate that the cushioning effect of the pier configuration should be considered together with the added mass effect, as will be discussed in Section 5.3.

Other ways in which energy is dissipated upon impact include deformation of the vessel's hull and induction of vibrations in the vessel's hull and pier structure, which ultimately are released as heat. The latter factor is considered negligible, but it is often assumed that the vessel's hull will absorb about 5% to 10% of the energy of impact—in other words, the softness coefficient C_s ranges from 0.9 to 0.95. In certain instances, values of C_s from 0.5 to 1.0 have been used (8), the lower value for cases of rigid fenders and flexible ships and the value of unity for very soft fenders and rigid ships. Unless a more detailed study is conducted, a value of $C_s = 1.0$ is recommended where soft fendering is used, and a value of C_s between 0.9 and 1.0 where hard fendering is used, with soft and hard fenders defined as deflecting more than and less than six inches, respectively, under the design impact energy (3).

5.3 THE CONCEPT OF ADDED MASS

A body in an accelerated (decelerated) fluid flow will be acted upon by inertia forces that resist changes in velocity. A vessel traveling at constant velocity is not opposed by inertia forces, but by resistance due to viscous effects and turbulent flow that create a pressure field around the vessel. When being slowed or suddenly stopped, the total force (F_R) resisting the change in velocity is given by:

$$F_R = M_s \frac{\partial V_s}{\partial t} + \sum \frac{\partial (M_w V_w)}{\partial t} + R_v \tag{5-9}$$

where M_s is the mass of the ship, M_w is the mass of water moving with the ship, V_s is the velocity of the ship, V_w is the velocity of entrained water particles,

142 DESIGN OF MARINE FACILITIES

and R_v is resistance due to velocity forces. The mass of water supposed to move along with the ship is termed the *added mass* or *hydrodynamic mass*. The total mass of the vessel plus entrained water is termed the *virtual mass*. Likewise, a hydrodynamic mass coefficient (C_h) and virtual mass coefficient (C_m) can be defined, where the total virtual mass ($M_s + M_w$) is given by $C_m \times M_s$ or $(1 + C_h) M_s$.

The reader should be wary, when reviewing published data, about which form of the added mass is used. For purposes herein, C_m will be used for convenience, allowing equation (5-9) to be written:

$$F_R = C_m M_s \frac{\partial V}{\partial t} + R_v \tag{5-10}$$

provided we also assume that $V_s = V_w$.

Virtually all published values of C_m are based upon empirical and semiempirical methods. From classical hydrodynamic theory, the added mass of a cylinder oscillating in an accelerated flow is given by:

$$M_w = \frac{\pi}{4} \rho D^2 LWL \tag{5-11}$$

where ρ is the mass density of water ($= \gamma/g$, where γ is the unit weight of water), D is the vessel draft, and LWL is the vessel's waterline length. This equation, which has been used for design purposes, usually will give corresponding values of C_m on the order of 1.5 to 1.6 for typical larger vessels such as tankers. Another commonly used formula for C_m, as recommended by Vasco Costa (6) after the work of Grim (9) is:

$$C_m = \left(1 + \frac{2D}{B}\right) \tag{5-12}$$

where B is the vessel's beam. This empirical equation usually yields values of C_m on the order of 1.3 to 1.8 for most oceangoing vessels. Neither of these formulas, however, considers the importance of relative water depth as given by the ratio d/D and other potentially important parameters such as the vessel's initial velocity, the proximity of solid structures, the deceleration distance, and fender system stiffness. Figure 5.4 illustrates the various parameters in the evaluation of added mass effects.

In open water without defined boundary conditions, the added mass of a vessel in forward motion is on the order of 10%, or $C_m = 1.10$ (8). Various investigators have determined values of C_m ranging from around unity to near 16.0. The higher values generally have been based on model experiments and

BERTHING LOADS AND FENDER SYSTEM DESIGN 143

Figure 5.4. Added mass definition sketch.

have not been reconciled with practical experience. Qualitatively, however, they are very instructive in terms of demonstrating the effects of under-keel clearance and other boundary conditions.

Notable investigations of added mass effects that are relevant to vessel berthing impacts are described in references (9) through (16). Mathematical models also have been developed to treat the berthing impact problem analytically, such as those presented by van Oortmerssen (17) and by Fontijn (18). Middendorp (19) developed an analytical solution using long wave theory to describe the movement of water around the berthing ship.

Although convenient for some purposes, the concept of added mass as entrained water lumped together with the vessel's mass is somewhat misleading. A lucid but simplified explanation has been given by Ball (20). For a vessel moving sideways (sway motion), the rate of displacement of water forward is balanced primarily by a return flow under its keel and around its ends. When the water depth is shallow relative to the vessel's draft, more water will be forced to flow around the vessel's ends. When the vessel is brought to rest more or less suddenly, the displacement flow is stopped, but the momentum of the moving water (i.e., the under-keel flow) will continue, creating in effect a negative pressure wave inboard and a consequent drop in water level and a positive pressure wave and rise in water level on the vessel's outboard side (see Figure 5.4). Ball refers to these as surge waves that result in a difference of water levels across the ship, forcing the vessel against the fender. Because the surge force is time-dependent, the peak surge lags the instant of impact, so that the total added mass felt by the fenders will depend upon their stiffness and rate of load application, the hydrodynamic boundary conditions, and, importantly, whether or not the vessel was accelerated or decelerated immediately before striking the fender. A seeming paradox is that for a given impact velocity a ship will produce more fender energy (E_F) if slowed down just prior to impact than

if accelerated beforehand. Reduced under-keel clearance results in an increased under-keel flow rate, and it has the effect of increasing the time over which the surge force acts and thus increasing E_F.

Ball and Markham (16) reported on scale model tests undertaken to evaluate the relative importance of various parameters and emphasized the need for full-scale verification. Some of their conclusions, based upon a 1/142 scale model 100,000 DWT tanker driven sideways onto a fender of average prototype stiffness and at a prototype velocity of 0.1 meter per second, are:

1. Under-keel clearance in the range of 2.5% to 5% of draft ($d/D = 1.025$ to 1.05) results in C_m values much higher than are usually applied.
2. The presence of a solid wall within distance B of the fender face may significantly increase C_m; a separation of approximately $B/4$ has the greatest effect.
3. The vessel does not have to be moved great distances to develop large C_m values; approach distances of approximately $B/2$ have the greatest effect.

The second item is contrary to the usual practice of reducing E_F by the configuration coefficient for solid structures. Another seeming contradiction is that soft fenders with large deflections actually increase the added mass for a given velocity of impact. Clearly there is a need for more definitive research and full-scale corroboration of results in this area of study. As will be seen in Chapter 6, added mass effects are important to the determination of mooring loads, and a vessel has different added mass relationships for movements or rotations about its various coordinate axes. This discussion has been limited to added mass in sway, although yawing, or rotation about a vertical axis through the vessel's c.g., also is important to the berthing ship.

Hayashi and Shirai (12) recognized the importance of velocity flow effects in their studies of tanker models, and also found very high values of C_m for low d/D ratios. They gave their results in terms of the nondimensional Froude number $N_F = V_N/\sqrt{gd}$ (see Section 6.7). Their results were presented for N_F ranging from 0.01 to 0.07 and d/D from 1.0 to 2.1. Beyond $d/D = 2.0$, this parameter appears to become insignificant. Hydrodynamic mass is inversely proportional to N_F and V_N; therefore, with a decrease in V_N, the N_F decreases, and consequently C_m will increase for any depth of water.

The evaluation of a vessel's virtual mass under given conditions is an area for further study, with field verification being most important. As for current design practice, it appears that the extreme values associated with $d/D \leq 1.1$ are unlikely to be attained. The configuration coefficient should perhaps be taken into account in the calculation of C_m and not applied separately. Studies in Japan (2), based upon actual measurements of tankers berthing at open sea berths, indicated higher values of E_F than predicted by calculation. The discrepancy

appeared to be in the evaluation of C_m, and it was determined that for a $d/D = 1.1$, the minimum practical design depth, C_m will range between 1.61 and 2.31. The designer is left with a choice of applying conventional methods of calculation such as equation (5-11) or (5-12) or reviewing the literature to find data most applicable to the particular problem at hand. Where under-keel clearances are small ($d/D \leq 1.5$) and large vessels are being considered, special consideration should be given to the evaluation of C_m. The PIANC report on fender systems (1) reviews other studies on this subject and provides additional recommendations for design.

5.4 BERTHING IMPACT ENERGY DETERMINATION

Berthing impact energy levels to be used for fender system selection and design can be determined by one or more of four basic approaches to the problem. In the conventional calculation procedure, the kinetic energy of the moving vessel is multiplied by various coefficients that account for specific assumed design conditions. This approach has already been introduced, and discussion of the coefficients serves to illustrate the many factors involved. Where sufficient field measurements exist, a statistical approach may be taken, making use of the fact that the frequency distribution of impact energies measured at various berths approximately follows a log-normal distribution. Such analysis generally is valid only for a specific harbor and vessel size and type. The empirical approach involves the application of simple formulas based upon past experience and certain assumptions relating the vessel's displacement or other characteristics directly to the berthing energy. Mathematical models have been developed that attempt to model the entire process and allow analysis of a wide range of input conditions. These models need to be calibrated through actual measurements under controlled conditions or via physical model experiments. The direct use of physical models to determine berthing energies cannot be applied with full confidence under the current state of the art. This section will review the application of these four methods.

Conventional Calculations

The conventional method of berthing energy calculation has been presented as equation (5-2), and various factors relating to the use of this equation have been explored. The remaining and perhaps key factor in its application is the selection of an appropriate velocity of approach (V_N). The velocity at impact is affected largely by vessel controllability and the skill and techniques of those in charge. The number, type, and use of tugs, the use of bow thrusters, weather conditions (including visibility, wind, sea state, and the presence of currents), the maneuvering space available, and the frequency of berthing affect design

146 DESIGN OF MARINE FACILITIES

velocity selection. Docking sonar or berthing aid systems are employed at some large vessel facilities, especially at exposed locations, to monitor the vessel's speed and distance off. Docking sonars may have large display panels mounted atop the berth structure or may transmit data directly to the vessel, or both. Measurements from more than one location can be used to determine the vessel's angle of approach and rate of rotation. Berthing aid systems should be required at all facilities that berth vessels of 200,000 DWT and over (21). A general discussion of their operation and advantages is presented by Tomlinson (22).

Normal velocities used in design generally are in the range of 0.3 to 1.0 foot per second (fps). Usually higher speeds are associated with smaller vessels, and conversely for larger vessels. Various criteria have been presented for the selection of design velocities with regard to exposure, approach conditions, vessel displacement, and so on, such as described in references (1) through (4) and (6). Brolsma et al. (23) demonstrated the use of probability laws in the determination of design velocities. Figure 5.5 illustrates V_N versus vessel displacement for varying exposure conditions and is somewhat representative of North American design practice. A mean value of V_N of from 0.3 to 0.5 fps is representative for the larger vessels. Table 5.1 summarizes actual measurements taken at several berths in Japan (24). Design velocities for ferry and Ro/Ro operations that feature high frequency berthings are generally much higher than

Figure 5.5. Vessel normal velocity vs. displacement. (After NAVFAC (4).)

Table 5.1. Berthing speed.*

Case 1: Berth located in harbor at which operating conditions are moderate.
Case 2: Berth facing open-sea at which operating conditions are relatively severe.
Case 3: Open-sea berth at which operating conditions are severest.

Operating conditions	Berths	Ship size (DWT)	Berthing	Berthing speed (cm/sec) Range	Berthing speed (cm/sec) Average	n
Case 1.	T company (K berth)	75,000–250,000	First impact	1–13	5.3	50
	T company (S berth)	50,000–245,000	"	1–10	4.3	59
	N company (No. 3)	21,000–280,000	"	1– 8	3.3	36
	N company (No. 4)	160,000–480,000	"	1– 5	2.3	12
	N company	60,000–170,000	"	2– 9	2.3	16
Case 2.	K company	155,000–250,000	First impact	1–10	4.3	8
	O company	100,000–250,000	"	1– 7	3.7	30
Case 3.	I company	100,000–220,000	First impact	2–10	4.5	19
				5–28	17.1	18

*Reproduced from reference (24), courtesy of the Bridgestone Tire Company. *Note:* 1 cm/sec = .0328 fps; n = number of data.

these values. Such vessels usually are guided into their slips by outer guide-in dolphins or fender racks and then by wharf-side fenders before finally contacting the head dolphins or end fender units end-on. Recommended design velocities for outer guide-in dolphins range from 2.0 to 3.0 meters per second (m/s) (1, 3), or approximately 6 to 10 fps, to 0.5 to 1.0 m/s (1.6 to 3.3 fps) for wharf-side fender units where the vessel makes a direct approach. These velocities usually are associated with a minimum approach angle of 10° to 15°. The actual velocity used should consider the vessel's stopping distance in relation to the slip length. Lower velocities apply where the vessel is essentially stopped and worked into the slip or alongside in a transverse mode (1, 3). The end fenders should be designed for approach velocities of from 0.5 to 1.0 m/s (1.6 to 3.3 fps) when a direct, end-on approach is made. It is customary to select a design velocity representative of usual conditions with some overload capacity for high energy berthings and anything beyond, which is considered an accident. Reference (21) notes that the extreme velocities from record data are on the order of four times the most favorable berthing values. In regard to accidents, economic analysis can be made of various fendering systems for different levels of energy if reasonable assumptions can be made about the probability of damaging accidents over the structure's life (6, 25).

The angle of approach is related to the normal velocity and is typically less

than about 5° for larger vessels being maneuvered by tugs, whereas angles of 10° to 15° generally apply to smaller vessels making an approach under their own power. Angles as high as 20° have been proposed for ferry slip design (26).

An alternative approach to the conventional calculation method is the force-acceleration method, whereby the time history of the vessel's deceleration is equated directly to the deflection of the fender system and/or berthing structure. Derucher (27) has applied this approach to the general evaluation of pier fendering systems, and Sherman (28) has demonstrated the approach in the design of a ferry fender rack.

Statistical Approach

The conventional method of energy calculation leaves many difficult choices to the designer, especially for berths frequented by vessels of different sizes. It should be intuitively obvious that, whatever the level of energy selected for design, there will always be some risk that the energy level will be exceeded during the structure's life (29). The decision then as to what E_F to design for must consider the usual range of E_F under normal berthings, the overload capacity required to accommodate some infrequent level of high energy berthings, and the probability and risk of damage caused by accidents associated with low probability levels. These constraints strongly imply the application of statistical techniques, but the main problem with this approach is with the availability of a sufficient number of actual measurements relevant to a given berthing situation. Actual measurements of berthing energies were reported by Goulston (30) and by Svendsen (31), the latter finding a direct correlation with ship size and current conditions as the most significant parameters.

Dent and Saurin (32) and Toppler and Weersma (33) both proposed a statistical approach based upon many direct measurements. Dent and Saurin recommended the following energy levels for design for a moderately exposed tanker berth:

- An E_F of 16.8 foot-kips per 1000 DWT of design vessel for fenders at the yield stress.
- An E_F of 11.2 foot-kips per 1000 DWT of design vessel under normal maximum conditions, corresponding to the working stress in fenders and berthing structures.
- An E_F that is 5/8 of the above for protected berths and also for loading terminals where vessels normally arrive in ballast condition.

Toppler and Weersma noted that the following conditions must be accepted in developing a rational statistical approach: berthing energies in excess of the ultimate design energy may occur, and their occurrence is related to the number

of berthings during the lifetime of the structure, the size distribution of vessels, in particular the maximum and average size, and their relative frequencies, as well as the site conditions. They further noted that once a sufficient volume of statistical data is available, a risk factor can be determined that can assign probabilities of damaging impact, which can be weighed against structure first cost and life cycle cost. They then proposed the following design energy levels for tankers greater than 100,000 DWT:

- An E_F of 2000 foot-kips per 100,000 DWT for low frequency berthing (i.e., less than 40% of all berthings).
- An E_F that is 1.25 times the above for high frequency berthing (i.e., greater than 60% of all berthings).
- An E_F equal to 85% of the above for loading terminals (vessels in ballast) and sheltered locations.
- An E_F equal to 110% of the above for very exposed locations.

Figure 5.6 summarizes their proposed design criteria.

The British Petroleum Company has taken an active role in conducting actual measurements at its marine terminals, as reported by Balfour et al. (34). Balfour and his colleagues grouped vessels into various size ranges and berth types and then produced log-normal plots of energy versus probability of exceedence, which plot as nearly straight lines. They applied a correction factor (C_f) for fender spacing, acknowledging the importance of the distance from the center of mass to the point of fender contact. The correction factor is given by:

$$C_f = \frac{C_e \text{ for standard }.15 \text{ } LOA \text{ eccentricity}}{C_e \text{ for eccentricity of half fender spacing}}$$

Balfour and colleagues further note, however, that extreme impacts are not necessarily associated with center-of-mass berthings. For design purposes, they suggest that the maximum impact to be resisted should be that corresponding to a return period equal to the structure design life. Their base case is assumed as a 30-year structure life with 100 berthings per year for a total of 3000 berthings. The associated probability of occurrence, then, is .033% or 1/3000. Figure 5.7 illustrates the application of their method to two particular piers in Rotterdam Harbor. Figure 5.8 is a generalized recommendation for design energies versus displacement for any enclosed harbor. Balfour also found that the berthing energy is related to the inverse square of the vessel displacement.

Empirical Methods

The empirical approach is best suited to specific sites where the relative berthing frequency, vessel characteristics, berthing conditions, and so on, remain fairly

150 DESIGN OF MARINE FACILITIES

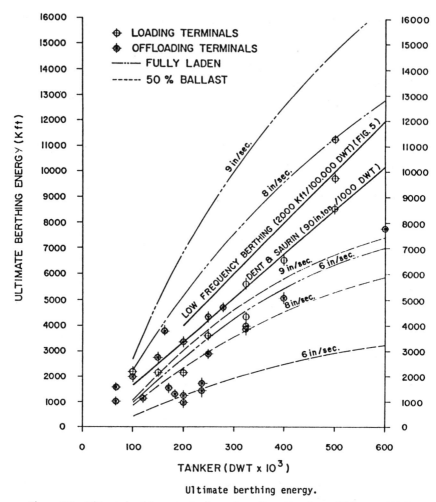

Ultimate berthing energy.

Figure 5.6. Ultimate berthing energies for tankers. (From Toppler and Weersma (33).)

constant. Empirical relationships or formulas can be developed from record data and past experience. Girgrah (35) proposed the following empirical relation based solely on vessel displacement:

$$E_F = \frac{\Delta}{120 + \sqrt{\Delta}} \text{ for } E_F \text{ in tonne-meters} \qquad (5\text{-}13)$$

This equation applies to fenders located at the point of first contact and can be reduced 50% for interior fenders. Piaseckyj (36) notes that the relationship can

Figure 5.7. Berthing energy vs. Displacement for 1:3000 probability of occurrence. (From Balfour et al. (34).)

Figure 5.8. Design energy values for an enclosed harbor. (From Balfour et al. (34).)

be adapted to suit a given designer's experience by varying the constant, here given as 120.

Mathematical Models

Mathematical models have been developed for predicting the motions and forces on a vessel moored in shallow water, as described in Section 6.7, but such models have not been so widely used in describing vessel impacts during berthing. Notable efforts in this field are described in references (11) and (17) through (19). Further description of the analytical treatment of berthing vessels can be found in various papers of the NATO Advanced Studies Institute proceedings (37, 38).

An example output from a numerical berthing model based on Fontijn's work (18) is given in Figure 5.9, for a 17,500 DT naval vessel berthing against a foam-filled fender unit. The initial approach velocity of the vessel was 0.5 foot per second, and the impact occurred at a point one-fourth of the length of the ship away from the vessel's center of gravity. The figure presents a load-deflection curve for the foam-filled fender, as well as time histories of the fender force, vessel sway, and vessel yaw during the berthing impact. It is important to note that because fender systems may have a major influence on mooring loads, their design should integrate both berthing and mooring conditions. Mathematical models could provide a detailed description of the interrelationship of berthing and mooring problems, but the practical problems of physically modeling the impact process under a wide range of conditions in order to calibrate the models have perhaps inhibited development in this area.

In summary, several methods are available for the determination of berthing impact energy and/or the resultant forces as they relate to marine fender system design. The conventional calculation method remains the most generally applicable approach that a designer can readily implement. Where a sufficient data base exists, and/or where there is a long record of experience such as within developed harbors with similar facilities, then the statistical or empirical methods offer important advantages. At remote and exposed sites for large vessels and/or where major facilities are planned, mathematical models offer the advantages of considering a wide variety of conditions and integrating berthing and mooring design requirements.

5.5 FENDER SYSTEM TYPES AND SELECTION

A wide variety of fender system types are available to the designer. In addition to commercially manufactured units, many other options are open to the designer developing custom-designed fender systems, depending upon specific facility requirements. Fender systems may dissipate impact energy in various

BERTHING LOADS AND FENDER SYSTEM DESIGN 153

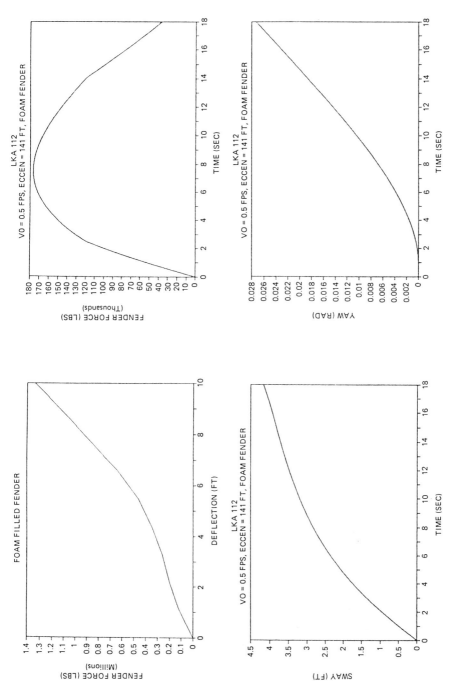

Figure 5.9. Load deflection curve and time histories of fender force and vessel sway and yaw for a foam-filled fender, based on numerical berthing model of Fontijn (18). (Provided by John R. Headland, P.E., NAVFAC.)

ways: through conversion of kinetic energy to potential energy, as in gravity and buoyancy-type fenders; by elastic deformation in compression, bending, shear, and/or torsion, as in elastomeric-type units; by conversion of energy into heat by friction, as in hydraulic units; and even by plastic deformation of certain expendable-type fenders. Various fender systems and their characteristics are described in some detail in reference (1), and some of the more common types are described later in this section.

The selection of the optimum fender type for a given application depends upon the following factors:

- Energy absorption requirements.
- Maximum reaction force.
- Maximum deflection and load deflection characteristics.
- Vessel deceleration rate and fender rebound characteristics.
- Effect of angular impact on performance.
- Allowable hull pressure on the vessel.
- Coefficient of friction and vertical and longitudinal rubbing forces.
- Range of vessel sizes and hull shapes.
- Vessel standoff distance requirements.
- Range of tide and exposure conditions.
- Environmental exposure effects.
- Frequency of berthing and wear considerations.
- Factor of safety and overload capacity.
- Cost and long-term maintenance/repair costs.
- Local availability, costs, and construction practices.

Fender system selection often is a compromise of first cost and/or life cycle cost versus vessel and pier safety and maintenance. For a given berthing energy (E_F), the terminal force or reaction (R) on the pier structure and vessel hull is a function of the load deflection characteristics of the pier and fender system. In most cases, except for certain dolphins that absorb the impact by their own deflection, the deflection and hence the energy absorption by the structure itself are neglected. Fenders usually are designed for some prescribed level of berthing impact energy, and in general it is desirable to minimize the reaction to energy ratio (R/E) of the fender system. The lower the R/E, the more "efficient" is the fender. Contact bearing pressures on the vessel's hull are kept within allowable limits by properly sizing fender face panel dimensions. It should be noted that fenders play an important role in the loads imposed upon moored vessels, and their characteristics must be checked against mooring requirements as discussed in Chapter 6 and in Section 5.6.

Fender systems can be classified as to their general load deflection characteristics, as shown in the generalized curves of Figure 5.10. The area under the

BERTHING LOADS AND FENDER SYSTEM DESIGN 155

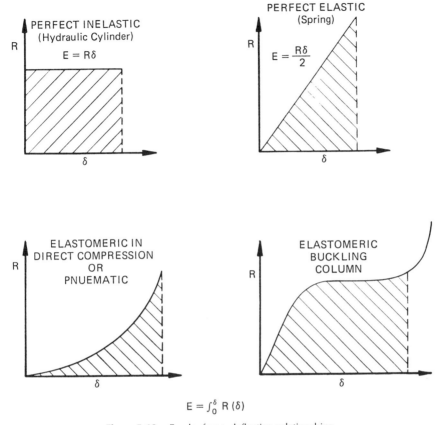

Figure 5.10. Fender force–deflection relationships.

curve is equal to the work done in deforming the fender, and hence to E_F. Equation (5-1) can be rewritten to express the relationship between R and E_F as:

$$E_F = \int_0^\delta R(x)\,dx \tag{5-14}$$

where x is the deflection, and δ is the deflection at the rated or desired energy level. For example, for the linear elastic case of a steel spring or cantilever pile, the energy is equal to one-half the reaction (R) times the deflection (δ). A spring constant, $K_F = R/\delta$, can be defined over the elastic range of the spring. When the pier or dolphin deflection is significant, the fender and structure act like springs in series, and the combined spring constant (K_c) can be found from:

$$\frac{1}{K_c} = \frac{1}{K_F} + \frac{1}{K_s}$$

where K_s is the equivalent spring constant of the structure. This equation can be used for any number of springs in series by summing the reciprocals of the linear spring constants. For springs in parallel, such as for a rigid wale backed by several units along its length, the spring constants are directly additive; that is, $K_c = K_{F1} + K_{F2} + \ldots + K_{FN}$. For a hydraulic-type fender, the load remains nearly constant with deflection, so that the maximum energy is extracted per unit deflection, and the fender does not recoil, reflecting energy back into the ship. For most compressible and buckling column-type resilient rubber units in common use today, however, the spring constant cannot be defined except along a limited length of the curve; so common practice is to obtain both the reaction and the energy at a given deflection from the manufacturer's published data. Care must be taken to correct published energy values for the angle of approach.

Fender Types

Perhaps the most common types of fender systems in the United States today are the rigid timber framework and/or a row of piles interconnected by timber wales that absorb energy by compressing and bending the wood fiber. These systems have a minimal energy absorption capability and sometimes are backed up at deck level by solid or hollow core rubber units. Figure 5.11 illustrates a typical timber fender pile installation, and Figure 5.12 shows a continuous fender rack as used at active ferry slips. Timber systems usually are economical on a first-cost basis, easy to repair, and adaptable to large tide ranges. Their disadvantages include high maintenance costs and limited protection. Many timber systems are being gradually replaced by more contemporary rubber units.

Camels are floating separators that may simply consist of a log or pile tethered to the pier or quay face or may be more substantial built-up timber or steel pontoon structures (see Section 9.1). Camels may be employed as the sole source of fendering, especially at solid-faced quays, or they may be used in conjunction with timber fender pile systems. Caution should be exercised when using camels with springing pile fenders, as they may exert a concentrated load at the piles' midspan, resulting in frequent pile breakage. Therefore, camels used in this way should be long enough to distribute berthing and mooring loads over a sufficient number of piles. *Pneumatic floating fenders* made from rubber or other elastomeric materials may double as fendering and camels to maintain a proper vessel standoff distance. Figure 5.13 shows a row of floating units alongside a rigid timber pile fender system. Such units also are used as separators between vessels and are available in a large range of sizes and energy absorption

BERTHING LOADS AND FENDER SYSTEM DESIGN 157

Figure 5.11. Timber fender pile system.

capacities. Pneumatic-type fenders also are available as fixed units to be mounted on the quay face.

Hanging or draped fender units consisting of hollow core rubber sections or foam-filled rubber units sometimes are employed at solid-face quays. In many European ports, especially inland ports and waterways, brushwood fenders consisting of cylindrical bundles of tree branches bound together , such as willow, are suspended along the quay face.

158 DESIGN OF MARINE FACILITIES

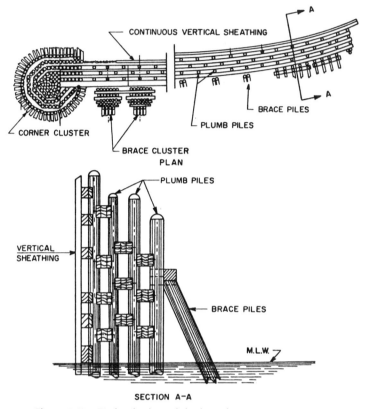

Figure 5.12. Timber fender rack for ferry slip. (From NAVFAC (4).)

Berthing beams typically consist of horizontal concrete or steel beams faced with resilient fendering and supported by cantilevered steel or prestressed concrete piles, which typically are raked at an angle toward the vessel. The rake or batter usually ranges from 1 on 12 to 1 on 5 horizontal to vertical, and it ideally results in a dead load condition of maximum working stress in the pile, in order to gain a greater working stress and energy capacity when the piles are bent back in the opposite direction. Berthing beams are used where a large range of vessel sizes is likely, usually to fend vessels from fragile structures that cannot support fender systems themselves.

Rotating fenders typically consist of either a pneumatic tire or a hollow rubber unit mounted on a vertical shaft. The tire or cylinder is free to rotate, thus reducing friction, and to absorb energy by its deformation. Such fenders are ideally suited to corner protection, especially where vessels frequently are warped about them. They are generally less tolerant of vertical motions.

Less common today are mechanical-type systems consisting of springs and

Figure 5.13. Pneumatic-type floating fenders alongside pier (Photo courtesy of Seaward International, Inc.)

linkages, and/or gravity systems consisting of suspended weights or buoyancy units. Some of these systems are described further in references (1), (39), and (40).

Hydraulic fenders may be preset to a given reaction force level. Bruun (41) has advocated their use at exposed locations because of their energy absorption efficiency and nonrecoiling characteristics. The fact that they do not immediately return to their original position after impact may be a problem at certain locations. Others have noted that hydraulic units are likely to require greater than average maintenance (1).

At most major marine terminals today, however, high-energy-capacity resilient rubber units are employed. There are a wide variety of shapes and sizes of elastomeric units on the market today. Figure 5.14, from reference (3), illustrates some of the more common generic types. The elastomer is of either a natural rubber or a synthetic rubber, usually of the butyl type, the properties of which can be varied to obtain different characteristics. Rubber units may be worked in direct compression, bending via buckling column action or in shear or torsion. More energy per unit volume can be obtained by shearing rubber versus compressing it. However, shear fenders are more difficult to construct and wear more quickly. The larger units are equipped with frontal frames or face panels to minimize contact pressure on the vessel's hull. Further discussion of their design follows in Section 5.6.

Padron and Han have presented the results of a study of fender system problems in U.S. ports (42), based upon a study conducted for MarAd (43) that is of interest to those evaluating alternative fender types with particular regard to maintenance costs and problems. Their findings are summarized in Table 5.2, and the most prevalent types are illustrated in Figure 5.15. Padron and Han found that timber systems generally had greater maintenance problems than rubber systems, chief among them damage due to high energy berthings, wear, and attack by marine organisms. Rubber fender units when properly sized and installed may have practical design lives of 10 to 20 years or more, depending upon the level of activity at the berth (36, 44). Replacement of deteriorated systems with contemporary resilient fenders can play an important role in the upgrading of existing facilities (45).

5.6 FENDER SYSTEM DESIGN

The design of marine fender systems usually begins with a determination of the fender energy absorption requirements and other fender type selection factors, as introduced in Section 5.5. Table 5.3, from reference (3), lists some additional design considerations for specific facility types. The need to consider the fender's performance while the vessel is moored cannot be overemphasized. This is especially true at exposed locations and for larger vessels subject to dynamic forces (46–48). The fender load/deflection properties should be compatible with the elasticity of the mooring lines at open sea berths (47), and berthing/mooring design requirements should be integrated.

The ultimate capacity of a fender unit and its supporting structure should be on the order of twice the nominal design energy level, subject to site-specific studies which may determine that a higher or lower figure is warranted (3). The overload factor also depends to some degree upon the type of fender system,

BERTHING LOADS AND FENDER SYSTEM DESIGN 161

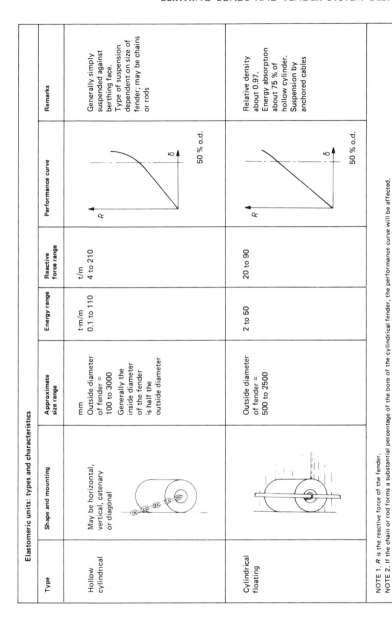

Figure 5.14. Elastomeric fender units: types and characteristics. (From British Standards Institution, BS 6349 (3).)

162 DESIGN OF MARINE FACILITIES

(continued)

Type	Shape and mounting	Approximate size range	Energy range	Reactive force range	Performance curve	Remarks
V-shaped	Closed leg / Open leg	mm H = 150 to 1300	t·m/m 0.3 to 60	t/m 6 to 110	(curve, 50% H)	Large variety of rubber grades made with single base plate (closed leg) or double base plate (open leg); also with resin board facing to decrease surge friction. It is also possible to use multiple units with a fender frame
Buckling column		H = 400 to 2500	3 to 160	20 to 180	(curve, 50% H)	Large variety of rubber grades. It is used with a fender frame. Reaction R at least 30% less than for V-shaped fenders. Fenders may need to be supported by piles

NOTE. R is the reactive force of the fender.

Figure 5.14. *(Continued)*

BERTHING LOADS AND FENDER SYSTEM DESIGN 163

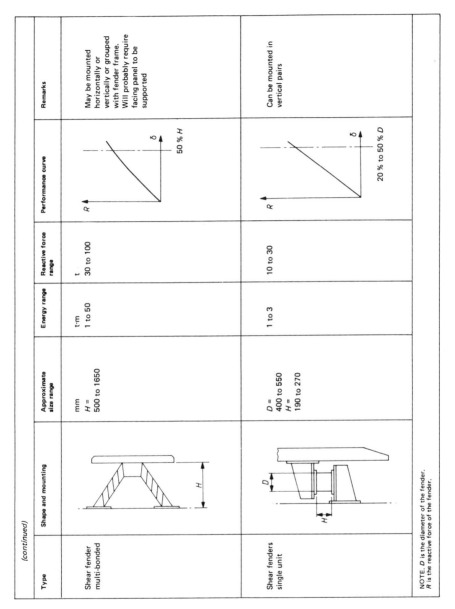

Figure 5.14. (Continued)

164 DESIGN OF MARINE FACILITIES

Type	Shape and mounting	Approximate size range	Energy range	Reactive force range	Performance curve	Remarks
Hollow cylindrical- axially loaded	(figure with H)	mm $D = 300$ to 3500	t·m 0.4 to 700	t 4 to 600	(curve, 40% to 50% H)	Large variety of rubber grades available
Rectangular square and 'D' hollow	Suspension system (figure) or (figure)	mm Cross-sectional dimensions vary from 150×125 to 305×305	t·m/m 1 to 4	t/m 30 to 90	(curve with *)	*Large variation in performance in relation to deflection limits

(concluded)

NOTE. D is the diameter of the fender.
R is the reactive force of the fender.

Figure 5.14. (Continued)

Table 5.2. Ranking of fender system problems and fender system types by prevalence/severity of fender system problems.*

	Rating of severity of fender system problems as percentage of total wharf length												Rank of fender
	Timber fender systems							Rubber fender systems					
Description of fender system problem (1)	A 34.3 (2)	B 9.1 (3)	C 4.9 (4)	D 3.3 (5)	E 5.9 (6)	F 1.2 (7)	G 14.1 (8)	H 22.9 (9)	I 2.1 (10)	J 1.7 (11)	K 0.5 (12)	Summation (13)	system problem (14)
High berthing energy	5	4	5	5	5	4	5	1	1	1	1	380.9	1
Wear	4	3	5	5	3	3	3	1	1	2	2	298.5	2
Deterioration by marine organisms	4	4	4	4	4	4	2	1	1	3	1	293.6	3
Securing lines to fender system	4	4	4	1	3	3	—	—	—	—	—	217.8	4
Performance adversely affected	2	2	2	2	3	3	3	1	1	2	1	195.7	5
Snagging by vessels	1	1	1	1	3	3	3	1	3	3	3	151.0	6
Corrosion of steel components	1	1	1	1	1	1	1	2	2	3	3	129.4	7
Summation	720.3	172.9	107.8	62.7	129.8	25.2	239.7	160.3	18.9	23.8	5.5		
Rank of fender system type	1	3	6	7	5	8	2	4	10	9	11		

Note: Summations are the sum of the products of wharf length and rating for each row or column.
*From: Padron and Han (42).

its mode of failure, and the consequences of such a failure. Steel pipe pile dolphins, which absorb energy via bending deflection, often are designed to be at up to 80% of the yield stress in the steel under nominal design conditions, corresponding to a factor of safety of approximately 1.5 on the design energy at the yield of the steel (1). Other aspects of fender system design to be discussed further here include: installation requirements with particular regard to vertical and longitudinal rubbing forces, face panel dimensions with regard to allowable hull pressures, dimensioning, layout and spacing requirements, and material specifications. Worked examples of the design of various types of fender systems are given in references (4), (24), (39), and (40), and in the product literature of certain fender manufacturers. The design of dolphins as fenders, such as timber pile clusters and cantilever pile arrangements, is discussed in Section 7.4.

166 DESIGN OF MARINE FACILITIES

Figure 5.15. Prevalent types of timber and rubber fender systems. (From Padron and Han (42).)

Mounting and Installation

Fender installations should be as simple and rugged as possible with maintenance and ease of replacement as key features. The use of chains, turnbuckles, and moving parts should be kept to a minimum. The ability of the pier or berthing structure to distribute and resist all berthing and mooring reaction forces must be carefully checked. Figure 5.16 shows a representative contemporary rubber, buckling-column action type, fender installation fitted with a low friction face panel and retaining chains. Alternatively, the face panel may be mounted on piles, where large tide ranges must be accommodated. The instal-

Table 5.3. Facility type fender design considerations.*

Vessel categories Type	Features to be taken into account in the design of the fendering system
Train and vehicle ferries	Quick turn round
	End berthing
	High docking velocities
	Intensive use of berth (see also notes 1, 2, and 4)
Roll on–roll off (Ro–Ro) vessels	Loading ramps, slewed or end loading (vessel mounted or shore based) end berthing (see also notes 1 and 2)
LNG/LPG carriers	Shallow draught even at full load
	Low berthing pressures on hull
	Single type vessels using dedicated berth
	Need to avoid fire hazards from sparking or friction (see also notes 1 and 3)
Coastal tankers	Very low amidships freeboard
	Intensive use of berth
	Need to avoid fire hazard from sparking or friction (see also notes 3 and 4)
Containerships	Flared clipper bows with liability to strike shoreside installations (see also note 1)
Bulk carriers	Need to be close to berth to minimize shiploader outreach
	Possible need to be warped along berth for shiploader to change holds
	Large change in draught between empty and fully laden conditions (see also note 1)
Passenger liners	Little change of draught between empty and fully laden condition (see also note 1)
Tankers	Need to be close to berth to reduce loading arm length
	Large change in draught between empty and fully laden conditions
	Need to avoid fire hazard from sparking or friction (see also note 1)
General cargo vessels	Need to be close to berth to minimize outreach of quayside cranes and/or ship's gear
	Large change of draught between empty and fully laden conditions
	Possible long occupancy of berths
Coastal cargo vessels	Short straight run of body (see also note 4)
Miscellaneous tugs, supply boats, barges, lighters, and fishing boats	Need for very substantial fendering for heavy use
	Timber fendering usually provided (see also notes 2 and 4)
Yachts	Need for soft fendering which is sometimes provided by the yachts themselves

Note 1. The vessels are possibly fitted with bulbous bows.
Note 2. The vessels are possibly fitted with belting.
Note 3. The vessels do not necessarily have manifolds at the amidships position.
Note 4. The vessels may often berth without the aid of tugs.
*From reference (3), reproduced with the permission of the British Standards Institution.

168 DESIGN OF MARINE FACILITIES

Figure 5.16. Rubber fender installation. (Courtesy Bridgestone Tire Co., Ltd. (24).)

lation of high-energy-capacity units on piers often requires the use of raised parapets or lowered skirt beams along the pier face.

The design of all rubber dock fenders that are self-supporting and have a frontal load distribution panel utilizes suspension, shear, and restraining chains for safety and additional support. In actual practice, chains are needed only in case of an accident. The fenders generally are capable of very large shear deflections without damage and under the design loads; there will be no damage due to the deflections. On the other hand, in an accident, having the chains act as stoppers may mean the difference between costly damage and an undamaged fender. In operational use, chains are used to keep the tension in the rubber to a minimum because after several years of continuous tension, the fender may sag. Another operational application of chains is in wale systems, where they are used as an equivalent structural joint or support.

Rubbing Forces

Because the velocity of impact does not act exactly normal to the fender face, there always will be some horizontal component of velocity acting parallel to

the face of the structure. The upper bound of the resulting force is determined by the coefficient of static friction between the vessel's hull and the fender face material. Various recommendations have been followed for design values for the rubbing forces, typically ranging from 0.20 to 0.50 or more of the normal fender reaction. There is wide scatter in the reported values of coefficients of friction. Table 5.4 gives a range of values suggested for design purposes from various sources. The actual coefficient of friction (μ) also depends upon whether the contact surfaces are wet or dry, and ice- or oil-covered, as well as the surface roughness, which will vary with the degree of wear and/or the condition of the vessel's hull. Several fender manufacturers produce low friction facing materials, usually of a high density polyethylene (HDPE) or of nylon, that also possesses good impact strength and abrasion resistance.

Maximum rubbing forces, both vertical and horizontal, usually are associated with mooring forces rather than berthing forces, although it is usual for designers to apply an assumed value of μ times the design berthing reaction force. Vertical forces, however, are more likely to occur while the vessel is moored rather than during berthing. NAVFAC (4) suggests designing for a longitudinal force of 0.5 times the maximum berthing load and for a vertical force of 0.3 times the maximum berthing load. My recommendation is that the vertical and longitudinal rubbing forces associated with the fender reaction forces under the maximum design mooring conditions also be checked and compared with the berthing conditions.

Allowable Hull Pressures and Fender Face Dimensions

Vessel parameters affecting fender layout and design include hull strength and allowable bearing pressures and hull geometry, especially hull curvature and length of parallel midbody. For smaller vessels, the local hull strength is not usually a problem because of the closer frame spacing, greater curvature, and inherently greater stiffness, compared to larger vessels. For larger vessels with large areas of vertical and parallel sides, the shell plating and stiffeners, which are designed for local hydrostatic pressures, are vulnerable to local point loads.

Svendsen and Jensen (49) provided a general solution in graphical form for tankers and bulk carriers based on the plastic moment capacity in bending of

Table 5.4. Coefficients of friction for fender face materials.*

Material	Extreme range	Typical design values
Timber	0.3–1.0	0.4–0.6
Rubber	0.5–0.7	0.5–0.6
Steel	0.15–0.74	0.25
HDPE	0.10–0.25	0.20

*Against steel hull. Static coefficient usually controls. Higher end of range applies to dry, worn, and roughened surfaces; lower end of range to wet, new, and smooth surfaces.

longitudinal and transverse stiffeners and frames for mild ship hull steel. The ratio between the fender panel width and the vessel transverse frame spacing should not be less than 0.5 to 0.65, and the ratio between the fender panel height and the side longitudinal spacing not less than about two. Piaseckyj (36) applied the work of Svendsen and Jensen using characteristics of contemporary fender systems and came up with the following maximum recommended hull pressures for design:

- LNG, LPG, modern bulk carriers: 2500 psf.
- Containerships: 5000 psf.
- Freighters: 5000 psf.
- Tankers: under 50,000 DWT—4000 psf; under 100,000 DWT—5500 psf; over 100,000 DWT—7000 psf.

The above values fall in the range of from 15 to 50 psi maximum. There is a correlation of increasing local strength with increasing draft. As a general rule of thumb, for large vessels such as tankers the maximum allowable hull pressure on side plating and longidutinals in tonnes per square meter is approximately equal to the fully loaded draft in meters (1). Caution should be applied in dealing with lightly constructed vessels such as certain contemporary naval warships; NAVFAC (4) gives limiting hull contact pressures ranging from 5 psi for vessels of 150 tons displacement to 20 psi at 20,000 tons displacement. For aircraft carriers, specific allowable hull pressures range from 15 to 44 psi. Where hull pressures may be critical, the naval architect or vessel owners should be consulted for specific requirements.

Fender Spacing and Layout

The vessel's geometry affects fender spacing in particular, as well as the location and number of fenders contacted. Fender spacing additionally depends upon the type of fender system and structural support, the range of vessel sizes to be accommodated, and the type and arrangement of berth and mooring loads. Typical berth arrangements are discussed in Chapters 3 and 6. A tanker often is moored against two discrete dolphins, spaced on the order of 25% to 50% of the vessel's *LOA* apart. Along a pier or quay individual fenders may be spaced from 30 to 120 feet apart, with 50 to 80 feet being most common. The fender spacing should allow for the smallest design vessel to safely lie at any location alongside. Spacing ranging from 8% (36) to 15% (3) of the vessel's *LOA* have been proposed. The distance between centers of fender units often is referred to as the pitch. Specific recommendations have been made for Ro/Ro and ferry-type vessels (1, 3). Figure 5.17 illustrates typical fender spacings for various berth types.

Technically, a vessel alongside only requires two points of contact while in berth, although three or more are recommended, which means that the absolute maximum spacing is controlled by the length of the vessel's parallel sides. In general, the ratio of a vessel's parallel midbody length to its overall length is on the order of 35% to 55% of its *LOA*, usually being larger for longer vessels. This ratio often determines the point of first contact with the vessel's hull, which

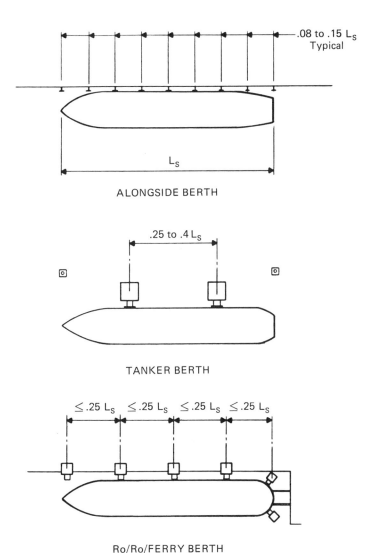

Figure 5.17. Fender spacings for typical berth arrangements.

172 DESIGN OF MARINE FACILITIES

is usually at the end of the parallel midbody, and also the length of vessel available to contact fenders under moored conditions.

The effect of the hull's curvature near the bow or stern on fender spacing is illustrated in Figure 5.18. Vertical curvature of the hull and hull flare and overhangs and/or projections such as bulbous bows also must be considered in fender system layout. The standoff distance, from the face of the pier to the face of the fender, should be minimized in the interest of increasing the effective reach of loading equipment, but should also provide a sufficient buffer zone to prevent contact of any parts of the vessel with the pier face with fenders at 50% compression. The standoff distance usually ranges from three to six feet for most seagoing terminal facilities. At offshore installations, this distance will be closer to ten feet or more.

Material Specifications

All materials for fender installations should be specified to meet minimum strength and durability requirements as described in Section 3.5. All metal hard-

Figure 5.18. Effect of hull shape on fender spacing.

Table 5.5. Properties of selected elastomeric fender and fender facing materials.*

Property	ASTM test method	Natural rubber	Butyl rubber	Neoprene rubber	HDPE
Density (S.G.)	—	0.93	0.92	1.23	0.93–0.96
Tensile strength (psi)	D-412	4000	2000	3000	3100–5500 (ASTM D-638)
Elongation (%)	D-412	700	600	800	600–800
Hardness (Shore A—durometer)	D-412	70 ± 5	70 ± 5	70 ± 5	50–60 (D—durometer)
Tear strength (pli)	D-624	400+	300	300	—
Compression Set (%)	D-395	—	25%	—	—
Water absorption (%)	D-471	—	0.2	—	0.01
Ozone resistance	—	Poor	Excellent	Good	—
Hydrocarbon resistance	—	Poor	Fair	Good	—

*Elastomeric materials may exhibit a wide range of properties. The above values are representative of typical fendering materials in new condition for general reference only. Elastomer properties in general are specified in accordance with ASTM D-2000.

ware items should be hot-dipped galvanized with a minimum of two ounces of zinc per square foot of coated surface. Elastomeric fender materials can be of natural or synthetic rubbers. Synthetic rubber preferably should be of the butyl or neoprene type. Latex rubbers generally are unacceptable because of their poor hydrocarbon resistance. Rubber properties can be varied widely to obtain the desired stiffness and resistance to sunlight, chemicals, and temperature effects. An important property of elastomeric materials is their rebound curve as observed in the hysteresis effect. Rubbers with a greater hysteresis effect will exhibit less recoil or tendency to bounce the vessel off.

Table 5.5 gives some important properties of representative elastomeric fender materials and HDPE fender facing materials for comparison. Young (50) provides further discussion on specifying marine fenders. Potential fender manufacturers for a given project should be contacted prior to the preparation of fender specifications in order to verify the optimum rubber properties available.

REFERENCES

1. PIANC (1984). "Report of the International Commission for Improving the Design of Fender Systems," PIANC, Supp. to Bull. No. 45, Brussels.
2. ——— (Mar. 1980). "Design of Fender Systems," Japanese National Section PIANC, Working Group on Fender System Design.
3. BSI (1985). *British Standard Code of Practice for Maritime Structures*. Part 4: "Design of Fendering and Mooring Systems," BS-6349. London.
4. NAVFAC (1980). *Piers and Wharves*, DM 25.1. Dept. of the Navy, Alexandria, Va.

5. Vasco Costa, F., (1981). "Berthing Maneuvers," in *Port Engineering*. Gulf Publishing, Houston, Tex.
6. Vasco Costa, F., (May, June, July 1964). "The Berthing Ship," *The Dock and Harbour Authority*, Vol. XLV, London.
7. Saurin, B. F. (June 1963). "Berthing Forces of Large Tankers," *Proc.* of Sixth World Petroleum Congress, Section VII, Paper No. 10.
8. Kray, C. J. (1982). "Safety of Ships and Structures During Berthing and Mooring," *Marine Technology Society Journal*, Vol. 16, No. 1, Washington, D.C.
9. Grim, O. (Sept. 1955). "Das Schiff und der Dalben," *Schiff and Haven*.
10. Giraudet, P. (Mar.–Apr. 1966). "Recherches experimentales sur l'energie d' accostage des navires," *Annales des Ponts et Chaussées*, No. 2.
11. Lee, T. T., Nagai, S., and Oda, K. (May 1975). "On the Determination of Impact Forces, Mooring Forces, and Motions of Supertankers at Marine Terminals," *Proc.* OTC, Houston, Tex.
12. Hayshi, T. and Shirai, M. (1976). "Added Masses of Large Tankers Berthing to Dolphins," *Proc. ASCE*, 15th Coastal Engineering Conf., Honolulu.
13. Blok, J. J. and Dekker, J. N. (May 1979). "On Hydrodynamic Aspects of Ship Collisions with Rigid or Non-rigid Structures," *Proc.* OTC, No. 3665, Houston, Tex.
14. Veda, S. (Dec. 1981). "Study on Berthing Impact Force of Very Large Crude Oil Carriers," *PHRI* of Japan, Vol. 20, No. 2.
15. Ball, D. J. and Hall, C. D. (Oct. 1982). *Ship Berthing Forces*, Vols. 1-6. Simon Engineering Laboratory, University of Manchester, U.K.
16. Ball, D. J. and Markham, A. (Oct. 1982). "Maximum Added Mass for a Berthing Tanker in Very Shallow Water," *The Dock and Harbour Authority*, London.
17. Van Oortmerssen, G. (May 1974). "The Berthing of a Large Tanker to a Jetty," *Proc.* OTC., paper No. 2100, Houston, Tex.
18. Fontijn, H. L. (May 1980). "The Berthing of a Ship to a Jetty," *Proc. ASCE*, WPCOE Div., WW-2, Vol. 106.
19. Middendorp, P. (Oct. 1981). "Fender Forces from Berthing Ships," *The Dock and Harbour Authority*, London.
20. Ball, D. J. (Mar. 1982). "The Concept of Added Mass," *The Dock and Harbour Authority*, London.
21. BSI (1984). *British Standard Code of Practice for Maritime Structures*. Part 1: "General Criteria," 856349. London.
22. Tomlinson, P. (July 1986). "Practical Advantages of Berthing Aid Systems," *The Dock and Harbour Authority*, London.
23. Brolsma, J. U. et al. (1977). "On Fender Design and Berthing Velocities," *Proc.* PIANC, 24th Int. Nav. Congress, 5-II-4, Leningrad.
24. ——— (n.d.). *Marine Fender Design Manual*, Catalog No. F100E-1. The Bridgestone Tire Co., Tokyo.
25. Wirzbitzki, B. et al. (1977). "Criteria for Economical Design of Fender Systems," *Proc.* PIANC, 24th Int. Nav. Congress, Leningrad.
26. NAVFAC (1981). *Ferry Terminals and Small Craft Berthing Facilities*, DM 25.5. Dept. of the Navy, Alexandria, Va.
27. Derucher, K. N. (June 1983). "Evaluation of Pier Fendering Systems," *The Dock and Harbour Authority*, London.
28. Sherman, Z. (1980). "Ferry Fender and Rack Design," American Wood Preservers Association, AWPI, Tech. Guidelines, 5-6, McLean, Va.
29. Eddie, A. G. F., (1981). "Risk Analysis of Berthing Accidents Involving Large Ships," *Proc.* PIANC, 25th Int. Nav. Congress.
30. Goulston, G., (1973). "Measured Berthing Energies and Use of Control Instrumentation," in

BERTHING LOADS AND FENDER SYSTEM DESIGN 175

"Analytical Treatment of Problems of Berthing and Mooring of Ships," *Proc.* NATO Advanced Study Institute, Wallingford, U.K.
31. Svendsen, I. A. (Sept., Oct. 1970). "Measurement of Impact Energies on Fenders," *The Dock and Harbour Authority*.
32. Dent, G. E. and Saurin, B. F. (Nov. 1969). "Tanker Terminals Berthing Structures," Conf. on Tanker and Bulk Carrier Terminals, *Proc.* ICE
33. Toppler, J. F. and Weersma, J. (May 1972). "Planning and Design of Fixed Berth Structures for 300,000 to 50,000 DWT Tankers," *Proc.* OTC, No. 1646, Houston, Tex.
34. Balfour, J. S., Feben, J. C., and Martin, D. L. (1980). "Fendering Requirements/Design Fender Impact Criteria," *Proc. ASCE*, Ports 80, Spec. Conf., Norfolk, Va.
35. Girgrah, M. (1977). "Practical Aspects of Dock Fender Design," *Proc.* PIANC, 24th Int. Nav. Congress, 5-II-4, Leningrad.
36. Piaseckyj, P. J (1977). Untitled paper 5-II-4, presented at 24th Int. Nav. Congress, PIANC, Brussels; incl. 1982 update By Duxbury Int., Duxbury, Mass.
37. NATO (1965). "Analytical Treatment of Problems in the Berthing and Mooring of Ships," *Proc.* NATO Advanced Study Institute, Lisbon, published by ASCE, New York.
38. NATO (1973). "Analytical Treatment of Problems in the Berthing and Mooring of Ships," *Proc.* NATO Advanced Study Institute, Wallingford, U.K.
39. Quinn, A. D. F. (1972). *Design and Construction of Ports and Marine Structures*. McGraw-Hill, New York.
40. Derucher, K. N. and Heins, C. P. (1979). *Bridge and Pier Protective Systems and Devices*. Marcel Dekker, New York.
41. Bruun, P. (1981). *Port Engineering*. Gulf Publishing, Houston, Tex.
42. Padron, D. V. and Han, H. Y. (Aug. 1983). "Fender System Problems in U.S. Ports," *Proc. ASCE*, JWWPCO Div., Vol. 109, No. 3.
43. Han, E. et al. (1978). "Improved Fender Systems for Shallow and Deep Draft Berths—Phase I," Report #MA-GEN-970-78046, prepared for U.S. Dept. of Commerce, MarAd, by Dravo Van Houten, Inc., New York.
44. Thoresen, C. A. (1988). *Port Design: Guidelines and Recommendations*. Tapir Publishers, Trondheim, Norway.
45. Asayama, K. and Ohtsuka, T. (1983). "The Role of Marine Fenders for Repairing, Upgrading and Rehabilitating Existing Facilities," *Proc. ASCE*, Ports '80 Spec. Conf., New Orleans, La.
46. Bruun, P. A. (Feb. 1984). "Mooring and Fendering Rational Principles in Design," *The Dock and Harbour Authority*, London.
47. Koman, B. (1980). "Fender System Requirements at Open Sea Berths," *Proc. ASCE*, Ports '80, Spec. Conf.
48. Khanna, J. and Birt, C. (1977). "Soft Mooring Systems at Exposed Terminals," *Proc. ASCE*, Ports '77, Spec. Conf., Long Beach, Calif.
49. Svendsen, I. A. and Jensen, J. F. (June 1970). "The Form and Dimensions of Fender Front Structures," *The Dock and Harbour Authority*.
50. Young, R. A. (1980): "Marine Fender Specifications," *Proc. ASCE*, Ports '80, Spec. Conf., Norfolk, Va.

6

Mooring Loads and Design Principles

Mooring loads usually govern the required lateral load capacity of a pier or berth structure. They are induced by the action of wind, current, and waves on the moored vessel and are transmitted to the structure via direct breasting or through the pull of mooring lines. Within protected harbors, wind and current loads usually are estimated for some assumed limiting wind and current velocities, and the effects of waves and harbor motions are neglected. Today with marine terminals in more exposed locations, wave loads cannot be neglected and, in fact, may govern design. With larger vessels transiting more crowded harbors, the effects of passing vessels on vessels already moored must be considered. Even in protected harbors, vessel movements in response to long period waves and harbor oscillations known as seiches may cause damage to vessel and pier alike, even when winds and currents are calm.

This chapter introduces some fundamental concepts of moored vessel behavior and the layout of mooring lines and hardware for vessels moored at fixed structures. Evaluations of wind, current, standoff forces, and wave loads are treated separately. Because of the complexities of calculating wave loads and other dynamic effects, mathematical and physical scale models are often used, and this subject is introduced at the end of the chapter. This discussion is oriented to vessels moored at fixed structures. The subject of freely moored vessels at single point moorings is addressed later, in Section 9.7.

6.1 MOORING SYSTEM PRINCIPLES

Mooring forces acting on a ship at berth arise from the following sources:

- Wind
- Current
- Passing vessel effects
- Hydrodynamic standoff forces
- Wave action
- Seiche and long period wave effects
- Tidal variations and vessel draft changes
- Vessel movements at berth
- Ice

MOORING LOADS AND DESIGN PRINCIPLES 177

In addition to staying secured in its berth, a moored vessel must remain as motionless as possible within a limited range of acceptable motions. At most harbor and near-shore locations it is usually sufficient to design the mooring system components for the maximum expected or prescribed limiting wind plus current loads. Mooring systems adequately designed for appropriate combined wind and current loads usually will possess sufficient capacity to cope with minor wave and water level changes, wind gust effects, and so on. Along river banks with strong currents or narrow channels, however, significant off-berth standoff forces may arise due to locally increased water velocities and resulting negative pressures caused by passing vessels or strong currents. At offshore berths and exposed near-shore sites, wave action also must be considered. Certain harbors are affected by long period wave action such as surging motions, sometimes referred to as range action or seiching, which can result in very large mooring line forces. In areas of large tidal range, especially when combined with rapid cargo transfer operations, mooring lines may be subject to rapidly changing lengths and angles, resulting in increased line loads. Large vertical forces may result from friction between the vessel and the fender facing. Warping and shifting of the vessel within its berth may impose longitudinal rubbing forces. Ice is seldom a factor in mooring design loads, but it should be considered in areas subject to moving ice floes.

The general approach to calculating mooring loads on piers, wharves, and berthing structures depends somewhat upon the vessel size and type, site exposure and environmental conditions, and amount of information available describing those factors. For example, for smaller vessels, and for general and bulk cargo ships up to around 20,000 tons displacement at protected harbor locations, it may be adequate to provide mooring bollards or bitts of specified capacity at suitable locations, usually between 30- to 60-foot centers along the wharf face. A similar approach may be taken for even larger vessels when little is known about the specific vessels to be berthed. The capacity of the mooring hardware can be presumed from some guideline standard such as references (1) through (4) or other accepted practice. British (1, 2) and German (3) code recommendations for bollard capacities are summarized in Table 6.1. Another method is to size the mooring hardware to exceed the breaking strength of the vessel's docking lines if this information can be reasonably attained.

The mooring point loads are distributed into the overall pier structure as described in Section 7.3. Equivalent distributed mooring loads for oceangoing vessels usually range from 1000 to 2500 pounds per lineal foot (plf) of pier face applied horizontally at deck level for vessels from 2000 to 20,000 tons displacement (DT), to over 3000 plf for the largest vessels. Longitudinal mooring forces are usually on the order of 25% to 75% of the lateral forces. Vertical uplift on bollards also must be considered.

In general, for vessels around and exceeding 20,000 DT, for special vessels

178 DESIGN OF MARINE FACILITIES

Table 6.1. Mooring point (bollard) loads for general and bulk cargo ships (after BSI, 1984, and EAU, 1980).

Vessel displacement (tons)	Bollard load (kn)*
Up to and incl. 2,000	100
10,000	300
20,000	600
50,000	800
100,000	1,000
200,000	1,500
> 200,000	2,000

*For exceptional wind, current, or other adverse effects, increase the above values by 25%. Per EAU (3), provide corner bollards of 2500 kN capacity for vessels > 100,000 T. Assumed bollard spacings of 15 m to 30 m. *Note:* 1 kN = 224.8 lb.; 1 m = 3.28 ft.

such as those with high superstructures, and/or where severe site conditions exist, the maximum probable loads should be calculated for each mooring point for the entire range of vessel sizes and types expected. This involves calculating the combined forces and moments produced by the design level wind plus current plus equivalent static wave force and/or standoff forces, as applicable and with regard to the elasticity and geometry of the mooring lines. The calculations may be simplified in some instances by assuming that only bow/stern and/or breast lines take the lateral loads, and spring lines take the longitudinal loads. Refer to Figure 6.1 for mooring line and layout definitions.

Another approach is to assume an unequal distribution of the total mooring force among the mooring points. According to BSI (2), when the vessel is moored at six points, one-third of the total force should be assumed at any one point; and where the vessel is moored at four points, one-half of the total force should be assumed to act at any single point. Calculation of line loads is discussed at the end of this section. In cases where wave loadings may be severe or long period waves or range action is present, a more rigorous analytical treatment and/or the use of physical models may be required, as described in Section 6.7.

A freely floating vessel possesses six degrees of freedom of motion—three translational and three rotational, as illustrated in Figure 6.2. These possible motions are defined as follows with respect to a cartesian coordinate system centered at the vessel's center of gravity (c.g.): *surge*, *sway*, and *heave* represent longitudinal, transverse, and vertical displacements in the x, y, and z directions, respectively, and *roll*, *pitch*, and *yaw* represent rotations about the x, y, and z axis, designated by ψ, ϕ, and θ, respectively. Heave, roll, and pitch motions are acted upon by the restoring force of gravity; so, a free floating vessel will possess natural periods of roll, heave, and pitch as described in

MOORING LOADS AND DESIGN PRINCIPLES 179

Figure 6.1. Vessel mooring lines and berth layout definitions.

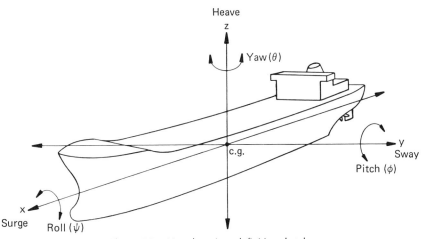

Figure 6.2. Vessel motions definition sketch.

Section 9.3. The periods of heave and pitch are usually on the order of one-half to two-thirds of the roll period. The heave response, or ratio of heave amplitude to wave amplitude, and the pitch response, or ratio of trim angle to sea surface slope, are always less than unity when the ratio of wavelength to ship length (L/L_s) is less than approximately 2. For $L/L_s > 2$, the ship basically will be in concert with the sea motion. In a seaway the various motions may interact via exchange of energy between modes, resulting in even more complex, coupled motions. This is especially true for coupled heave and pitch and for surge and yaw. Surge, sway, and yaw motions only exhibit a natural period when acted upon by some restoring force such as mooring lines and fenders; and it is these motions that are of principal interest herein. The importance of the dynamic behavior of the ship and mooring system as a spring-mass oscillating system will become apparent in the following sections.

Environmental disturbing forces manifest themselves in a spectrum of periodic forces covering a wide range of periods, as introduced in Section 3.4. A moored ship may respond selectively even to very subtle influences such as the wave drift force discussed in Section 6.6. When the ratio of the ship/mooring system natural period (T_n) nearly coincides with the period of the disturbing force (T_f), then resonance results in a dynamic amplification that theoretically would tend to infinity without damping. Typical natural periods of moored ships range from around 20 seconds for vessels of 3000 DT to 60 seconds and longer for vessels of 100,000 DT and larger.

Mooring Arrangements

Figure 6.1 illustrates the general arrangement and nomenclature for a vessel moored alongside a pier or quay, breasting against a fender system and being restrained by mooring lines. Bow and stern lines are attached to the respective ends of the vessel and usually make an angle of around 45° with the face of the pier (larger angles are not recommended), so that they provide some degree of both lateral and longitudinal restraint. Breast lines are nearly normal to the pier face and provide only lateral restraint, whereas spring lines run fore and aft, usually at an angle of 5° to 10° to the pier face, and provide only longitudinal restraint. The total length of a mooring line, or hawser, from connection point on the pier to connection point on the vessel is denoted by l. As a practical matter, mooring lines should not be less than 100 feet or more in length, to allow for change in tide and ballast conditions and to provide sufficient elasticity. The horizontal angle with a line normal to the vessel is denoted by α, and the vertical angle in the plane of the hawser by β. Vertical angles should not exceed 25° to 30° maximum and will vary with the stage of tide and the vessel ballast condition. In general, a minimum of four lines are required to safely moor a vessel, and a dozen or more lines may be required to secure larger

MOORING LOADS AND DESIGN PRINCIPLES 181

Figure 6.3. Tanker berth mooring line layout. (After OCIMF (8).)

vessels. Further description of typical pier-side mooring line layouts can be found in references (5) through (7).

A tanker berth type of mooring arrangement is illustrated in Figure 6.3. The vessel is breasted against two or more breasting dolphins with mooring lines running to smaller mooring dolphins, usually interconnected by catwalks for personnel access. Bow, stern, and breast lines should preferably have a maximum angle α of 15° and spring lines a minimum α of 80°, in accordance with the Oil Companies International Marine Forum (OCIMF) criteria (8). In general, this type of berth should be laid out symmetrically. Note that because the dolphins are set back from the berth line, the bow and stern lines are more nearly normal to the vessel's hull than they otherwise would be, allowing for greater lateral restraint. A vessel may be safely moored within its own length provided that the dolphins and hardware are properly designed. Bulk cargo terminals also may use this concept, frequently employing a greater number of dolphins, especially in cases where the ship needs to be shifted in position for access to all its hatches by loading/unloading equipment. Specific guidelines for tanker berth layouts and recommended mooring practice can be found in reference (8).

The overall adequacy of a vessel's mooring is determined not only by its structural integrity and safety under storm conditions, but also by the limitation of vessel movements at berth to prevent possible damage to the ship, pier, and loading equipment. Table 6.2, from Bruun (9), suggests approximate ranges of limiting motions for safe operation of various large ship types. Further discussion of allowable vessel motions can be found in references (9) through (13). Special care must be exercised in the location and orientation of a berth to avoid exposure to potentially damaging waves and/or nodal points in harbors subject to seiche or ranging action (see Section 3.4). In the planning of a facility in an exposed location, a statistical approach usually is taken, in which the percent frequency of time the berth is occupied and subject to conditions exceeding certain prescribed limits (e.g., limiting wave height by direction) can be compared with the economics of vessel turnaround time.

Figure 6.4 shows a variable orientation berth for an exposed offshore bulk terminal. The arrangement of dolphins and shiploader characteristics allow for a choice of six orientations, increasing the total berth availability to 90% as opposed to 70% for the most favorable single orientation. This facility and other multiorientation berth arrangements have been described by Sugin (14) and by Soros and Koman (15).

Table 6.2. Ranges for allowable maximum movements for large vessels > 200 meters at berth for unloading periods of oscillations of 60 to 120 sec (9).

	Surge (m)	Sway (m)	Yaw (degrees)	Heave (m)	Note
Tanker	±2[1]	+0.5 (away from berth)	1	±0.5	Surge most important.
Ore bulk (crane unloading)	±1.5	1.0 (away from berth)	1/2 not important	±0.5	Surge most important.
Grain bulk	±0.5	+0.5 (away from berth)	1/2 not important dep equipment	±0.3	Surge, sway most important.
LNG	very small (v.s.) ±0.2	v.s.	v.s.	v.s.	All movements risky.
Container[2]	±0.5	+0.3 (away from berth)	-nil	±0.3	
RO/RO (side)	±0.3	+0.2 (away from berth)	-nil	±0.1	For most effective operation,
RO/RO (bow or stern)	±0.1	nil	-nil	±0.1	all movements nil.

1. Depending upon mooring forces this movement could be even larger, e.g., accepted at Antifer, France.
2. In article by P. J. B. Slinn published in the *Dock and Harbour Authority*, August 1979, the results of field tests on loading of containers in a moving cell was that loading rates of 26 containers per hour were obtained with stiff moorings for surge movements of 0.92 m (101.5 sec) and sway movements of 0.48 m (40 sec).
From: Brunn (9). *Note:* 1 m = 3.28 ft.

MOORING LOADS AND DESIGN PRINCIPLES 183

Figure 6.4. Variable orientation, offshore loading terminal at Punta Colorada, Argentina. (Photo courtesy of Soros Associates, Consulting Engineers, New York).

In general, berths should be oriented as much as possible with the prevailing current direction in order to minimize current forces and increase controllability during berthing.

Design Mooring Conditions

It is generally considered unnecessary to design a mooring facility for the most extreme (e.g., 100-year) storm condition with the maximum vessel alongside. Hunley and Lemley (6) point out that, based upon available studies, 90% of the winds experienced at commercial ports are below 35 knots, discounting gusts of less than five minutes' duration. Standard practice in many instances is to assume that vessels will put to sea in extreme weather conditions when winds exceed certain threshold values, often taken to be around 50 to 60 knots. Where such assumptions are not considered valid, the 50-year recurrence level of extreme wind conditions has been adopted by many authorities (1, 6). For permanently moored vessels such as floating dry docks, floating piers or storage vessels, and mothballed vessels, the maximum winds associated with 100-year return periods should be used. Conversely, for temporary facilities with short design lives, say of one to ten years' maximum, the extreme winds associated with 10- to 25-year return periods may be applied with the designer's discretion. At terminals serving a mix of vessel sizes, and especially for those with rapid

vessel turnaround times, the joint probability of the largest vessel being alongside during the occurrence of 50- to 100-year storm conditions for the site generally will be acceptably low. For most active marine terminals relatively protected from wave action, a minimum design wind speed of around 60 knots from any direction combined with the maximum spring tidal current acting on the largest, most frequently calling vessel seems to be a reasonable minimum design condition in the absence of other information. Design wind speeds must be referenced to specified durations and elevations, as described in Section 6.3. Static wind and current analysis may be applied to vessels moored at "protected" locations.

The U. S. Navy, NAVFAC (16), requires that special consideration (beyond static wind and current analysis) be given when the following conditions obtain:

- Wave heights exceed: 1.5 feet for small craft or 4 feet for larger vessels.
- Winds exceed 60 knots.
- Currents exceed 3 knots.
- Tide range exceeds 8 feet.
- The site is exposed to hurricanes or typhoons, there is any evidence of seiche action, and/or free-floating ice is known to be a hazard.

Dynamic analysis is required for the above wave, seiche, and/or hurricane conditions. The wave period may be just as important a criterion as wave height for determining its potential effect on moored vessels. In particular, the peak spectral period (T_P, the period at which most of the sea's energy is concentrated), is most critical. For oceangoing vessels when T_P exceeds approximately 4 seconds, waves will begin to have a significant effect, depending upon the vessel's orientation to the sea. Wave conditions will be dealt with further in Section 6.6.

The OCIMF (8) recommends the following minimum design criteria for VLCC berths not exposed to wave action:

- A 60-knot wind (as measured at 10 meters and corresponding to a 30-second gust duration) from any direction plus one of the following:
 - 3-knot current from 0° and 180°
 - 2-knot current from 10° and 170°
 - 3/4-knot current from abeam
- All current forces based upon a $d/D = 1.1$ to 1.0. For combined loading, $d/D = 1.1$ to 3.0 must be investigated; for a vessel in full load to light condition, the current speed is average over the vessel's draft.

Under the above conditions, the load in the most heavily loaded line should not exceed 55% of the mean breaking load (MBL) of the line. Mooring lines should

be between 35 meters and 50 meters in length and preferably pretensioned to 10% MBL (8).

At offshore locations, design mooring conditions must be specified under which vessels of a given size would remain alongside. Such design criteria often are presented in terms of prescribed limiting conditions for berthing operations and terminal operations, and/or maximum conditions under which the vessel would remain alongside, as well as survival conditions for the terminal facilities without the ship in berth. For example, design criteria for the offshore berth depicted in Figure 6.4 are described as follows, after Sugin (14):

- Maximum ship size: 60,000 DWT.
- Water depth at berth: 42 feet with 30-foot tide range.
- For design berthing conditions:
 - Wind = 5 to 30 knots.
 - Wave height = 4 to 6 feet (head seas).
 - Current = N.A.
- For design operating conditions:
 - Wind = 10 to 35 knots.
 - Wave Height = 6 to 8 feet (head seas).
 - Current = N.A.
- For survival condition without ship in berth:
 - Wind = 80 knots.
 - Wave height = 12 feet significant wave (Hs)/29-foot maximum wave (Hmax).
- Current = 2 knots.

The above conditions must be analyzed over a range of tide levels and wind and sea directions to obtain the worst load combinations. Sugin (14) also has summarized design criteria for several other offshore bulk loading terminals.

Mooring System Design

Once the berth layout and design mooring conditions have been determined, then wind, current, and equivalent static wave forces, if applicable, can be calculated, in accordance with discussions in the following sections, and combined vectorially to obtain the maximum total longitudinal and lateral forces and yaw moment. The calculations may need to be carried out over a range of water depths, but only for the combined maximum conditions that are likely to occur simultaneously. When dynamic wave loads are involved, they cannot in general be accurately combined with steady wind and current loads by linear superposition (17); such calculations must be regarded as approximate and subject to dynamic analysis, as described in Sections 6.6 and 6.7. Passing vessels,

hydrodynamic standoff forces, and long period waves may produce large mooring loads; but in general these conditions do not act simultaneously with the maximum design wind, wave, and current conditions.

For a given site, berth orientation, and mooring arrangement, the design process generally proceeds as follows:

1. Define environmental conditions.
2. Calculate environmental loadings.
3. Calculate mooring system response and component loads.
4. Determine factors of safety.

Steps 2 through 4 may have to be reiterated for alternative berth orientations, and steps 3 and 4 for alternative mooring arrangements. Factors of safety on mooring system components generally range from 1.5 to 3.0 times ultimate strength, depending upon the severity and statistical probability of design environmental conditions, the level of confidence in design assumptions and calculation methods, and the nature and consequences of a failure. Redundancy should be provided as much as is practical.

Individual mooring line loads are calculated by balancing the sum of the static wind plus current longitudinal (F_x) and lateral (F_y) force components and yaw moments (M_y) about the vessel's geometric center and considering the geometry and elasticity of the mooring lines. If α is the horizontal angle in plan view between a line normal to the vessel and the vessel's chock, and β is the vertical angle in the plane of the mooring line between the pier mooring point and the vessel's chock (see Figure 6.1), then the tension (T_l) of the mooring line is related to the longitudinal (F_x), lateral (F_y), and vertical (F_z) force components as:

$$T_l = \frac{F_x}{\sin \alpha \cos \beta} = \frac{F_y}{\cos \alpha \cos \beta} = \frac{F_z}{\sin \beta} \qquad (6\text{-}1)$$

As an aid to determining the component line loads and identifying critical line loads, it may be useful to plot the line loads and total elastic restoring force (F_R) versus vessel surge and sway excursions (ϵ_x, ϵ_y), as illustrated in Figure 6.5. The mooring line force is then a function of its stretched length, assuming some initial at-rest length. The effect of any pre-tension in the line also should be accounted for. Normally a pre-tension of approximately 10% MBL will remove the initial sag from a synthetic line. The total elastic restoring force (F_R) then can be determined and balanced against the applied forces. Calculations may be simplified by neglecting the effect of spring lines on lateral restraint and of breast lines on longitudinal restraint. Individual line loads should be tabu-

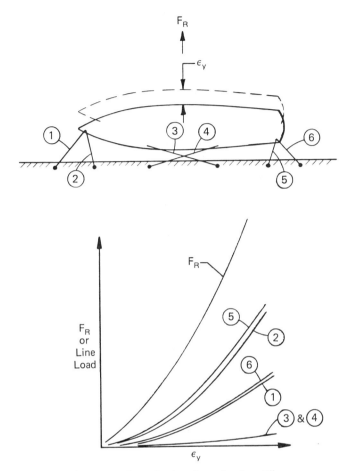

Figure 6.5. Example plot of mooring line stiffness.

lated and compared with the line's allowable load so that critical, highly loaded lines can be rearranged or resized.

The OCIMF (8) guidelines for tanker berths gives step-by-step instructions for calculating line loads and identifying critically loaded lines. In order to calculate mooring line loads with reasonable accuracy and thus provide a high degree of confidence in the security of a mooring layout, the OCIMF emphasizes the following good mooring principles:

- Lateral and longitudinal restraint functions are separated.
- Layouts should be as symmetrical as possible.
- Vertical line angles (β) should be less than $25°$.

- Bow and stern line deviations from true breast lines (α) should be less than 45°.
- All lines should be of similar material, construction, makeup, and size.
- Lateral lines should be effectively grouped at the bow and the stern.

The tension in lateral restraint developed in any line is proportional to cos α/l, and the component of tension effective in resisting the lateral load is proportional to $\cos^2 \alpha/l$. The maximum restraint capacity of a mooring layout is limited by the allowable or ultimate tension capacity of the most heavily loaded line. The most critical line in a group will be the one with the highest value of cos α/l; so the efficiency of any line in resisting an imposed load is related to the geometry of the critical line as well as its own.

Chernjawski (18) has presented a matrix method of solving for mooring line forces in response to wind, current, and equivalent static wave loads, which also considers the hydrostatic restoring forces in response to vessel trim, list, and immersion and the nonlinear modulus of elasticity of the mooring lines. NAVFAC (16) describes a microcomputer program for calculating mooring line loads.

The general design of marine structures to accommodate vessel mooring loads is addressed in the general port and harbor engineering literature cited in Chapter 1 and in references (1) through (5), (8), and (16). A review of design case histories can be most informative. Many such case histories can be found among the various technical papers of the conference proceedings listed in Appendix 2. Bruun (19) has summarized the contents of several important papers covering rational principles in mooring and fendering design presented at the Eighth International Harbour Congress.

6.2 MOORING LINES, HARDWARE, AND EQUIPMENT

Mooring lines provide the critical link between the vessel and the berth structure, although the importance of their proper application to mooring safety often is overlooked. Mooring lines usually consist of synthetic fiber rope and/or steel wire rope. In general maritime usage, *rope* becomes a *line* when put to some specific use, and in turn mooring lines are often referred to as *hawsers*. Fiber ropes may be constructed of three (plain-lay) or four (shroud-lay) bundles of wound fibers called strands, which are twisted together; alternatively, plaited or braided rope construction may be used. Plaited rope is made up of pairs of strands that twist in opposite directions and thus eliminate the problem of kinking known as a hockle. Braided rope construction is similar to plaited rope, but is made up of many more smaller yarns rather than strands. Rope may be made from natural fibers such as manila, sisal, jute, or hemp, or more commonly today of synthetics such as nylon, dacron, polypropylene, polyethylene, and newer low-stretch materials such as mylar and kevlar. Although manila line is

Table 6.3. Sizes of synthetic mooring ropes normally carried by vessels, and the breaking load of these ropes.

Type	Gross registered tonnage (grt)	Rope diameter (mm)	Breaking load of eight-stranded plaited polypropylene (kN)
Dry cargo ships	Up to 2,000	40 to 48	190.2 to 266.7
	Over 2,000 up to and including 4,000	48 to 56	266.7 to 353.0
	Over 4,000 up to and including 8,000	52 to 60	308.9 to 404.0
	Over 8,000 up to and including 15,000	56 to 64	353.0 to 457.0
	Over 15,000	64 to 72	457.0 to 573.7
Ferries	Up to 2,000	40 to 48	190.2 to 266.7
Liners	Over 2,000 up to and including 6,000	48 to 60	266.7 to 404.0
	Over 6,000 up to and including 10,000	56 to 64	353.0 to 457.0
	Over 10,000	64 to 72	457.0 to 573.7
	Deadweight tonnage (dwt)		
Tankers	15,000 up to and including 20,000	56 to 60	353.0 to 404.0
Bulk carriers	Over 20,000 up to and including 40,000	60 to 64	404.0 to 457.0
	Over 40,000 up to and including 70,000	64 to 72	457.0 to 573.7
	Over 70,000 up to and including 120,000	72 to 80	573.7 to 706.1

Reproduced from reference (2) with permission of the British Standards Institution.
Note: 1 kN = 224.8 lb.; 1 mm = .0394 inch.

still in occasional use, eight-strand or double-braid nylon or polypropylene is the most common rope material on oceangoing vessels.

Table 6.3 gives sizes and strength properties of synthetic ropes normally carried by vessels. The minimum bending radii for all fiber lines should not be less than four rope diameters. The elastic stretch at break ranges from around 10% to 15% elongation for manila to around 50% elongation for nylon. Nominal working loads are on the order of 10% to 15% of tensile strength for manila and 10% to 20% for braided nylon where the elongation can be tolerated. The tensile strength of a line can be characterized in terms of its mean breaking load (MBL) based upon full-scale test results. The strength of a line of a given material and construction is proportional to the square of the diameter times a constant. Figure 6.6 illustrates approximate load/elongation characteristics for selected mooring lines.

The elastic properties of a line are of particular interest herein. A fiber line when initially loaded to around 50% or more MBL will not return to its original length when the load is released. Under repeated loading to a given stress level, the line will develop a stable hysteresis loop with relatively predictable elastic properties. Unfortunately a modulus of elasticity for fiber lines cannot be accurately defined, and the designer must refer to the manufacturer's test data when such information is important. Nylon line in general exhibits stiffer behavior under cyclic loading, and greater elongation at a given load when wet than when dry. The dynamic load elongation behavior of nylon line is dependent upon the initial tension and its magnitude in relation to the line's ultimate

Figure 6.6. Approximate load/elongation characteristics for selected mooring lines.

strength. Ropes usually are terminated by splices or knots, or by belaying to bitts or other mooring hardware. An eye splice is about 95% efficient, whereas the use of knots results in around 50% to 70% efficiency.

Wire rope is constructed of bundles of strands that are made up of individual wires twisted together. The rope construction is designated by the number of strands times the number of wires within each strand. Most wire rope used in mooring line applications consists of six strands, with 6 × 24, 6 × 36, and 6 × 37 construction being most common for hawsers. In the center of the twisted strands is a fiber or steel wire rope core. Table 6.4 gives sizes and mean break-

Table 6.4. Sizes of galvanized steel wire ropes normally carried by vessels, and the breaking load of these ropes

Type	Gross registered tonnage	6 × 24 (15/9/Fibre) Fibre main core		6 × 36 (14/7 and 7/7/1) Fibre main core		6 × 36 (14/7 and 7/7/1) Steel core		6 × 37 (18/12/6/1) Fibre main core	
		Nominal diameter (mm)	Minimum breaking load at 145 grade (kN)	Nominal diameter (mm)	Minimum breaking load at 180 grade (kN)	Nominal diameter (mm)	Minimum breaking load at 180 grade (kN)	Nominal diameter (mm)	Minimum breaking load at 145 grade (kN)
Dry-cargo ships	Up to 2,000	22 to 26	193.2 to 269.7	18 to 22	188.3 to 281.4	X	X	22 to 26	202.9 to 283.4
	Above 2,000 up to and including 4,000	26 to 30	269.7 to 358.9	22 to 26	281.4 to 393.2	22 to 24	304.0 to 361.9	26 to 32	283.4 to 429.7
	Above 4,000 up to and including 8,000	28 to 32	311.8 to 407.9	24 to 28	335.3 to 456.9	22 to 26	304.0 to 424.6	28 to 32	328.5 to 429.7
	Above 8,000 up to and including 15,000	X	X	26 to 28	393.2 to 456.9	24 to 28	361.9 to 493.3	X	X
	Over 15,000	X	X	28 to 32	456.9 to 596.2	28 to 32	493.2 to 644.2	X	X
Ferries	Up to 2,000	22 to 26	193.2 to 269.7	18 to 22	188.2 to 281.4	X	X	22 to 36	202.9 to 283.4
Liners	Above 2,000 up to and including 6,000	26 to 32	269.7 to 407.9	22 to 28	281.4 to 456.9	22 to 26	304.0 to 424.6	26 to 32	283.4 to 429.7
	Above 6,000 up to and including 10,000	X	X	26 to 28	393.2 to 456.9	24 to 28	361.9 to 493.3	X	X
	Over 10,000	X	X	28 to 32	456.9 to 596.2	28 to 32	493.3 to 644.3	X	X
	Deadweight tonnage								
Tankers	15,000 up to and including 20,000	X		26 to 28	393.2 to 456.9	24 to 26	361.8 to 424.6		
	Above 20,000 up to and including 40,000	X		28	456.9	26 to 28	424.6 to 493.2		
	Above 40,000 up to and including 70,000	X		28 to 32	456.9 to 596.2	28 to 32	493.2 to 644.3		
	Above 70,000 up to and including 120,000	X		32 to 35	596.2 to 712.9	32 to 36	644.3 to 815.9		
Bulk carriers								32	429.6
								X	X
								X	X
								X	X

X denotes not available for these vessel sizes.
Reproduced from reference (2) with permission of the British Standards Institution. *Note*: 1 kN = 224.8 lb.; 1 mm = 0.394 inch.

ing strengths for various classes of construction of wire rope normally carried by vessels. The standard practice of the wire rope industry is to use a factor of safety of 5 times the MBL for nominal working loads. Wires usually are made of a mild steel such as an improved-plow-grade steel, and are often galvanized for corrosion protection. The yield point of wire rope is not well defined, but usually is on the order of two-thirds of the ultimate break strength. The elongation at break ranges from 2% to 5%. Approximate load elongation characteristics of a representative wire rope are shown in Figure 6.6 for comparison with synthetic lines.

After repeated loading to around 30% of its ultimate tensile strength, wire rope becomes relatively elastic, and a modulus of elasticity (E_{wr}) can be defined within the elastic limit. Typical values of E_{wr} range from 10×10^6 to 22×10^6 psi, depending upon the rope construction and usage (20). Wire rope end connections may be made up by swaging on end fittings, splicing, or using wire rope clips. Swaging and splicing are from 75% to 90% efficient, and the use of clips is from 75% to 80% efficient. The use of wire rope clips is the most common means of connection because they can be readily applied in the field for rope of up to $1\frac{5}{8}$-inch diameter. A minimum of three clips, ranging up to six clips with increasing size, usually are required. Clips are spaced from 3 inches to $9\frac{3}{4}$ inches apart. Weight and strength characteristics and details of wire rope and synthetic rope construction and hardware can be found in the various manufacturers' product literature. General treatment of fiber and wire rope for mooring line applications can be found in references (20) and (21). Wilson (22) has provided an in-depth study of the elastic characteristics of moorings.

In general, it is undesirable to mix wire and fiber lines as independent mooring lines because wire is much stiffer than fiber and will carry most of the load. The total length of a line also affects its relative elongation. Wire rope and nylon, for example, can be effectively mixed in the makeup of individual lines, with the nylon used as a tail for ease of tying up and to take advantage of its elastic properties. In a multicomponent mooring arrangement, the length of nylon tail can be varied in order to maintain approximately equal loads among lines under anticipated movements. The nylon also aids in absorbing shock loads. The change in length (Δl) of a line under load is given by:

$$\Delta l = \frac{T_l l_0}{A E_{(T)}} \qquad (6\text{-}2)$$

where T_l is the tension, l_0 is the initial unstretched length, A is the cross-sectional area of the wire or fiber, and $E_{(T)}$ is the modulus of elasticity as a function of the tension. As fiber rope has no well-defined value for $E_{(T)}$, it should be determined from actual stress–strain curves for the specific use conditions. The

elastic restoring force versus vessel excursion then can be determined from the geometry of the layout and elastic properties, as illustrated in Figure 6.5.

Fenders are an important part of the overall mooring system. The compression of fender systems in general helps to reduce peak loads from sudden short-duration excitations such as waves and wind gusts, although under certain circumstances (because resilient fenders act like loaded springs) they may increase periodic loads if the combined spring stiffness of the total mooring system approaches that of the excitation. Bruun (23) has made a case for the use of non-recoiling-type fenders to avoid the problem of resonance and reduce vessel movements. Fenders must be able to resist peak mooring loads and to distribute them effectively into the pier structure. As a general rule, the compression of a resilient fender unit under maximum design mooring loads should not be more than about 50% of the maximum rated deflection in order to allow for unequal load distribution among fender units and additional capacity to absorb peak overloads.

Mooring hardware includes: bollards, bitts, cleats, chocks, fairleads, pad eyes, hooks, and so on. Bollards, bitts, and cleats are used for securing mooring lines, and normally are made of cast steel, such as ASTM A-27 or ductile iron (ASTM A-536) and secured to the structure deck with high-strength steel bolts. Larger bollards and bitts may be filled with concrete after installation. Bolts preferably should be installed through pipe sleeves cast through the pier deck, and their top exposed ends run in with lead after proper tightening. Figure 6.7 illustrates some typical mooring hardware.

Bollards are primarily used for snubbing and checking a vessel's motion, but may also be used for tying up. Bollard capacities are usually on the order of 40 to 100 tons horizontal load, but they are available up to 300 tons or more. Often larger bollards, called corner posts, are placed at the extreme ends of piers and wharves. Smaller bollards and/or bitts may be spaced approximately 40 to 100 feet apart, alternating with cleats in between. *Bitts* may be single or double, referring to the number of posts, which usually have horns to keep lines from riding up. Bitts are used for tying up a vessel and have typical capacities on the order of 35 to 60 tons horizontal. *Cleats* are used for tying up, usually for smaller vessels and lighter lines. They are not usually load-rated and typically range in overall length from 16 inches to 42 inches or more. *Pad eyes* of various dimensions sometimes are provided for the attachment of chain or wire rope. *Mooring rings* may be provided along the face of quays where small craft are likely to tie up. Rings should be of at least one-inch-diameter solid steel rod and let in flush with the quay face where possible. *Chocks* and *fairleads* are devices used to change direction or guide lines between points of attachment. They may be open at the top, closed, or fitted with an opening latch bar. Some also are equipped with roller guides mounted either vertically or horizontally, or both, to reduce friction and chafe.

194 DESIGN OF MARINE FACILITIES

Figure 6.7. Typical mooring hardware.

Quick release hooks most often are employed at isolated mooring dolphins, especially at large tanker berths. They provide added safety and security in releasing heavily loaded lines quickly. Release hooks are available with multiple hooks, and capacities range from 60 tons to 150 tons per hook with total capacities up to 600 tons. Figure 6.8 shows a high-capacity, multiple-hook in-

MOORING LOADS AND DESIGN PRINCIPLES 195

Figure 6.8. A multiple quick release mooring hook with a capacity of 150 tonnes per hook. (Photo by Dennis Coutts; courtesy of Mampaey International, B. V. and W. B. Arnold Co., Inc.)

stallation. These devices are swivel-mounted with varying degrees of movement, allowing the hook to align itself with the load. Load monitoring devices equipped with threshold alarm systems also are available and can be wired to a central console so that proper adjustments can be made if required.

A variety of equipment is used for mooring. *Capstans* are powered devices consisting largely of a rotating drum used to haul in a vessel (see Figure 6.9). They are common equipment at dry docks where a vessel must be carefully hauled into a controlled position over keel blocks. Capstans typically have hauling capacities on the order of 5 to 20 tons, with hauling speeds of 20 to 40 feet per minute (fpm) being common. *Winches* also are rotating drum devices, most commonly used with wire rope, which is spooled onto the winch drum. Winches are common shipboard equipment, and many larger vessels are equipped with *constant tension winches* that can be preset to pay out or take up line when a limiting tension value is exceeded. This is especially useful to vessels such as tankers and bulk carriers that change draft dramatically during loading and unloading. Additional description of shipboard line-handling equipment can be found in references (6) and (20).

Electric-powered winches also may be located ashore or on berthing structures to provide shore augmentation. In this instance, the shore-based mooring line normally is hove out to the vessel by messenger after the vessel has been

Figure 6.9. Electric-powered capstans of 6 tonnes capacity. (Photo courtesy of NEI Clark Chapman, Ltd. and W. B. Arnold Co., Inc.)

safely berthed. Shore-based winches may offer greater control of vessel mooring lines while in berth, but also may result in an undesirable division of responsibility for the tending of lines (19).

6.3 WIND FORCES

Wind and current loads result primarily from velocity forces, which are calculated from the well-known drag force equation:

$$F_D = \frac{\gamma}{2g} C_D V^2 A_P \qquad (6\text{-}3)$$

where F_D is the drag force, γ is the unit weight of fluid, g is the acceleration of gravity, V is the velocity of the fluid relative to the object, A_P is the projected area normal to the direction of flow, and C_D is the drag coefficient, which itself is dependent upon the object's shape and orientation to flow, surface roughness, Reynolds number (N_R), proximity to other objects, and boundary conditions. Often, the problem of calculating static wind and current loads for design purposes can be reduced to one of selecting an appropriate drag coefficient.

For the windspeed (V_w) in knots, F in pounds, and A_P in square feet, equation (6-3) reduces to:

$$F_w = .0034\, C_D V_w^2 A_P \qquad (6\text{-}4)$$

for $\gamma = 0.0765$ pcf at standard temperature of 59°F and sea level pressure. Note that air density increases almost linearly with decreasing temperature, which results in a nearly 13% increase at 0°F. Variations with moisture content and pressure are less noticeable.

The moisture content per se does not have an important effect on air density as it relates to most drag force calculations, but some investigators have questioned the effect of entrained water under storm spray and heavy rain conditions. Even a small volume of entrained water has the potential to increase theoretical wind loads by an order of magnitude. However, other investigators have argued that the water would slow down the air, thus maintaining a constant level of kinetic energy. The issue remains somewhat unresolved at this time, but it is not customary to consider entrained water effects in standard drag force calculations. Wind force calculations that are carefully carried out in accordance with the following procedures generally can be considered as accurate to within 10% to 15%.

Wind force calculations normally are carried out by resolving the wind force into longitudinal (F_{wx}) and lateral (F_{wy}) force components and a yawing moment (M_{yw}), which tends to rotate the vessel about a vertical axis through its center. Accordingly, we can write the following expressions:

$$F_{wx} = .0034 \, C_{Dx} V_w^2 A_x \qquad (6\text{-}5)$$

$$F_{wy} = .0034 \, C_{Dy} V_w^2 A_y \qquad (6\text{-}6)$$

$$M_{yw} = F_{wy} LOA \, C_{ym} \qquad (6\text{-}7)$$

for F in pounds, V_w in knots, A in square feet, and LOA in feet, and where C_{Dx} and C_{Dy} are the drag force coefficients in the respective directions, and C_{ym} is the yaw moment coefficient—all of which are functions of the relative wind direction, or angle of attack (θ) (see Figure 6.1). A_x and A_y are the end-on and side projected areas of the vessel, respectively, and include the area of masts, stacks, rigging, deck cargoes, and so on. The yaw moment is given in terms of the lateral force times the vessel's length overall (LOA) and C_{ym}. The total resultant force for wind from any direction ($F_w(\theta)$) is found from this equation:

$$F_w(\theta) = .0034 \, C_D(\theta) V_w^2 (A_x \cos^2 \theta + A_y \sin^2 \theta) \qquad (6\text{-}8)$$

Figures 6.10 and 6.11 illustrate the usual range of values and variation with wind direction of the coefficients C_{Dx}, C_{Dy}, and C_{ym} for typical merchant vessels and for vessels with boxlike superstructures, respectively. Numerous wind tunnel studies have been undertaken on a wide range of vessel types, and useful design data can be found in references (24) through (35). Note well that the typical values for C_{Dx}, C_{Dy}, and C_{ym} given herein are provided for general in-

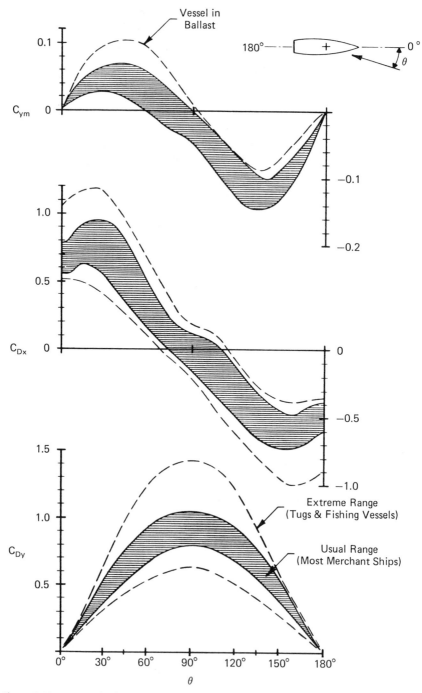

Figure 6.10. Range of values and variation with wind direction of the lateral and longitudinal force coefficients and yaw moment coefficient for typical merchant vessels.

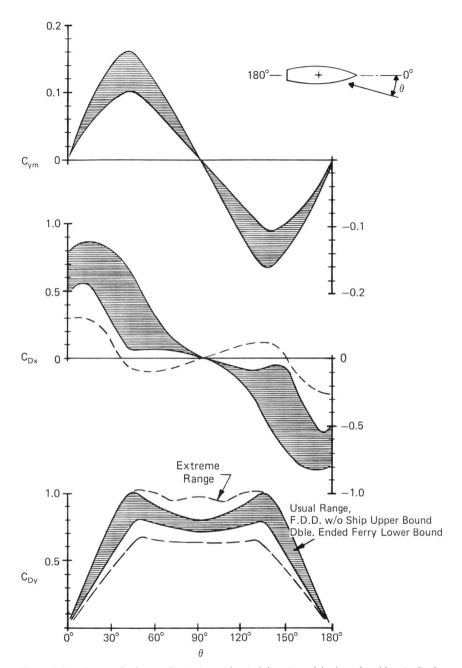

Figure 6.11. Range of values and variation with wind direction of the lateral and longitudinal force coefficients and yaw moment coefficient for typical ferries, liners, and floating dry docks.

formation purposes, and the designer should pursue the literature to determine appropriate values for specific applications. Wind coefficient data also may be presented in different formats, which must be verified before substitution into equations (6-5), (6-6), and (6-7).

For typical oceangoing vessels values of C_{Dy} for the wind abeam condition ($\theta = 90°$) range from 0.6 to 1.4, with 0.8 to 1.0 being most representative. Occasional higher values have been reported. For wind ahead or astern conditions, C_{Dx} typically ranges from 0.4 to 1.2, with 0.7 to 0.9 being representative for most vessels. In general, C_{Dx} tends to be slightly less with the wind astern than for the wind ahead. In the absence of other data, it can be assumed that C_D values vary sinusoidally between maximums at 0° and 180°, with maximums at 90° and 270° for F_{wx} and F_{wy}, respectively. Note, however, that the maximum F_{wx} and F_{wy} do not necessarily occur at these positions. For example, for most vessels with boxlike superstructures, such as ferries and passenger liners and empty floating dry docks, the maximum F_{wy} occurs at $\theta = 45°$ to 60° and $\theta = 120°$ to 135°. For floating dry docks occupied with capacity vessels, the C_D values tend to resemble those of the docked vessel. Values of C_D vary with a vessel's overall hull shape and the extent, location, and configuration of the superstructure (including masts, stacks, rigging, deck cargo, etc.). For a given vessel, C_D will likely also vary noticeably with the ballast condition. For the case of vessels moored nearly alongside one another such that one is directly downwind, it is often assumed for design purposes that the wind load on the downwind, shielded, vessel is reduced by 50%. If the shielded vessel is much larger than the upwind vessel, then this assumption should not be considered valid. Wind tunnel model tests have been carried out for two and more rafted vessels of various types (26, 27).

$M_{yw}(\theta)$ typically has maximum values for θ at 30° to 60° and 120° to 150°, which range from around 0.05 to 0.15. In the absence of other information, for preliminary estimates, the maximum M_{yw} can be taken at around 0.10 at 45° and 135°, varying sinusoidally in between. Note that for vessels with large superstructures located aft or for other asymmetrical configurations the value of $M_{yw}(\theta)$ may be much less for wind forward of the beam than for aft of the beam. The yaw moment is the result of the distance between the center of applied wind pressure and the vessel's center of lateral resistance, which are constantly changing with the angle of attack (6, 32). Another important factor affecting the vessel's relative response to wind and current forces and the separations of their lines of action is the ratio of above-water to below-water areas. This ratio typically varies from around 0.4 for fully laden tankers to greater than 3.0 for passenger and Ro/Ro vessels with high deck cargoes as applied to their lateral profiles (32).

The lateral wind force will cause a vessel to list away from the wind. The wind heeling moment (M_{hw}), which is resisted by hydrostatic restoring forces

and/or mooring system components, is given by:

$$M_{hw} = F_{wy} h C_{hm} \cos^2 \psi \qquad (6\text{-}9)$$

where h is the vertical distance from the center of wind pressure to the center of lateral resistance (CLR) of the vessel's underwater profile, and ψ is the angle of heel. Values of C_{hm} for most vessels of interest herein range from 0.85 to 1.40. The wind heeling moment is not normally considered a significant factor in most mooring load calculations, although it can be readily accounted for in computer programs (18).

Accurate wind area information for specific vessels is often difficult to obtain. Wind profile areas for selected vessels can be found in references (11), (21), and (36), and approximate areas for selected vessel types and sizes are presented in Table 6.5. For larger vessels such as tankers and bulk carriers, wind areas can be reasonably well correlated with displacement or DWT (1, 37). Thoresen (11) cites approximate values for design estimates ranging from 22 square feet per linear foot (s.f./ft.) to 36 s.f./ft. for a 30,000 DWT general cargo vessel in laden and light condition, respectively, to 63 s.f./ft. or more for large passenger vessels.

Wind loads also may be evaluated directly from the results of scale model tests, such as the example curves shown in Figure 6.12. Such data can be scaled to different wind velocities or to estimate loads on smaller or larger vessels of

Table 6.5. Approximate lateral wind areas for selected vessels.

Vessel type	Size (DWT)	Projected side area (A_y) (S.F.)
Tanker	5,000	7,668*
	20,000	16,712*
	50,000	27,968*
	100,000	41,290*
	200,000	60,956*
	300,000	76,557*
General cargo (class)	5,000	9,716*
	10,000	14,059*
	12,900	24,000+
	20,000	20,342*
	30,000	25,249*
Containership	10,000	18,292[x]
	20,000	33,356[x]
	30,000	47,344[x]
	50,000	65,636[x]

*Vessel in ballast condition, calculated after PHRI (37).
+Vessel in light ship condition, after Quinn (35).
[x]Vessel with containers on deck, after Thoresen (11).

202 DESIGN OF MARINE FACILITIES

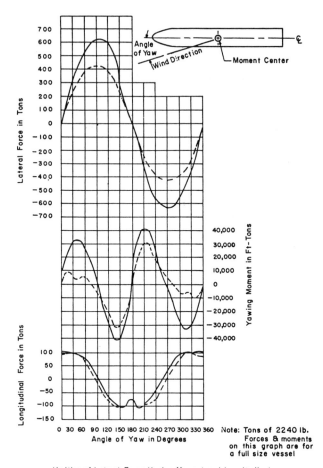

Varition of Lateral Force, Yawing Moment and Longitudinal Force with Angle of Yaw for a Single AO-143 (T5 Tanker) Wind Speed 125 Knots, Dashed Line-Laden Condition, Solid Line-Unladen Condition.

Figure 6.12. Example model test results, for determination of longitudinal and lateral forces and yaw moment due to wind on a loaded and ballasted tanker. (Reproduced from NAVFAC (26) after Ayers and Stokes (27).)

similar type (26). When test results are presented as actual measured loads on a model, care must be taken in applying the laws of dimensional analysis and similitude and in accounting for other possible scale effects when converting data to scaled prototype forms.

Reliable data from actual field measurements are relatively scarce. Fang (38) reported on an experimental study to verify predicted wind and current loads

on VLCCs, and the test results demonstrated the accuracy of the OCIMF (24) methodology. Fang notes that the shortest gust that fully excited the moored tanker was 20 seconds, and the maximum response was induced for an approximately 60-second-duration gust.

As with berthing loads, wind forces are proportional to the square of the velocity, and the determination of appropriate wind velocity is a very important factor. Design wind speed usually is determined by some assumed threshold velocity, beyond which it is assumed that the vessel would not remain alongside, or on a statistical return period basis as described in Sections 6.1 and 3.4. Design wind speeds must be corrected for height and duration. If the design wind speed is based upon records at an inland site (airports are a common source of such data), then it must be further corrected for overland–overwater effects. This correction usually will be on the order of a 10% increase for the harbor site, depending upon the proximity of the measurement site and exposure of the harbor. Overwater, the vertical profile of wind speed varies roughly as the one-seventh power of the height above water. The standard reference height for anemometer records is 10 meters (33 feet) above the surface. Wind speeds at other heights may be corrected to the velocity at 33 feet (V_{33}) from the velocity at elevation z, $V(z)$, according to:

$$V(z) = V_{33} \left(\frac{z}{33}\right)^n \quad (6\text{-}10)$$

where the value of the exponent (n) can be varied to account for surface roughness, and is usually taken as $1/7$ for coastal sites. For most vessels of interest herein, the center of wind pressure will be near the standard reference height, and applying the reference height wind velocity as an average over the vessel's height above water should be sufficiently accurate. For small craft and low-profile vessels such as submarines, resulting wind force estimates will be conservative. For vessels with high profiles, say 50 feet or more above the water surface, the height correction should be considered.

Maximum wind speed data often are recorded as peak gust or fastest mile wind speeds, the latter referring to wind speeds sustained long enough to cover a gust length of one mile. According to OCIMF (8), a gust of 30 seconds is required to force a response of a VLCC in light condition on an all-wire mooring, and a gust of 60 seconds duration to force a response of a fully loaded VLCC. NAVFAC (16) has adopted the 30-second wind speed duration for design. Table 6.6 gives the approximate conversion factor, C_{tw}, for converting wind speeds of any duration to the 30-second equivalent. Additional discussion of wind data, design wind speed criteria, and bluff body aerodynamics can be found in references (39) and (40).

Table 6.6 Wind duration correction factor relative to 30-sec gust speed.

Wind duration	Correction factor (C_{tw})
1 sec	1.22
3 sec	1.16
10 sec	1.08
30 sec	1.00
60 sec	0.95
10 min	0.87
30 min	0.80
60 min	0.75

6.4 CURRENT FORCES

Steady current forces, like wind forces, may be calculated using the drag force equation (6-3), and similarly can be resolved into longitudinal (F_{cx}) and lateral (F_{cy}) force components and a yaw moment (M_{yc}). Using foot-pound units and taking the unit weight of seawater as 64 pcf and with the current velocity (U_c) in knots, we can write:

$$F_{cx} = 2.86 C_{cx} U_c^2 A_{cx} \quad (6\text{-}11)$$

$$F_{cy} = 2.86 C_{cy} U_c^2 A_{cy} \quad (6\text{-}12)$$

$$M_{yc} = F_{cy} LWL\, C_{ymc} \quad (6\text{-}13)$$

where C_{cx}, C_{cy}, and C_{ymc} are the respective forces and yaw moment coefficients, and A_{cx} are the projected underwater areas normal to the flow for head currents and beam currents, respectively. For design purposes, A_{cx} can be taken as the product of the draft times the beam at the waterline ($D \times B$), and A_{cy} as the product of the draft and length on the waterline ($D \times LWL$). Figure 6.13 illustrates the usual range of drag and moment coefficients for deep water where the water depth to draft ratio (d/D) is greater than approximately six. The lateral force coefficient is dramatically increased by shallow water depths, and a correction factor must be applied as given by Figure 6.14.

Current forces calculated using equations (6-11) through (6-13) may be subject to greater uncertainty than wind loads calculated in the same manner. Several important factors make accurate calculation of current forces more problematical. The Reynolds number (N_R) of flow (see Section 6.7) for ships in currents normally is within the range of turbulent flow but near enough to the transition zone that a wider scatter of the force coefficients is observed (6). Flow regime effects also affect the extrapolation of scale model test results. Full-scale

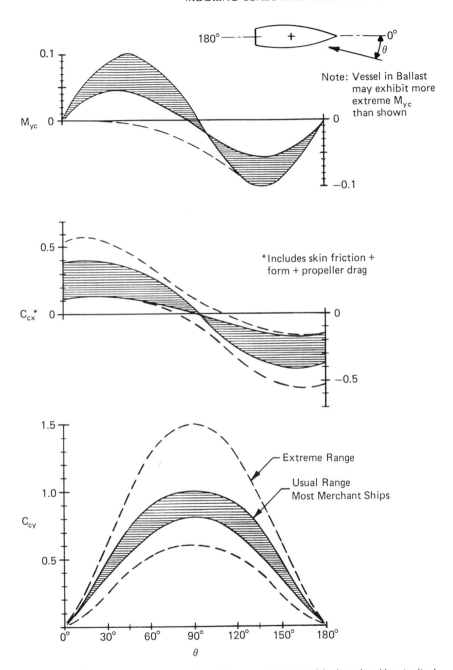

Figure 6.13. Range of values and variation with current direction of the lateral and longitudinal force coefficients and yaw moment coefficient for typical vessels in deep water.

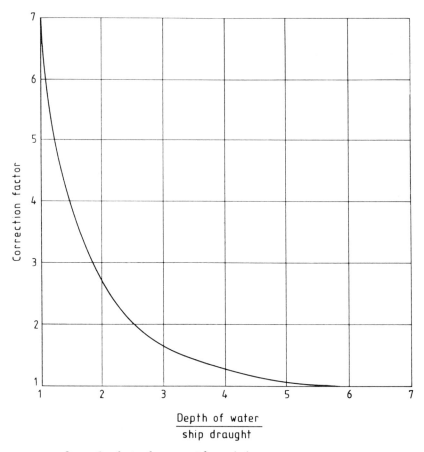

Figure 6.14. Current force correction for lateral force in shallow water. (After BSI (1).)

measurements of current forces on a moored tanker and destroyer, as described by Palo (41), have challenged the adequacy of conventional methods of calculating current loads. Some important conclusions from Palo's work are that the lateral force coefficients are relatively insensitive to hull shape but show a significant dependence upon the vertical distribution of current velocity, also referred to as current shear. Palo further notes that the usual shallow water correction factors do not adequately represent actual shallow water behavior, but without further actual prototype testing an improved design methodology cannot be devised. Other factors also make current force determination difficult: the longitudinal force cannot be accurately calculated using a simple form drag coefficient; the velocity profile around the entire vessel is difficult to assess and

may be significantly affected by bottom conditions and proximity effects; current velocities are not necessarily steady, and dynamic behavior may need to be considered.

For the time being, port engineers must apply conventionally accepted methods but should be aware of their limitations and of particular situations where they are more or less applicable. Design guidance and current coefficient data for evaluating current loads can be found in references (1), (8), (16), (24) through (27), (29), and (34). An in-depth treatment of hydrodynamic forces on ships' hulls and of scale model applications is provided by Newman (42).

In deep water, overall lateral force coefficients, C_{cy}, generally range from 0.5 to 1.5, with 1.0 being the approximate mean value. The lateral force varies nearly sinusoidally with the angle of attack but exhibits greater variation than wind forces in this regard (refer to Figure 6.13). As previously noted, measured lateral force coefficients show little sensitivity to hull form and may increase on the order of six times their deepwater value ($d/D > 6$) when d/D is reduced to 1.1. The maximum lateral force may occur at current attack angles other than 90°. Palo (41) noted maximum lateral forces for $\theta = 110°$.

The lateral force and yaw moment are closely related, and Palo notes maximum values of C_{ymc} of about 0.1 with a zero crossing near $\theta = 90°$ as reasonably representative in uniform current fields. Current yaw moments vary more symmetrically with direction than wind moments, and like the lateral force are dramatically affected by the relative water depth. The ballast condition and the vessel trim also significantly affect the yaw moment and lateral force.

Longitudinal current forces are the result of both form (pressure) drag and surface friction drag. Propeller drag often is added separately to the hull form plus friction components. In reality, form, skin friction, and appendage drag interaction cannot be effectively separated, and many designers choose to apply a single overall drag coefficient for a given vessel type. Overall form drag coefficients for F_{cx} range from around 0.10 to 1.0 (0.20 to 0.60 are more representative), depending upon the hull configuration, appendages, surface roughness (e.g., fouling can greatly increase longitudinal drag forces), angle of attack, and so on. Interestingly, vessels with twin propellers tend to have lower overall form drag coefficients than those with single screws. It should be noted also that the maximum longitudinal force may occur for angles of attack other than 0° and 180°, and in fact it often remains near its maximum to within 10° to 15° of lateral. Skin friction factors are dependent upon the Reynolds number (N_R) and for the range of interest herein may range from .001 to .01, with the higher value being more appropriate for design. The skin friction term is similar to the calculation of equation (6-11) except that the area used is the total surface area of the underwater portion of the hull (A_s).

Specific information on wetted surface areas for a given vessel may be difficult to obtain. NAVFAC (16) presents a simple formula for calculating vessel

surface areas. The designer with a need to accurately determine a vessel's wetted surface and appropriate friction factors should consult the general literature on naval architecture as referenced in Chapter 2. The form drag of a vessel's hull alone facing the current direction is relatively small. Hunley and Lemley (6) note an average drag coefficient corresponding to C_{cx} for the hull alone of 0.088 with a standard deviation of 0.039 based upon some 60 ships from SNAME resistance data sheets.

The longitudinal resistance of locked propellers can easily exceed that of the vessel's hull itself. Propeller drag can be estimated by applying a drag coefficient of unity to the projected blade area of the propeller, which generally can be approximated as one-half the propeller diameter squared. Alternative means for calculating propeller drag separately can be found in reference (16). It should be further noted that for small craft in very strong currents, say exceeding 4 to 5 knots, wave making resistance may become important. The wave making resistance of a vessel's hull usually becomes dominant when its speed–length ratio, as given by V_s/\sqrt{LWL}, exceeds a value of around 0.6 where V_s is in knots and LWL in feet. In such cases, the powering and resistance data published in most naval architecture texts or model testing must be utilized.

For design purposes, the value of U_c used should be the average over the vessel's draft for the maximum current expected at the site. Where current velocities are significant, say in excess of 2 to 3 knots, site-specific measurements should be taken. Current velocities may vary greatly with distance from shore, water depth, and local boundary conditions. It also should be noted that the construction of a new structure may alter existing current patterns. In shallow open water, the vertical velocity profile usually is assumed to vary (like wind) with the one-seventh power of the height above bottom and is maximum at the surface. (Refer to the discussion of currents in Section 3.4.) As previously noted, the actual vertical distribution or current shear may have a significant impact upon the forces realized. For large deep-draft vessels in relatively shallow water, the variation of current with depth must be considered. Figure 6.15 illustrates

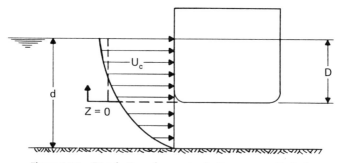

Figure 6.15. Distribution of current velocity over vessel's draft.

the distribution of current velocity normal to a ship's hull in shallow water. The averaged squared velocity of the current can be found from:

$$U_c^2 = \frac{1}{D} \int_0^D (U_z)^2 \, dz \qquad (6\text{-}14)$$

where U_z is the velocity at depth z. Assuming the current velocity varies according to the one-seventh power law, then for a vessel with small under-keel clearance ($d/D = 1.1$) the average velocity over the vessel's draft would be about 90% of the surface current velocity (24).

6.5 STANDOFF FORCES AND PASSING VESSEL EFFECTS

A vessel moored alongside a pier or quay located in a river or narrow tidal estuary with strong currents may be subject to hydrostatic pressure forces arising from an unequal velocity distribution about the vessel. A manifestation of this force is illustrated by the case of a vessel transiting a channel. If the vessel is too near a bank, the flow of water will essentially channel itself around the outside of the vessel, resulting in an accelerated flow around one side, which in turn causes a hydraulic gradient across the vessel's width with higher water near the bank, thrusting the vessel away from the bank. As the vessel moves toward the center, the redistribution of flow results in a suction back toward the bank. At the center of the channel, side forces cancel, and no net suction force acts on the vessel. In the case of a vessel moored alongside in a relatively strong current, the slowing down of water passing through, or being totally obstructed by, the pier creates a velocity-induced pressure head against the vessel's side, resulting in an outward pull on the mooring lines, known as a standoff force. Figure 6.16 illustrates this effect.

Khanna and Sorensen (43) reported on a design case history where model experiments were used to predict the standoff forces, which were manifested in a net sway force away from the structure, a longitudinal surge force, and a yawing moment. They also noted dynamic forces in terms of coupled surge–yaw and sway motions that took place at the natural periods of the moored ship. The differential head across the vessel can be expressed by the Bernoulli equation as a coefficient (C_{sy}) times the velocity head $U_v^2/2g$, where U_v is the undisturbed current velocity. Actual field measurements originally reported by Jackson (44) indicate that this head can be as high as $0.42 \, U_v^2/2g$. Khanna and Sorensen, however, found values of C_{sy} ranging from 0.10 to 0.15 and 0.07 to 0.12 for scaled current velocities of 5 knots and 7 knots, respectively, depending upon the berth location. Their experiments were carried out on a 1:100 scale model of a 300,000 DWT tanker with $d/D = 1.5$. They found that a

210 DESIGN OF MARINE FACILITIES

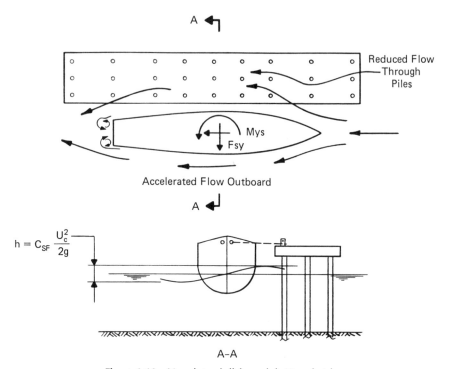

Figure 6.16. Vessel standoff force definition sketch.

reduction in d/D to 1.32 resulted in an approximately 35% increase in the standoff force. Motions of the vessel started when the current exceeded a threshold velocity of around 2 knots, and dynamic mooring line loads were on the order of two to three times that due to the static head. The use of more resilient, "softer," moorings has been shown to reduce peak dynamic loads (45, 46). The longitudinal force also was increased by the standoff effect. There was some correlation with pile area density, but a 16% increase in pile density did not alter their results significantly. Ball and Wilcock (47) reported on follow-up studies to determine the effects of pile density, spacing, and configuration.

A similar phenomenon is the effect of passing vessels on moored ships. Experiments have shown that the forces and moments induced on a moored vessel by a passing vessel, which can be very significant, are more importantly related to the pressure gradient surrounding the vessel than to the waves generated by the moving vessel. In general, the forces associated with surge, sway, and yaw motions are proportional to the square of the speed of the passing vessel (V_s) and the separation distance. Other factors affecting the maximum mooring line loads are: length of time for passing vessel to transit, ratio of length of moored vessel (L_{s1}) to length of passing vessel (L_{s2}), under-keel clearance, vessel bal-

last condition, and mooring line pre-tension. In general, stiffer moorings result in smaller forces. According to Wang (48), the passing vessel forces are also dependent upon the ratio of the midship section areas of the moored and passing vessels. Forces on the moored vessel are greatest when $L_{s1} < L_{s2}$ and are only noticeable for very close separation distances when $L_{s1} > L_{s2}$. The effect of water depth is minimal when d is greater than the center-to-center vessel separation distance. When currents are present, the speed of the vessel is that relative to the moving water; so an opposing current will increase the force, and following currents will decrease it.

Figure 6.17 shows a schematic plot of the variation of forces and moments on a moored vessel due to a passing vessel. Wang has presented graphs of nondimensional surge and sway forces and yaw moments for variable vessel length and midship area ratios based upon theoretical analysis. Scale model test results have been reported by Remery (49) and by Lean and Price (50). Muga and Fang (51) have compared theory and experiment. Standoff forces due to currents and to passing ships potentially may be very large, and a dynamic response of moored vessels is possible. At sites where predisposing conditions exist, the designer should investigate the problem further, as no simplified equations or design method have been developed for the general case.

6.6 WAVE LOADS AND VESSEL MOTIONS

There is no standard procedure for calculating wave loads on fixed moored vessels. Wave forces and mooring system reactions arise from a complex interaction of the water particle kinematics, vessel motion, and mooring system response. Determinations of added mass, damping, and higher-order nonlinear effects add to the difficulty. Because of this complexity, an analytical treatment is most difficult, and scale model testing is often used. Section 6.7 will provide a brief introduction to these subjects. This section discusses certain approaches that have been taken to the problem and presents some general findings of various investigations.

Wave action can be generally classified according to the frequency of motion as short period wind waves and swell and long period wave action, which is generally a result of harbor or basin oscillations known variously as seiche, surge, and ranging action. Wind waves and swell are of central importance at offshore terminals, and typically have periods of less than 20 seconds. An important second-order nonlinear effect arising from continuous wave action against a moored object is the *drift force*; it consists of a steady state component and an oscillating component (in irregular seas) associated with wave grouping action and having a periodicity on the order of 20 to 100 seconds, which is within the range of natural periods of large moored vessels. Long period wave action manifests itself typically in periods ranging from around half a minute

212 DESIGN OF MARINE FACILITIES

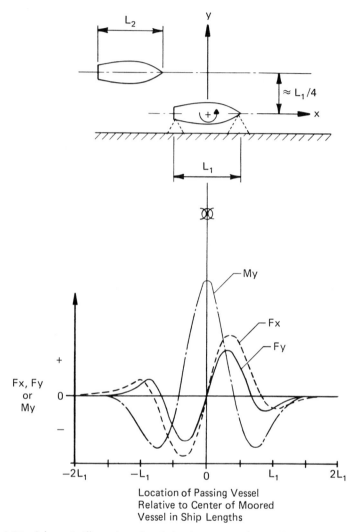

Figure 6.17. Schematic illustration of variation in surge and sway force and yaw moment acting on a moored vessel due to a passing vessel. (Adapted from: Wang (48).)

to several minutes or more and can be a serious menace even in protected harbors when the natural period of the mooring system and the seiche period are nearly coincident, and/or when a vessel is located near a nodal point (see Section 3.4) in the harbor where the horizontal flow of water associated with the seiche action is at its greatest. Notable investigations of vessel surging problems due to long period waves have been presented by Wilson (52, 53) and by O'Brien and Kuchenreuther (54).

If we consider a free floating unrestrained object, it will respond to the passage of waves with some degree of motion in each of its six degrees of freedom. The force imposed upon a mooring structure is largely a function of the degree of restraint or the effective spring constant of the fenders and mooring lines. Because waves are a time-dependent phenomenon, dynamic analysis must be applied to obtain meaningful results. In general, the force is proportional to the wave height and vessel draft for a given wavelength and water depth. The behavior of moored vessels as oscillating systems was introduced in Section 6.1 and has been treated in more depth by Vasco Costa (55). The natural period of a moored vessel can be increased by the use of longer and softer hawsers and softer fenders, and can be decreased by the use of shorter, stiffer hawsers and stiffer fenders. This fact can be used to reduce the potential for mooring systems resonance, depending upon the prevailing wave climate at the site.

Wave forces on moored vessels manifest themselves in two components. The first is a linear oscillatory force at the frequency of the waves. This force can be found by integrating the varying water pressures around the submerged portion of the hull (see Figure 6.18). Diffraction theory must be employed in order to do this, as the presence of the vessel modifies the incident wave train. The oscillating vessel scatters waves, which propagate away from the hull and contribute to the damping of the oscillations. The hydrodynamic coefficients of added mass and damping are different for each of the six degrees of freedom and are dependent upon the frequency of the oscillation. In shallow water, the water depth to draft ratio (d/D) or the vessel under-keel clearance $(d - D)$ has a significant effect on these coefficients. Buoyant restoring forces may result in heaving and/or pitching oscillations at some natural period of the system. The restoring forces of mooring lines and fenders are nonlinear and may result in resonant horizontal motions. Resonant response may also occur at some subharmonic of the wave frequency, that is, the wave period divided by an integer.

The second type of wave force is nonlinear in nature and a result of the irregular sea state. This force, known as the drift force, is primarily a consequence of wave grouping and set-down effects. The momentum flux and radiation stress of progressive groups of higher and then lower waves result in a net force in the direction of wave propagation. The drift force is a steady force in regular waves and a slow-varying/low frequency force in irregular waves. Because damping is relatively low at a lower frequency and because the usual range of periods of 20 to 100 seconds is also within the range of natural periods of moored ships, the slow-varying drift force may cause overstressing of mooring lines and large fender forces. Analytical and computational and physical modeling techniques used to approach the general moored ship problem are reviewed in the next section.

Goda and Yoshimura (56, 57) presented a method of determining the static wave force on a vessel rigidly moored at offshore (wave-transparent) dolphins by solving the general diffraction problem for an elliptical prism in shallow

214 DESIGN OF MARINE FACILITIES

Figure 6.18. Schematic Illustration of wave force on fixed moored vessel in beam sea.

water. The method has been adopted in the *Technical Standards for Port and Harbour Facilities in Japan* (4). According to Goda and Yoshimura, the maximum value of the total longitudinal force (F_{wax}) and total lateral force (F_{way}) can be found from:

$$F_{wax} = C_{mx} \gamma H D_0^2 \frac{\pi^2}{8} \cos\theta \left[\frac{\sinh kd - \sinh k(d-D)}{\cosh kd} \right] \quad (6\text{-}15)$$

$$F_{way} = C_{my} \gamma H D_0^2 \frac{\pi^2}{8} \sin\theta \left[\frac{\sinh kd - \sinh k(d-D)}{\cosh kd} \right] \quad (6\text{-}16)$$

where:

H = Incident wave height.
D_0 = Projected width of the vessel normal to the direction of wave approach as given by: $D_0 = (L_s - B)\sin\theta + B$ where L_s is ship length and B is ship beam.
k = Wave number = $2\pi/L$ where L is the wavelength.

The terms C_{mx} and C_{my} are added mass coefficients in the two principal directions, for which Goda and Yoshimura have presented nondimensional graphs as functions of D_0/L and B/L_s. For usual large vessel proportions, B/L_s = .13 to .18, the longitudinal virtual mass coefficient remains relatively constant with θ and D_0/L, with a maximum value of about 0.25 for D_0/L less than about 0.25. The lateral force coefficients show much greater variations with D_0/L and θ. C_{my} has a maximum value of approximately 1.45 for $D_0/L = 0.25$ and B/L_s = .15 in the beam sea condition (i.e., $\theta = 90°$). For $\theta = 60°$, the value of C_{my} increases to approximately 1.75 at $D_0/L = 0.25$. The above method does not account for the vessel relative motions or mooring stiffness but can be readily applied to estimate the magnitude of total wave force to be resisted by the dolphins. Textbook treatment of the general approach to solving wave–structure interaction problems for wave loads on floating bodies can be found in references (42), (58), and (59).

The drift force is generally much smaller than the oscillating wave force. In fact, van Oorshot (60) notes that for tests carried out on a model ore carrier the steady drift forces in regular waves were found to be on the order of 1% or less in the head sea and 2% to 4% in the beam sea condition of the harmonic oscillatory wave forces. Because of its importance in the mooring of large vessels, however, the slow-varying drift force has been studied by several investigators, such as described in references (60) through (64). Chakrabarti (61) has reviewed the work of various investigators and the theoretical development on steady and oscillatory drift forces on floating objects. The drift force is generally propor-

tional to the square of the wave height. According to Sarpkaya and Issacson (58), where no energy is dissipated, the magnitude of the drift force as derived from small amplitude wave theory can be calculated from:

$$F_{dr} = \gamma \frac{H_r^2}{8}\left(1 + \frac{2kd}{\sinh kd}\right) \tag{6-17}$$

where H_r is the reflected wave height. It is therefore necessary to determine a reflection coefficient ($C_r = H_r/H_i$) for the given sea conditions. Note that the incident, reflected, and transmitted wave heights are related by $H_i^2 = H_r^2 + H_t^2$, as described in Chapter 9.

Where wave-induced vessel motions and mooring forces can be assumed to be linear and frequency-dependent, their statistical values can be determined by spectral response analysis techniques. This requires that a transfer function, or response amplitude operator (RAO), which describes the vessel unit response in a given mode and at a given frequency, be determined. RAOs for specified conditions may be determined from theory or experiment as described in the following section. For a unidirectional sea spectrum, the response spectral density ($S_r(\omega)$) is obtained by multiplying each ordinate of the input wave energy spectrum ($S(\omega)$) by the square of the RAO, as given by:

$$S_r(\omega) = \left[T_j(\omega, \theta_1)\right]^2 S(\omega) \tag{6-18}$$

Here the subscript j indicates the jth degree of freedom (DOF), the subscript l indicates the direction of the jth DOF, ω is the circular frequency $= 2\pi/L$, θ is the angle indicating the direction of motion, and $T_j(\omega, \theta_1)$ represents the transfer function or RAO.

Figure 6.19 schematically illustrates the response spectrum analysis procedure. A simple example applied to the heave motion of a rectangular barge is given in Section 9.3. Figure 6.20 shows computed mooring line force spectra for a 7348-ton cargo ship moored alongside an open pier. The spectra were generated by a computer program based upon a mathematical model as described by Bomze (65). Figure 6.21 shows the vessel motion response spectra for surge, sway, and yaw, and comparison with spectra developed from actual field measurements conducted during field validation studies.

Important conclusions of Bomze's study are that mooring line forces are induced almost entirely by waves of longer than 20 seconds with major contributions occurring in the period range of 40 to 200 seconds. Rolling is excited primarily by swells in the 13- to 19-second range. Pitching is small and insensitive to wave period, whereas heave motions tend to equal the wave height at long periods and are smaller with short period waves. As has been previously noted, the evaluation of a vessel's motion response to given sea conditions is

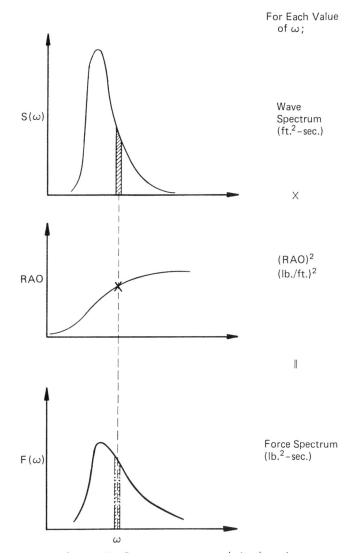

Figure 6.19. Response spectrum analysis schematic.

important to the planning, design, and operations of a terminal facility. A vessel's response depends importantly on its orientation to the sea as well as to the wave heights and periods.

Figure 6.22, after Sugin (14), shows the relative motion response (ship motion amplitude/wave amplitude) in the six DOFs for a 250,000 DWT ore carrier as a function of wave period and wave direction. Such information, developed

218 DESIGN OF MARINE FACILITIES

Figure 6.20. Example of mooring line force spectra. (After Bomze (65).)

for specific site conditions, is essential in optimizing the berth orientation. Other ship motion studies of interest that present data on hydrodynamic coefficients and RAOs include references (66) through (73).

It is interesting to note that larger vessels moored alongside in harbors and subject to short period wind waves (generally less than three to four seconds), and wave heights of less than three to four feet, do not seem to experience any great difficulty. In general, the short pulse duration and the large vessel inertia combine to greatly reduce the local peak wave pressure on the vessel as felt by

MOORING LOADS AND DESIGN PRINCIPLES 219

Figure 6.21. Example of vessel motion response spectra. (After Bomze (65).)

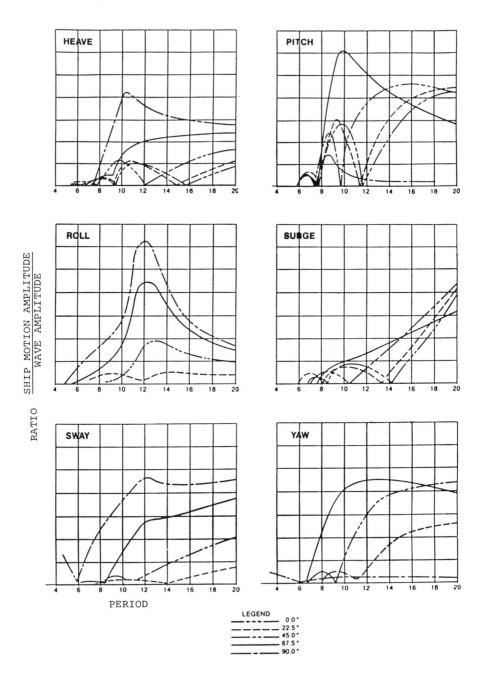

Figure 6.22. Example of relative vessel motions as a function of wave period and direction. (After Sugin (14).)

the mooring system. The short-crestedness of the waves generally also results in peak loads that only act over a short length of vessel at any given instant.

6.7 ANALYTICAL TREATMENT AND MODELING*

Rigorous mathematical treatment of the problem of combined wind, wave, and current loads, vessel motions, and mooring system response is most complicated. Recent advances in the application of high speed computers and computational techniques have allowed the use of mathematical models in solving the general problem. Early attempts at analytical solutions to the problems of moored ships were presented by Wilson (53, 74). Progress in the development of the analytical treatment of moored ships can be followed in the proceedings of the NATO Advanced Studies Institute's specialty conferences on the subject (75–77). Recent computer applications are described by van Oortmerssen et al. (78).

There are two general approaches to the mathematical simulation of ship moorings: (1) frequency domain analysis and (2) time domain analysis. Frequency domain analysis idealizes the moored ship as a linear system characterized by a mass, spring, and dashpot. The analysis is based on mechanical vibrations theory and involves solution of an equation of the following form:

$$(M + a(\omega))\ddot{x} + b(\omega)\dot{x} + cx = F(t) \qquad (6\text{-}19)$$

where M is the vessel mass, $a(\omega)$ is an added mass coefficient, x is vessel displacement, \dot{x} is vessel velocity, \ddot{x} is vessel acceleration, b is the damping coefficient, c is the linear spring constant, and $F(t)$ is a time-varying forcing function. In reality, there is a separate equation of the form of equation (6-19) for each of the six modes of vessel motion.

Terms on the left-hand side (LHS) of equation (6-19) account for forces imposed on the vessel as it moves in still water (i.e., in the absence of waves). The added mass, $a(\omega)$, accounts for those hydrodynamic forces proportional to acceleration of the vessel in still water. Similarly, the damping coefficient, $b(\omega)$, represents hydrodynamic forces proportional to the velocity of the ship in still water. The added mass, $a(\omega)$, and the damping coefficient, $b(\omega)$, are collectively called the hydrodynamic coefficients and are a function of the vessel motion frequency, ω. The spring constant, c, accounts for forces proportional to the vessel displacement and represents either hydrostatic restoring forces or a linearized mooring restraint, or both. $F(t)$, on the right-hand side (RHS) of equation (6-19), is a sinusoidally varying wave force and is a function of wave

*Portions of this section were contributed by John R. Headland, P.E., Design Manager, Harbor and Coastal Facilities, NAVFAC.

amplitude, frequency, and direction. $F(t)$ normally is computed by assuming that the vessel hull is held rigidly. For example, equations (6-15) and (6-16) are approximate expressions for the amplitude of the sinusoidally varying wave force on a hull in surge and sway, respectively.

Hydrostatic restoring forces are computed from basic principles of naval architecture, and are discussed in Chapter 9. The hydrodynamic coefficients, $a(\omega)$ and $b(\omega)$, and the wave forcing function, $F(t)$, usually are computed by using either slender body or diffraction hydrodynamic theory as discussed in reference (67) for a variety of motion frequencies. Similarly, the wave forcing function is computed for a variety of wave frequencies and directions. Equation (6-19) is then solved for the vessel displacement, velocity, and acceleration for each wave frequency and direction of interest. A consequence of this approach is that the vessel responds sinusoidally at a frequency equal to the wave frequency. Because an irregular wave field is characterized by a variety of frequencies and directions, equation (6-19) is solved for the range of wave frequencies and directions present in the wave field. Results of such analysis normally are computed for waves of unit amplitude and summarized in terms of RAOs such as those presented in Figure 6.22. Because the moored-ship system is linear, the motions of the vessel in an irregular wave field can be determined by using the spectral response techniques described in the previous section.

The frequency domain approach has been, and continues to be, widely used, as it is simpler and requires less computational effort than time domain analysis. Its shortcomings, however, are fundamental, and frequency domain techniques can, in many cases, provide misleading results. First of all, the mooring restraints must be linear; that is, the mooring restraint load must be a linear function of displacement. Unfortunately, mooring lines and fenders often are highly nonlinear. When the mooring restraints are nonlinear, nonnegligible subharmonic motions of the vessel can occur at frequencies that differ from the forcing wave frequency (see reference 67). Such motions are not simulated by frequency domain analysis. Second, as has already been discussed, vessels are subject to a low frequency, slowly varying wave drift force. Many moored-ship systems are characterized by relatively low natural frequencies in surge, sway, and yaw. This fact, coupled with the fact that damping is small at low frequencies, makes ship moorings prone to low frequency excitation. Soft mooring systems, such as spread moorings and single point moorings, are particularly susceptible to slowly varying drift forces. In fact, the first-order forces at wave frequency often are neglected in the analysis of soft single point mooring systems. Obviously, the above-described frequency domain analysis cannot be used to evaluate vessel response from slowly varying wave drift forces.

The shortcomings of frequency domain analysis are overcome by time domain analysis at the expense of added computational effort. Time domain analysis was developed by Cummins (79) and has been described in detail by van

Oortmerssen (67); it consists of solving equations of the following form:

$$(M + m')\ddot{x} + \int_{-\infty}^{t} K(t - \tau)\dot{x}d\tau + cx = F(t) \qquad (6\text{-}20)$$

where m' is a constant inertial coefficient, $K(t - \tau)$ is an impulse–response function where τ is a variable integration time denoting an earlier vessel position, $F(t)$ is an arbitrary forcing function, c is a hydrostatic restoring force coefficient, and other terms are as previously defined for equation (6-19). Like the frequency domain approach, terms on the LHS of equation (6-20) account for forces on the ship when moving in still water. In contrast to equation (6-19), however, the constant inertial coefficient, m', and the impulse—response function, $K(t - \tau)$, are independent of motion frequency and represent hydrodynamic forces on the vessel for any arbitrary vessel motion. The RHS of equation (6-20) is an arbitrary force–time function that may include first-order wave frequency forces, second-order wave drift forces, nonlinear mooring restraint forces, wind forces, current forces, and forces associated with passing ships.

The constant inertial coefficient, m', and the impulse–response coefficients cannot be determined directly. Instead, they are determined from the frequency-dependent hydrodynamic coefficients, as follows:

$$K(t) = \frac{2}{\pi} \int_{0}^{\infty} b(\omega) \cos \omega t \, d\omega \qquad (6\text{-}21)$$

$$m' = a(\omega) + \frac{1}{\omega} \int_{0}^{\infty} K(t) \sin \omega t \, dt \qquad (6\text{-}22)$$

Van Oortmerssen (67) provides a complete description of the numerical solution of equation (6-20). It suffices to say that the solution to equation (6-20) provides time histories of vessel motions and mooring restraint loads. Example time domain analysis results are provided in Figure 6.23 for a spread moored floating drydock. The mooring geometry and mooring line load deflection curve are shown in the figure. Figure 6.23 also provides a probability of exceedence curve for the load in mooring line 4.

Further description of the mathematical development and solutions to equations (6-20) through (6-22) is beyond the scope of this text; the reader is referred to the previously cited literature and to references (79) and (80) for discussion of the applications of mathematical models to vessel mooring problems. General description of the use of computer simulations can be found in references (81) and (82). Analytical treatment of berthing impacts has been presented by Fontijn (83), by Lee, Nagai, and Oda (66), and in the various papers of the NATO conferences proceedings.

224 DESIGN OF MARINE FACILITIES

Mooring Geometry

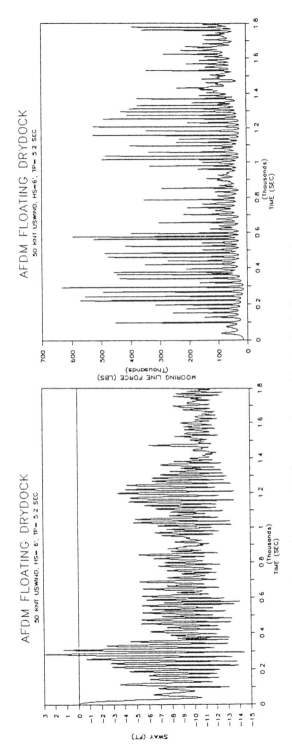

Figure 6.23. Example time domain analysis for a spread moored floating dry dock. (Provided by John R. Headland, P.E., NAVFAC.)

Figure 6.23. (Continued)

Hydraulic Models

Physical scale modeling may be resorted to on large-scale projects of complex geometry or to validate the predictions of mathematical models. Physical hydraulic models are expensive and time-consuming, and thus usually are employed to model the final design configuration and to examine specific effects. Hydraulic models are used in a wide variety of port and harbor problems such as breakwater, wave, and water circulation studies and to obtain mooring forces and vessel movements under variable environmental conditions. These models are based upon the principles of dimensional analysis and mechanical similitude as described in almost any fluid mechanics texts and in references (42) and (58). Usually only one or two types of force or behavior can be modeled in a given model because of relative scale effects. Certain nondimensional numbers that express the relationship between natural forces must remain constant between the model and the prototype in order for the forces to scale properly. For example, in determining mooring forces due to wave and current action, gravity and inertia forces govern, and their relationship is derived from the Froude model law, wherein the Froude number (N_F) must remain constant between model and prototype and is given by:

$$N_F = \frac{V}{\sqrt{gd}} \qquad (6\text{-}23)$$

where V represents the fluid velocity. The relationships between any linear dimension (L) for a linear scale of $1:L$ and other dimensional units in a Froude model are:

Acceleration:	a	$\propto 1:1$
Velocity:	V	$\propto L^{1/2}$
Time:	t	$\propto L^{1/2}$
Force:	F	$\propto L^3$
Force/unit length (or stiffness):	F/L	$\propto L^2$

Model scales are usually greater than 1:100 in order to reduce scale effects from viscous and friction forces and other factors. Figure 6.24 shows a model test setup for obtaining the motion response and mooring forces on a 200,000 DWT prototype tanker as used by van Oortmerssen (67) to confirm his theo-

Figure 6.24. Scale model test setup to evaluate vessel motions and mooring forces for vessel moored alongside pier. (Photo courtesy of the Netherlands Ship Model Basin (NSMB).)

retical analysis and mathematical model. The scale of the model was 1:82.5, wave heights were measured with an electrical resistance probe, and forces were measured by six strain-gage-type force transducers. The results for hydrodynamic coefficients and transfer functions often are plotted against the nondimensional frequency $\omega' = \omega \sqrt{L/g}$ for such tests.

In accordance with the principles of dimensional analysis, the force F due to wave action, for example, is in general a function of various other dimensions as given by:

$$F = f(\rho, g, H, L, D, d)$$

In this case, D is any characteristic width dimension, and ρ is the mass density (note that the product ρg = the unit weight of fluid γ). We can write the non-dimensional forces as:

$$\frac{F}{\rho g H D^2} = f(d/L, D/L, H/L) \qquad (6\text{-}24)$$

which is a common form of presenting wave force data. Where viscous forces are important, the Reynolds number (N_R), as given by:

$$N_R = \frac{VD}{\nu} \qquad (6\text{-}25)$$

where ν is the kinetmatic viscosity, expresses the relationship between inertia and viscous forces. In the general case of a time-invariant in-line force, such as a steady current drag force, N_R should be included as a constant parameter on the right in equation (6-24). In the linear diffraction case with $D/L > 0.2$, flow separation effects are negligible, and N_R may be neglected. Also for the linear diffraction case (see Section 4.5) the wave steepness usually is small, and the H/L term can be dropped from equation (6-24). Thus, the nondimensional force $F/\rho g H D^2$ can be evaluated in terms of constant d/L, D/L, or d/D values.

Kirkegaard et al. (84) demonstrated the importance of considering directional seas in model testing. Their model test results on a fixed moored tanker subject to both two-dimensional and three-dimensional seas revealed important differences in mooring forces. For the head sea condition, the 3-D representation results in twice the mooring force of the 2-D representation. In a quartering sea, there was little difference except for greater roll motion in 3-D. In the beam sea, the 3-D representation resulted in one-half the mooring forces and roll motion of the 2-D case. The greatest forces were generated in the beam sea condition.

REFERENCES

1. BSI (1984). *British Standard Code of Practice for Maritime Structures*, Part 1: "General Criteria," BS 64349, London (incl. 1987 amd.).
2. BSI (1985). *British Standard Code of Practice for Maritime Structures*, Part 4: "Design of Fendering and Mooring Systems," BS 6349. London.
3. EAU (1985). *Recommendations of the Committee for Waterfront Structures; EAU, 1985.* Society for Harbour Engineering and The German Society for Soil Mechanics and Foundation Engineering; Wilhelm Ernst & Sohn, Berlin–Munich.
4. PHRI (1980). *Technical Standards for Port and Harbour Facilities in Japan*. Ministry of Transport, The Overseas Coastal Area Dev. Inst. of Japan, Tokyo.
5. NAVFAC (1980). *Piers and Wharves*, DM-25.1. Dept. of the Navy, Alexandria, Va.
6. Hunley, W. H. and Lemley, N. W. (1980). "Ship Maneuvering, Navigation and Motion Control," in *Ship Design and Construction*, Taggert, ed. SNAME, New York.
7. Crenshaw, R. S. (1976). *Naval Shiphandling*, 4th ed. Naval Institute Press, Annapolis, Md.
8. OCIMF (1978). *Guidelines and Recommendations for Safe Mooring of Large Ships at Piers and Sea Islands*. OCIMF, Witherby & Sons, London.
9. Bruun, P. (1981). *Port Engineering*. Gulf Publishing, Houston, Tex.
10. Slinn, P. J. B. (Aug. 1979). "Effect of Ship Movement on Container Handling Rates," *The Dock and Harbour Authority*.
11. Thoresen, C. A. (1988). *Port Design: Guidelines and Recommendations*. Tapir Publishers, Trondhiem, Norway.
12. LeMéhauté, B. L. (1977). "Wave Agitation Criteria for Harbors," *Proc. ASCE*, Ports '77, Spec. Conf., Long Beach, Calif.
13. Bratteland, E. (1974). "A Survey on Acceptable Ship Movements in Harbours," *The Dock and Harbour Authority*, Vol. LV, No. 647, London.
14. Sugin, L. (1983). "Conventional vs. Offshore Loading and Unloading," Indonesian Coal/Lignite Dev. and U.S. Tech. Symp., Jakarta, Indonesia.
15. Soros, P. and Koman, B. (1982). "Variable-Orientation Berth for Large Carriers," *Proc. OTC*, Paper No. 4399, Houston, Tex.
16. NAVFAC (1986). *Fixed Moorings*, DM-26.4. Dept. of the Navy, Alexandria, Va.
17. Khanna, J. and Birt, C. (1977). "Two Mooring Dolphin Concept for Exposed Tanker Terminals," *Proc. ASCE*, Ports '77, Spec. Conf., Long Beach, Calif.
18. Chernjawski, M. (Jan. 1980). "Mooring of Surface Vessels to Piers," *Marine Technology*, Vol. 17, No. 1, SNAME, New York.
19. Bruun, P. (Feb. 1984). "Mooring and Fendering—Rational Principles in Design," *Dock and Harbour Authority*, London.
20. Meals, W. D. (1969). "Rigging, Tackle and Techniques," in *Handbook of Ocean and Underwater Engineering*, Myers et al., eds. McGraw-Hill, New York.
21. NAVFAC (1986). *Mooring Design Physical and Empirical Data*. DM-26.6. Dept. of the Navy, Alexandria, Va.
22. Wilson, B. W. (Nov. 1967). "Elastic Characteristics of Moorings," *Proc. ASCE*, Vol. 93, No. WW-4.
23. Bruun, P. (Sept. 1981). "Breakwaters vs. Mooring," *The Dock and Harbour Authority*, Vol. XLII, No. 730.
24. OCIMF (1977). *Prediction of Wind and Current Loads on VLCC's*. Witherby & Sons, London.
25. Benham, F. A. et al. (1977). "Wind and Current Shape Coefficients for Very Large Crude Carriers," *Proc. OTC*, Paper No. 2739, Houston, Tex.
26. NAVFAC (1968). *Harbors and Coastal Facilities*, DM-26. Dept. of the Navy, Washington, D.C. (Superseded by new series of point manuals.)

27. Ayers, J. R. and Stokes, R. C. (1958). "Berthing of U.S. Navy Reserve Fleet," *Proc.* PIANC, XVIII Int. Nav. Congress, Section II, Rome.
28. Aage, C. (1971). "Wind Coefficients for Nine Ship Models," Aerodynamics Dept., Danish Ship Research Lab.
29. Altman, R. (July 1971). "Forces on Ships Moored in Protected Waters," Hydronautics, Inc., T.R. 7096-1.
30. Gould, R. W. F. (1967). "Measurements of the Wind Forces on a series of Model Merchant Ships," National Physical Laboratory, Aerodynamics Div., Report No. 1233, U.K.
31. Isherwood, R. M. (1973). "Wind Resistance of Merchant Ships," *Trans. RINA*.
32. Martin, L. L. (1980). "Ship Manuevering and Control in Wind," *Trans. SNAME*, Vol. 88, New York.
33. Owens, R. and Palo, P. A. (1981). "Wind Induced Steady Loads on Moored Ships," Tech. Memo M-44-81-7, NCEL, Dept. of the Navy, Port Hueneme, Calif.
34. Palo, P. A. (1983). "Steady Wind and Current Induced Loads on Moored Vessels," *Proc.* OTC, Paper No. 4530, Houston, Tex.
35. Quinn, A. D. F. (1972). *Design and Construction of Ports and Marine Structures.* McGraw-Hill, New York.
36. Shearer, H. D. A. and Lynn, W. M. (Mar. 1960). "Wind Tunnel Tests on Models of Merchant Ships," *Trans. NECI*, Vol. 76, Part 5.
37. PHRI (1984). Technical Note No. 348, Japanese Ministry of Transport, Tokyo.
38. Fang, S. (1979). "Field Measurement of VLCC mooring Loads," *Proc.* OTC, Paper No. 3383, Houston, Tex.
39. Simiu, E. and Scanlan, R. H. (1978). *Wind Effects on Structures.* Wiley, New York.
40. ASCE (1987). "Wind Loading and Wind-Induced Structural Response," State of the Art Report by the Committee on Wind Effects, Structural Division, ASCE, New York.
41. Palo, P. A. (1986). "Full Scale Vessel Current Loads Data and the Impact on Design Methodologies and Similitude," *Proc.* OTC, Paper No. 5205, Houston, Tex.
42. Newman, J. N. (1977). *Marine Hydrodynamics.* M.I.T. Press, Cambridge, Mass.
43. Khanna, J. and Sorensen, T. (1980). "Stand-off force on Ships Moored in Strong Current," *Proc. ASCE*, Ports '80, Spec. Conf., New York.
44. Jackson, W. H. (1973). "Some Problems Associated with Jetties in Fast Flowing Water," PIANC, 23d Int. Nav. Congress, Section II, Ottawa.
45. Khanna, J. and Ottesen-Hansen, N. E. (1977). "Moored Ships Exposed to Combined Waves and Current," *Proc.* POAC '77, Conference, St. Johns's, Newfoundland.
46. Wood, J. S. et al. (1980). "Design of an Offshore Berth for Mooring Large Ships in a Strong Current," *Proc.* PIANC, 25th Nav. Congress, Section II, Vol. I.
47. Ball, D. J. and Wilcock, J. A. (June 1981). "Ship and Jetty Interaction," *The Dock and Harbour Authority*, Vol. LXII, #727, London.
48. Wang, S. (Aug. 1975). "Dynamic Effects of Ship Passage on Moored Vessels," *Proc. ASCE*, JWPCOE Div., Vol. 101, No. WW-3.
49. Remery, G. F. M. (1974). "Mooring Forces Induced by Passing Ships," *Proc.* OTC, Paper No. 2066, Houston, Tex.
50. Lean, G. H. and Price, W. A. (Nov. 1979). "The Effect of Passing Vessels on a Moored Ship," *The Dock and Harbour Authority*, London.
51. Muga, B. J. and Fang, S. S. (1975). "Passing Ship Effects from Theory and Experiment," *Proc.* OTC, Paper No. 2368, Houston, Tex.
52. Wilson, B. W. (1951). "Ship Response to Ranging in Harbor Basins," *Trans. ASCE*, New York.
53. Wilson, B. W. (Dec. 1959). "A Case of Critical Surging of a Moored Ship," *Proc. ASCE*, Vol. 85, No. WW-3.

54. O'Brien, J. T. and Kuchenreuther, D. I. (Mar. 1958). "Forces Induced on a Large Vessel by Surge," *Proc. ASCE*, Vol. 84, WW-2.
55. Vasco Costa, F. (Feb. 1983). "Moored Ships as Oscillating Systems," *The Dock and Harbour Authority*, London.
56. Goda, Y. and Yoshimura, T. (1971). "Wave Force Computation for Structures of Large Diameter, Isolated in the Offshore," PHRI Report, Vol. 10, No. 4, Nagase, Japan.
57. Ibid. (1972). "Wave Force on a Vessel Tied at Offshore Dolphins," *Proc.* 13th Coastal Engineering Conf., ASCE, New York.
58. Sarpkaya, T. and Isaacson, M. (1981). *Mechanics of Wave Forces on Offshore Structures*. Van Nostrand Reinhold, New York.
59. McCormick, M. E. (1973). *Ocean Engineering Wave Mechanics*. Wiley, New York.
60. Van Oorshot, J. H. (1975). "Subharmonic Components in Hawser and Fender Forces," *Proc.* 15th Coastal Engineering Conf., ASCE, New York.
61. Chakrabarti, S. K. (May 1980). "Steady and Oscillating Drift Forces on Floating Objects," *Proc. ASCE*, Vol. 106, WW-2.
62. Løken, A. E. and Olsen, O. A. (May 1979). "The Influence of Slowly Varying Wave Forces on Mooring Systems," *Proc.* OTC, Paper No. 3626, Houston, Tex.
63. Hsu, F. A. and Blenkarn, K. A. (May 1970). "Analysis of Peak Mooring Forces Caused by Slow Vessel Drift Oscillations in Random Seas," *Proc.* OTC, Paper No. 1159, Houston, Tex.
64. Faltinsen, O. M. and Michelsen, F. C. (1974). "Motions of Large Structures in Waves at Zero Froude Number," *Proc.* Int. Symp. on the Dynamics of Marine Vehicles and Structures in Waves, University College, London.
65. Bomze, H. (1980). "Math Model for Harbor Waves/Ship Motions/Moorings," *Proc. ASCE*, Spec. Conf., Ports '80, Norfolk, Va.
66. Lee, T. T., Nagai, S., and Oda, K. (May 1975). "On the Determination Impact Forces, Mooring Forces, and Motions of Super Tankers at Marine Terminals," *Proc.* OTC, Paper No. 2211, Houston, Tex.
67. van Oortmerssen, G. (1976). "The Motions of a Moored Ship in Waves," Ph.D. thesis Univ. of Tech. of Delft, NSME No. 510, Delft, The Netherlands.
68. Stammers, A. J. and Wennink, C. J. (Apr. 1977). "Investigation of Vessel Mooring Subject to Wave Action," *The Dock and Harbour Authority*, London.
69. Khanna, J. and Birt, C. (1977). "Soft Mooring Systems at Exposed Terminals," *Proc. ASCE*, Spec. Conf., Ports '77, Long Beach, Calif.
70. Eryuzlu, N. E. and Boivin, R. (1978). "An Experimental Program on Mooring Forces of Large Vessels Berthed at Offshore Terminals," *Proc.* of 4th Conf., POAC, St. John's, Newfoundland.
71. Koman, B. and Seidl, L. H. (Sept. 1979). "Mooring and Breasting Forces at an Offshore Berth," *Proc. ASCE*, Spec. Conf. C.E.O. IV, Vol. II, San Francisco.
72. Fang, S. and Scheider, R. M. (1980). "Design of a Tanker Mooring System for Dynamic Loads," *Proc. ASCE*, Spec. Conf. Ports '80, Norfolk, Va.
73. Hwang, Y. L. and Bando, K. (1987). "Motion Analysis of a Berthed Container Ship in Frequency and Time Domains," *Proc. ASCE*, Coastal Hydrodynamics, Spec. Conf., New York.
74. Wilson, B. W. (Oct. 1958). "The Energy Problem in the Mooring of Ships Exposed to Waves," The Princeton Univ. Conf. on Berthing and Cargo Handling in Exposed Locations, Princeton, N.J.
75. NATO (1965). *Analytical Treatment of Problems in the Berthing and Mooring of Ships*. NATO Advanced Study Institute, ASCE, New York.
76. NATO (1973). *Analytical Treatment of Problems in the Berthing and Mooring of Ships*. NATO Advanced Study Institute, Wallingford, U.K.
77. NATO (1987). *Advances in Berthing and Mooring of Ships and Offshore Structures*, Advanced Study Institute, Trondhiem, Norway.

78. van Oortmerssen, G. et al. (Mar. 1986). "Computer Simulation of Moored Ship Behaviour," *Proc. ASCE*, JWPCOE Div., Vol. 112, No. 2.
79. Cummins, W. E. (1962). "The Impulse-Response Function and Ship Motions," David Taylor Model Basin Report No. 1661, Washington, D.C.
80. Bomze, H. (May 1974). "Analytical Determination of Ship Motions and Mooring Forces," *Proc.* OTC, Paper No. 2072, Houston, Tex.
81. van Oortmerssen, G. et al. (1984). *Computer Simulation as an Aid for Offshore Operations*. WEMT, Paris.
82. Hooft, J. P. (Apr. 1986). "Computer Simulations of the Behavior of Maritime Structures," *Marine Technology*, SNAME, Vol. 23, No. 2, New York.
83. Fontijn, H. L. (May 1980). "The Berthing of a Ship to a Jetty," *Proc. ASCE*, Vol. 106, WW-2.
84. Kirkegaard, J. et al. (1980). "Effects of Directional Sea in Model Testing," *Proc. ASCE*, Spec. Conf., Ports '80, Norfolk, Va.

7
Design of Fixed Structures

Fixed structures to which vessels are secured directly include piers, wharves, bulkheads, quays, and dolphins. Ancillary structures, which serve to access piers or provide support for cargo handling equipment, product lines, and so on, include support platforms, trestles, catwalks, and moles. Structures may be generally categorized as being of open, pile-supported construction versus closed, fill-type construction. This chapter reviews general design principles common to both types of construction and the basis of selection between them, as well as particular aspects of open, pile-supported structure design. Aspects of closed, fill-type construction are addressed in Chapter 8. Important features of piers and wharves include crane trackage, ship services and utility systems, and such details as expansion joints, deck drains, handrails, curbs, and so forth. Fender systems are discussed in Chapter 5, and mooring hardware is described in Chapter 6. It is assumed that the reader is familiar with the general principles of structural and building code requirements for bridge and building construction. The following discussion is intended to emphasize the particular requirements of marine construction as they differ from those of traditional civil engineering works.

7.1 STRUCTURAL TYPES AND CONFIGURATIONS

A *dock* is the most general designation for a structure or place at which a vessel can be moored. A *pier* is a dock structure that typically projects outward nearly perpendicular to or at some skew angle with the shoreline. Outside of the United States, the word *jetty* is often used for a pier, whereas within the United States jetty generally refers to a coastal engineering structure used to stabilize coastal inlets. A pier is essentially a free-standing structure, shore-connected at one end, which allows the berthing of vessels along both sides. In some instances where deep water is not available near the shore end, the pier may be constructed in a T-head or L-head configuration, or less commonly a U-pier configuration, with the shore-connected stem or stems serving primarily as an access road and with vessels berthed alongside the T, L, or U portion, which is nearly parallel to the shoreline. In this case, the distinction between a pier and a wharf is imprecise, and the structures may also be referred to as T or L wharves. Other pier configurations include: breakwater piers, which provide wave protection and allow vessel berthing only along the protected edge; jack-

up type piers (De-Long piers) which consist of barge-like hulls floated into place and jacked up on free-standing legs spudded into the bottom; and floating piers, as discussed in Chapter 9.

A *wharf*, in contrast to a pier, usually is built nearly parallel to and contiguous with the shoreline, so that it also performs as a soil-retaining structure. A wharf may also be referred to as a *quay*, especially when it is of solid fill vertical wall construction and is long and continuous. Both piers and wharves may be of either *open*, pile- or column-supported construction or *closed*, solid fill-type construction. Figure 7.1 shows simplified cross sections of typical pier and wharf structural types.

The most common pier construction consists of a pile foundation of timber, concrete, or steel piles supporting a timber or concrete deck. The pile bents may be designed to resist lateral loads via the use of *batter* (or raker) piles, or by being rigidly cross-braced, or via cantilever action with moment-resistant deck connections. Batter piles that are secured to the lower level of bracing of a cross-braced pier sometimes are referred to as *spur* piles. Piles also may be driven through prefabricated tubular steel jackets, which are cross-braced and anchored by the piles. This type of construction, called *jacket/template*, is common in deepwater offshore operations. Piers also may be founded on large-diameter cylinder piles, or belled cylinders may be supported by an array of piles below the mudline. Open piers also may be supported by columns or pillars bearing directly on bedrock or a very firm substratum. Lateral stability is provided by rigid cross bracing, or sometimes by vertical shear walls, called *lamella walls*, similar in nature to full-width bridge piers. Closed pier construction may be of filled *sheet pile cells*, *diaphragm cells*, *concrete caissons*, or timber or concrete *cribs*, usually with stone fill.

Wharves may be of open pile construction with low-height retaining structures or relieving slabs along their inshore margins. *Relieving platform* construction consists of a low-level pile-supported deck that is filled over in order to gain stability and relieve soil pressures behind the wall in weak soil conditions. A sheet pile bulkhead or cutoff wall usually is required along its inshore edge. Relieving platforms alternatively may be constructed along the inside of sheet pile bulkheads. The exposed bottom slope below open-type wharves usually must be protected against erosion from wave, current, and vessel wash by placement of stone or concrete mattress slope protection. *Bulkheads* are vertical wall structures typically constructed of steel, concrete, or timber sheet piles anchored via tie rods to discrete *deadmen* or to continuous anchor beams or walls. Cantilever bulkhead walls may be employed for walls of relatively low height and low soil surcharge loadings. Cantilever walls often are made of large-diameter cylinder piles or heavy steel sheet pile Z-sections. Bulkheads also may be supported by batter piles or rock anchors.

Gravity walls gain stability against sliding and overturning via their mass,

DESIGN OF FIXED STRUCTURES 235

Figure 7.1. Pier, wharf, and dolphin structural types. (Prepared by Goldberg Zoino Associates, Inc.)

proportions, and soil friction. The simplest and earliest form of gravity wall consists of cut stone blocks, usually bedded in a foundation mat of smaller stones. Block walls also may be pile-supported by platforms, which usually are buried below the mudline. Stone masonry may be laid ''open,'' or the gaps between stones may be bedded with smaller crushed stone or filled with grout.

Figure 7.1. (Continued)

Precast concrete blocks are commonly used for contemporary block walls. Stone masonry blocks typically weigh from one to five tons, whereas solid concrete blocks often weigh from five to 200 tons as limited by placing equipment. Open, cellular blocks of precast concrete that are filled with earth or stone after place-

DESIGN OF FIXED STRUCTURES 237

Figure 7.1. (Continued)

ment can be used to reduce handling weights. Blocks on firm bearing material may be placed in an interlocking pattern with vertical joints staggered; or where settlements are possible, they should be placed in a sloping bond pattern with the nearly vertical joints aligned. Gravity walls also may be of massive "cyclopean concrete" construction poured in place, or of precast concrete caissons floated or slid into position and then filled with earth or stone. Precast *L-walls*

utilize the weight of earth fill to gain stability against sliding and overturning. Sheet pile cells are common in wharf construction, often capped with a continuous cast-in-place concrete edge beam or deck system.

Moles and *trestles* are primarily pier or platform access structures used for vehicular, pipeline or conveyor, and/or personnel access. Moles are of solid fill construction, and trestles are typically of free-standing pile bents or pile groups with a bridging structure spanning them.

Dolphins are solitary structures used primarily to absorb the impact of berthing ships, referred to as berthing or breasting dolphins, and/or to serve as a point for securing a vessel's mooring lines, referred to as mooring dolphins. Dolphins range from simple timber pile clusters to deepwater steel jacket structures, as described further in Section 7.4 along with special-purpose platform structures used to support cargo handling equipment. Additional description of various pier and wharf construction types can be found in references (1) through (7).

7.2 SELECTION OF OPTIMUM STRUCTURE TYPE

The most important factors in determining the optimum structure type at a given location generally include the subsurface soil conditions, in particular the depth to bedrock or firm bearing material, and the water depth. Both these factors usually have an important impact on the relative cost of open versus closed-type construction. As a general rule, closed-type construction, such as sheet pile cells, is favored where water depths are shallow to moderate and the depth to rock is shallow. Sheet pile cells typically are between 30 and 60 feet in diameter, and it is interesting to note that the weight of steel (hence its cost) per lineal foot of berth face is very nearly independent of the cell diameter. Cells become increasingly competitive with bulkheads and other fill-type structures with increasing water depth and wall height.

Anchored bulkheads require some minimum embedment depth into competent soils and may otherwise be cost-competitive at shallow water sites; these bulkheads usually are practical and economical for wall heights up to about 25 to 30 feet, although higher ones are built. At deepwater locations with soft soils extending relatively deep below the bottom pile foundations usually are required. Although pile or open column-type construction may be used at shallow rock locations, the cost to anchor the piles or columns to the bedrock and to provide adequate horizontal stability usually exceeds that of a suitable fill-type structure. Relieving platforms, which are a combination of open and fill-type construction, may be used to provide uplift resistance and thus lateral load resistance to pile foundations that cannot develop it within the soil. A relieving platform also reduces the required bulkhead wall height, thus extending the water depth capability of the system.

Other important factors that affect the relative cost of structure types include the local cost and availability of construction materials and labor and the accessibility of construction equipment. Factors that are not related to cost but may heavily impact the choice between open and closed-type construction include: the magnitude and nature of loadings; hydraulic conditions, such as wave action and currents; fire hazard and safety-related requirements; damage susceptibility and ease of repairs; construction schedule and "weather windows"; local construction practices; and environmental and regulatory concerns over water circulation and habitat loss, which almost always favor open-construction alternatives. Closed, solid-fill construction generally offers greater horizontal and vertical load capacity and impact resistance than does open pile construction. Closed-type construction typically presents a vertical to near-vertical structure face that reflects wave energy and may be a problem if the structure face is exposed to significant wave action. Possible scouring at the face of builheads due to prop wash or current action may strongly influence the choice of structure type.

The use of precast or prefabricated structural elements often results in overall cost savings and may be essential to meet critical construction schedules. The cost of form work, for example, is usually on the order of 10% to 15% of total construction cost for a typical pier, and additional savings in labor and potential down time due to weather or operations make precast concrete alternatives most attractive. Similarly, in deep water, prefabricated steel jacket/template structures or precast concrete caisson units often are mandatory to reduce construction exposure time. Precast concrete caisson or L-wall construction is more common on large-scale projects where many repetitive units are required and is a viable alternative where on-site construction time must be minimized and large lateral loads resisted. Caisson and L-wall construction are generally practical for wall heights up to around 60 feet.

In considering the lowest-cost alternative, the initial construction cost often is the sole determining factor, as average annual maintenance costs usually are relatively low, on the order of 1% or less of first cost per annum (8). Steel and open-type structures usually require more maintenance than massive concrete or stone construction, and, if all else is equal, long-term maintenance may be the determining factor. It is interesting to note, however, that marine structures tend to become obsolete or require upgrading within 15 to 25 years, as opposed to a usual physical structure life of 40 to 50 years; so annual maintenance cost comparisons may be projected over the shorter run. If a large degree of uncertainty exists over material costs and contractor efficiency with a given construction type, alternative final design plans may be prepared to generate competitive construction bids.

In the design of certain facilities, it may be useful to know and compare the incremental costs associated with greater berth depth, live load capacity, or

240 DESIGN OF MARINE FACILITIES

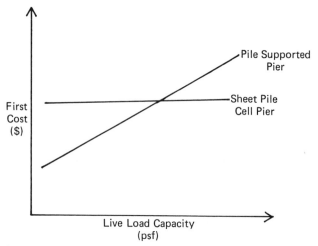

Figure 7.2. Relative cost comparison of pier and wharf structural types.

other variables that impose certain operational constraints. Figure 7.2 gives a relative cost comparison of basic structural types, in terms of cost per lineal foot of berth versus live load capacity. Such plots may be most useful in decision making where future expansion options are likely but uncertain. For example, although the cost of a sheet pile cell or filled caisson structure may be greater than the cost of a pile-supported pier designed to meet current live load requirements, the greater cost of the solid construction may be considered worthwhile when compared with the cost and uncertainty of designing the pile structure option for some assumed future capacity requirement. The cost of any operational restraints imposed by live load limits, berth depths, and so on, may be compared with the incremental first cost of the facility.

7.3 PIER AND WHARF STRUCTURAL DESIGN

Final design of a pier or wharf structure evolves from a preliminary design process based upon design vessel requirements, which in turn dictate the water depth and site layout, the general overall dimensions, cargo handling and deck loads, and berthing and mooring load requirements. Design criteria for deck equipment and cargoes also must include backing and turning radii, operating tolerances such as crane capacity versus reach and swing, minimum deck storage area, space for handling a vessel's lines, shore connections, services, utilities, lighting, fire protection, and so on. Proper and ample deck drainage is critical from both an operations and a maintenance viewpoint. In determining the structure's final layout and configuration, it is important to consider the possibility of future expansion or upgrading. The design of bulkheads and pile

foundations alike should be adequate for any foreseeable future dredging. The selection of appropriate construction materials with regard to design life and maintenance costs also must be given careful consideration. Marine structures in general are subjected to impact and accidental overloads, more rapid rates of material deterioration, and general neglect. General design criteria and considerations are discussed in Chapter 3.

The optimization of foundation systems and structural systems often involves a design spiral process whereby pile spacings and capacities, for example, are optimized against deck framing spans and loads. The structural design of piers and wharves generally follows the standard building code practices for concrete, steel, and timber (9–11), where the major differences between landside and waterfront construction are found in the determination of loads and related safety factors and in the selection of materials for durability. As pier construction is similar in many ways to bridge construction, the standard highway bridge design practices, as given by the American Association of State Highway and Transportation Officials (AASHTO) (12), are often referred to in pier design. Load factors and combinations for marine structures may differ from those used on land. The nature of marine work necessitates that different, usually more conservative, allowable stresses, larger minimum member sizes, and special material quality requirements be applied according to the designer's judgment. A corrosion allowance usually is added to required member sizes in steel design.

The U.S. Navy (7) requires the maximum allowable stresses shown in Table 7.1 for given load combinations based upon normal code allowable stresses for piers and wharves. Load combinations for berthing facilities located in frigid climates, where ice loadings are critical, should be developed after a review of site conditions. Note that for Group (e) the applied deck live load ranges between 10% and 50% of the total design live load, whatever yields the worst conditions. Other live load reduction factors may be applied as described in

Table 7.1. Allowable stresses for piers and wharves (7).

Load group	Loading	Percentage of unit stress
a	$D + L + I + E + B + Wa + Be$	100
b	Group (a) $+ S + T$	125
c	$D + W + Wa + B + E + S + T$	140
d	Group (a) $+ 30\% Wi + Ws + S + T$	140
e	$D + Wa + B + E + EQ$	133
f	Group (a) $+ Ice$	140
g	$D + Wa + B + Wi + E + Ice$	150

Note: D = dead load, L = live load, I = impact, E = earth pressure, B = buoyancy, Wa = water pressure, Be = berthing load, S = shrinkage, T = temperature, Wi = wind on structure, Ws = wind on ship, EQ = earthquake, Ice = ice.

subsequent sections. Other relevant load combination criteria can be found in reference (12). Higher permissible stresses and lower factors of safety (F.S.) may be allowable under extreme environmental conditions, as described in Section 6.1. It is important to note that piers and wharves built near the shore and along urban waterfronts may also be subject to state and local building codes. Such codes often are more restrictive—in terms of such as things as ultimate pile load capacities versus settlements, for example—than criteria that would apply to offshore construction.

Useful codes, standards, and guideline documents that pertain directly to the general and structural design of port and harbor structures include: the U.S. Naval Facilities Engineering Command (NAVFAC) design manuals (7, 13), the British Standards Institution (BSI) *British Standard Code of Practice for Maritime Structures* (14, 15), the *Recommendations of the Committee for Waterfront Structures* (EAU) (16), and the *Technical Standards for Port and Harbor Facilities in Japan* (PHRI) (17). Other foreign countries may have their own codes of practice. For the design of offshore terminal structures, useful guidelines can be found in references (18) through (21).

The preceding chapters of this book dealt with the determination of design criteria and loadings and the layout of fender systems and mooring hardware. With this background information, an experienced civil engineer should be able to proceed with the design of a marine terminal structure using the basic principles of civil/structural engineering practice. The remainder of this chapter is devoted to aspects of structural design that are not so common to land-based civil engineering structures, and additional design details and features peculiar to waterfront construction. All fixed structures ultimately behave as a soil/structure system, but open, pile-supported structures emphasize the structural engineering aspects of design, whereas solid, fill-type structures emphasize the geotechnical engineering aspects. The remainder of this chapter is concerned with structural and functional design features, with geotechnical aspects of fixed structure design covered in Chapter 8. Figures 7.3 and 7.4 present general nomenclature of typical timber and concrete pier construction that will be referred to in the following discussions.

Pile Foundation Design

The determination of pile axial and lateral load capacities is covered in Section 8.7 and in the general geotechnical engineering literature. The structural design of a pile foundation system includes: the selection of a pile type compatible with subsurface conditions; the determination of axial and lateral loads and associated shears and moments; the pier deck system vertical and lateral load distribution to individual piles; the attachment of the pile head to the deck framing; and the determination of effective column length and end fixity conditions.

DESIGN OF FIXED STRUCTURES 243

Figure 7.3. Typical timber pier nomenclature.

Pile types most common to marine construction include: timber (usually Douglas fir, southern yellow pine, or greenheart in the United States); steel H-piles and pipe piles; and precast/prestressed concrete solid section and hollow cylinder piles. Figure 7.5 illustrates these pile types and their usual dimensions and ranges of application. General criteria for pile type selection can be found in references (22) and (23). Guidelines for pile structural design and allowable stresses can be found in the following references: for timber piles, references (11) and (24); for concrete piles and concrete-filled steel pipe piles, reference (25); for prestressed concrete piles, reference (26); for large-diameter cylinder piles, reference (27); and for steel H-piles, reference (28). General design of marine piles is treated in references (5), (29), and (30). It is common practice to reduce the allowable stresses in piles supporting marine structures to account for environmental exposure and the possibility of accidental overloads. For example, allowable stresses in concrete piles should be reduced by 10% in the marine environment (25). Stresses in steel H-piles traditionally are limited to 9000 psi and may be further reduced for column action. Maximum loads on timber piles frequently are limited to 20 tons regardless of their calculated capacities.

From a purely structural engineering point of view, piles of circular cross section generally are preferred because of their efficiency as long columns; they

244 DESIGN OF MARINE FACILITIES

Figure 7.4. Contemporary concrete pier nomenclature.

have excellent torsional strength and exhibit the same strength properties in all directions, they are easy to clean and inspect, and they generally offer less surface area for coating and corrosion than other types of equal capacity. However, cost and subsurface conditions may favor other cross sections. Steel H-piles, for example, which are often favored for end bearing on rock conditions, offer other advantages; for example, they provide better penetration through rubble and subsurface obstructions, they are relatively easy to splice and secure to the bracing and superstructure, and they usually are readily available and cost-effective. Splicing of marine piles should be avoided in general, and should be restricted or prohibited within the tide zone or zones of high bending moments. Steel piles can be adequately spliced using full-penetration butt welds all around, but there is no wholly satisfactory means of splicing timber or concrete piles unless the splices are deeply buried below the point of contraflexure within the soil.

The pile to cap connection is a very important detail. Marine piles are often put into tension in order for the structure to resist lateral loads, and the resulting

DESIGN OF FIXED STRUCTURES 245

PILE TYPE	TYPICAL DIMENSIONS AND LENGTHS	USUAL RANGE OF CAPACITIES AND COLUMN LENGTHS	REMARKS
Timber	6″–8″ Tip diameter 12″–20″ Butt diameter Douglas fir to 80′ Southern yellow pine to 65′ lengths Special order up to 125′	Typically limited to between 15 and 20 tons for all column lengths	Either Southern yellow pine or Douglas fir with ready available lengths of up to 60 feet. Piles are usually pressure treated with creosote or CCA. Greenheart piles are usually untreated.
Steel H–Pile	Section depth 8″ to 14″ Unspliced– 60′–80′ lengths Splices– unlimited length	40 to 120 ton capacity with effective lengths of up to 60′.	Low displacement, able to penetrate through some obstructions, and easily spliced. The pile is vulnerable to corrosion and may be damaged or deflected when encountering obstructions. Favored for end bearing on rock.
Concrete Filled Steel Pipe	8″ to 48″ ϕ 5/16″ to 3/4 wall thickness. Unspliced– 60′–80′ lengths Splices– unlimited lengths	40 to 200 tons with effective lengths up to 100′	Displacement type piles may be driven either open or close ended, easily spliced, and provides good bending resistance
Precast Concrete	12″ to 24″ round, octagonal or square 60′ to 120′ lengths unspliced	20 to 120 tons with effective lengths up to 80′	High displacement pile may be provided with good corrosion resistance, tolerable of hard driving stresses and vulnerable to large handling stresses.
Concrete Cylinder	30″ to 54″ diameter 150′–200′ + lengths	120 to 240 tons with effective lengths up to 250′	Prestressing allows for large handling stresses and capable of tolerating high bending stress induced by lateral loading and long unsupported lengths.

Figure 7.5. Common marine pile types. (Prepared by Goldberg Zoino Associates, Inc.)

uplift forces may develop large pullout loads in the pile head. Further, even relatively rigid piers and wharves with batter piles to resist lateral loads will move under extreme lateral loads, resulting in moments at the pile head/cap connection. In general, minimum eccentricity factors of 10% of the pile di-

ameter should be applied to the pile design load to give the design moment unless pile head moments are calculated directly. Residual stresses due to jacking piles into place during construction also should be considered. The degree of moment restraint or fixity at the pile head also affects the column length factor and the critical buckling load of the pile. Timber pile to timber cap connections and piles with shallow embedment (less than three to four inches) into concrete caps should be considered as rotation-free, pin-end-connected. The degree of fixity increases greatly with the depth of embedment into a cast-in-place concrete cap. In general, timber piles should be embedded a minimum of six inches into concrete caps. The use of anchor bars with or without hooks well embedded within the concrete cap also greatly increases uplift resistance (31). Research on steel H-piles embedded in concrete caps (32) has shown that the tensile capacity is directly related to the depth of the anchor bars into the cap. End plates are not particularly effective in increasing either bearing capacity or pullout capacity, if adequate embedment is provided. For compressive loading of steel piles, the minimum embedment should be six inches, and the minimum edge distance from the center of the pile to the face of the caps should be 12 to 21 inches for pile capacities of up to 110 tons and over 140 tons, respectively. If this criterion is met, then ultimate bearing stresses in the concrete of $8f'_c$ (where f'_c is the 28-day compressive strength) can be developed between the pile end area and the concrete. When supplementary pile cap reinforcement is used (circling the embedded pile), the ultimate bearing strength may be increased to approximately $10f'_c$. Figure 7.6 illustrates some representative pile to cap connection details. Additional details can be found in references (5), (7), and (30).

The determination of the pile unsupported column length (l_u) and its lateral load resisting capability is a relatively complex process involving calculation of pile/soil relative stiffness properties with depth below the mudline under specified loading conditions. As a practical design approach, load/deflection properties of the given soil profile with depth are plotted (p–y curves) and related to the pile stiffness and applied load, such that an equilibrium position is found yielding shears, moments, slopes, and deflections of the pile/soil system along the length of the pile. This procedure, which is employed in the design of offshore structures, is described further in Section 8.7 and in references (33) and (34). In the design of near-shore and waterfront structures, however, a more simplified approach usually is taken, whereby an equivalent pile embedment or depth to fixity is assumed or calculated. The method is described further in references (25) and (35). The depth to fixity so found allows the pier, wharf, or individual pile to be analyzed as a simple structure where the maximum pile unsupported length and shears, moments, and deflections at the mudline sufficiently approximate the values found in a more rigorous soil analysis. The results generally are adequate for structures in water depths of less than 75

DESIGN OF FIXED STRUCTURES 247

*Deeper embedment increases pull-out capacity & pile head fixity.
Must also maintain minimum edge distance within narrow pile caps.

Figure 7.6. Pile head connection details.

to 100 feet. The approach assumes that the piles are sufficiently embedded in soils of uniform characteristics.

The depth to fixity (\overline{D}), based upon theoretical relationships introduced in Chapter 8, can be calculated from the following equations:

For granular soils, silts and normally loaded clays:

$$\overline{D} = 1.8 \sqrt[5]{\frac{EI}{n_h}} \qquad (7\text{-}1)$$

For preloaded clays:

$$\overline{D} = 1.4 \sqrt[4]{\frac{EI}{k_s}} \tag{7-2}$$

where E and I are the modulus of elasticity and moment of inertia of the pile, respectively; n_h is the coefficient of the horizontal subgrade modulus; and k_s is the subgrade modulus for clay, which can be taken as equal to 67 times the undrained shear strength of the soil (25). The value of n_h for normally loaded clay is equal to k_s divided by the depth below the ground surface, and can be taken as the best triangular fit of the top 10 to 15 feet on the k_s versus depth plot. Values of n_h of submerged sands range from approximately 1.5 pounds per cubic inch (pci), loose, to 30 pci, dense. In order for equations (7-1) and (7-2) to apply, the total pile embedment must be equal to or greater than 3 times \overline{D} (see Figure 7.7). For piles that "fetch-up" in end bearing on rock or another hard stratum above a level of $3\overline{D}$, the pile should be considered as pin-ended at its tip. One should measure \overline{D} from the top of the competent soil stratum, dis-

Figure 7.7. Pile depth to fixity definition sketch.

Figure 7.8. Pile fixity for sloping bottom and batter piles.

counting any loose, recently deposited, organic material or debris. Where the bottom slopes steeply below the structure, \overline{D} should be measured downward from a virtual ground line defined by a line drawn halfway between the soil surface at the given pile and the bottom (or dredge depth) at the face of the structure (17) (see Figure 7.8). For batter piles, \overline{D} should be measured vertically downward, not along the pile length. The value of \overline{D} usually lies within the range of 3.5 to 8.5 pile diameters (D). For typical near-shore waterfront structures of interest, this is usually on the order of 5 to 15 feet.

In figuring the pile unsupported length (l_u), consideration must be given to future dredging or the removal of material by scour or prop wash. The axial capacity of a pile is reduced as a function of its slenderness ratio, as given by: $K_c l_u / r$, where K_c is an effective column length factor that accounts for the pile's end fixity conditions, and r is the least radius of gyration of the pile cross section. In general, it is desirable that $K_c l_u / r$ values be kept below 120 and preferably below 90 to 100. For values of $K_c l_u / r$ within the intermediate column range of 30 to 100, the pile behavior is relatively insensitive to variations in K_c. For $K_c l_u / r$ greater than 100, such "long columns" become very sensitive to K_c; and for "short columns" with $K_c l_u / r$ less than approximately 30, the column strength reduction factor is relatively small.

Theoretical column length factors (K_c) are as follows:

$K_c = 0.5$, both ends fully fixed
$K_c = 0.7$, one end fixed/one hinged
$K_c = 1.0$, both ends hinged

The above values of K_c apply to structures that are rigidly braced or supported against horizontal forces by batter piles, and thus are essentially translation-fixed. Structures supported on unbraced cantilevered piles should be considered translation-free, and the appropriate value should be confirmed by analysis. Note that for one end fixed in the soil and the other rotation-fixed at the pile cap but subject to translation, the theoretical K_c value equals 1.0, as for two pin ends. Considering that full fixity at cap and soil cannot necessarily be guaranteed in most marine structures, a minimum design value of K_c equal to 0.75 generally is recommended (5) even though fixed/fixed end conditions are assumed. A minimum K_c of 1.2 is recommended for flexible cantilever pile piers. For piles that cannot be considered fixed below ground, some support against buckling will be provided even by relatively soft soils. Appropriate values of K_c may be found by rational analysis and applied to the total pile length.

Batter piles are employed to resist lateral loads and movements. When batter piles are used, the lateral load capacity of the vertical piles in the group usually is neglected. A batter pile in compression exerts an upward force that must be resisted by the dead load of the structure or by tension in other piles to resist lateral loads. Pile batters usually range from one horizontal to 12 vertical (1:12) to 5:12, which is usually considered the flattest batter that can be driven without special diving equipment. It usually is assumed that a batter pile has the same axial load capacity that a similar vertical pile has when driven to the same depth within a given stratum. The use of batter piles generally results in a dramatic reduction in pier deflections. Even a relatively shallow batter of 1:12 may reduce overall deflections some 40% or more over only vertical piles. A few piers have been designed on all vertical piles so that the deflection of the structure absorbs some of the impact energy of a berthing vessel (36). Lateral deflections at deck level of about six inches may be allowed under such circumstances, but pier deflections generally should be minimized, especially where vehicular equipment or cranes are operating.

Analysis of the distribution of both vertical and horizontal loads to individual piles often is simplified by reducing the general three-dimensional problem to the analysis of planar pile bents. This approach is generally valid, as pier and wharf structures usually consist of a great number of uniformly spaced pile bents. Longitudinal forces acting normal to the pile bent then are dealt with separately by providing independent batter pile groups, or in the case of a pier on cantilever piles by designing piles for biaxial shear and bending. Vertical loads usually are reduced to bent loads by assuming a tributary area equal to the bent spacing for uniformly distributed live loads. Live load reduction factors on the order of 25% sometimes are applied to the total uniform deck load in determining individual pile loads, depending upon the probability of the entire deck area being loaded to its design capacity at once. The distribution of concentrated loads is carried out by elastic analysis, according to the nature of the

loading. Wheel loads, for example, should be placed directly over a given pile to obtain maximum individual pile loads. It may be conservatively assumed that the entire wheel load is taken by a single pile although a more rigorous elastic analysis usually will demonstrate some degree of distribution along the pile cap, depending upon its depth and rigidity. Impact factors often are neglected in calculating maximum pile loads due to vehicles and moving equipment. Group action effects within the soil normally do not need to be checked when the pile spacing exceeds two to three pile diameters.

For the general case of a stationary concentrated load (F_v), the vertical load on an individual pile (P_v) may be calculated from:

$$P_v = \frac{\sum F_v}{n} + \frac{\sum F_v e_x c_x}{I_x} \qquad (7\text{-}3)$$

where n is number of piles in the bent, e_x is the eccentricity of the concentrated load measured from the center of gravity (c.g.) of the pile group, c_x is the distance of the individual pile from the c.g. of the group, and I_x is the moment of inertia of the pile group about its c.g. (Refer to Figure 7.9.) The three-dimensional case can be accommodated by adding a third term to equation (7-3) for e_y, c_y, and I_y. The above method assumes a rigid pile cap and deck system, and that all piles have the same elastic properties.

According to the theory of continuous beams on elastic supports, a "short beam" can be considered as rigid when its length is equal to or less than $1/\beta$, where β is the relative stiffness factor, given by:

$$\beta = \sqrt[4]{\frac{K_e}{4EI}} \qquad (7\text{-}4)$$

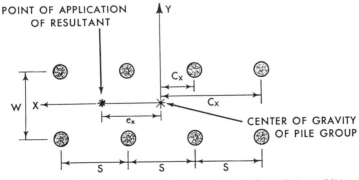

Figure 7.9. Pile load distribution definition sketch. (After U.S. Army (30).)

where K_e is the elastic support constant, which for a series of equally spaced elastic supports is equal to the support spring constant divided by the spacing, and E and I are the beam modulus of elasticity and moment of inertia, respectively. Equation (7-4) also assumes that the piles are linearly elastic in both directions, which is not strictly true for heavily loaded friction piles. Beams can be considered short when their length is less than approximately $5/\beta$ (37). For rigid beams, the load distribution to the piles can be treated as a statically determinate problem with a linear load distribution as reflected in equation (7-3); otherwise, a more rigorous elastic analysis must be performed.

When batter piles are used, the load distribution problem becomes largely indeterminate, as the lateral reactions must be balanced. When counteropposing batter piles are placed in A-frame fashion, the lateral forces will cancel (provided that the uplift piles can develop adequate tension), and the analysis can proceed. Where batter piles are used only in compression and depend upon the weight of the deck system or vertical piles in tension for uplift resistance, a different approach must be taken. Graphical methods such as the force "funicular polygon" sometimes are used, or conservative simplifying assumptions regarding load distribution are made. More accurate plane frame and three-dimensional analysis of pile bents can be readily carried out on one of the many structural analysis programs available for microcomputers. Such programs yield pile shears, moments, and deflections, whereas the above manual analysis employed in preliminary design usually considers only pile axial loads.

Figure 7.10 illustrates cantilever, braced, and batter pile bents and their lateral load resisting systems, idealized by using the equivalent pile length/depth to fixity method. Note that where batter piles are used, the depth to fixity method cannot be employed to accurately predict deflections and lateral load/deflection behavior because the structure's lateral deflection depends primarily upon the axial deflection of the pile. Therefore, the pile's elastic shortening and movement in the soil must be considered over the entire length of the pile. Pile axial stiffness is in general different for tension and compression. Axial load/deflection properties of the pile can be presented in the form of $t-z$ curves, similar to the $p-y$ curves described earlier for pile lateral deflections. Structural analysis then can proceed by replacing the portion of the pile below the mudline with an equivalent dummy pile having soil-spring reaction forces and moment as described in Section 8.7 and in references (33) and (34).

Wharf structures may gain lateral stability by rigid connection to retaining structures or anchoring structures such as friction slabs that are continuous along the wharf's inshore margin (see Figure 7.11). Alternatively, they may be shore-connected at discrete anchor points, an arrangement that requires careful analysis of the horizontal load distribution in both directions, or they may be constructed independent of their shoreside boundary and gain lateral stability via any of the methods shown in Figure 7.10 for pier construction. A critical aspect

DESIGN OF FIXED STRUCTURES 253

Figure 7.10. Pile bent configurations for resisting lateral loads.

254 DESIGN OF MARINE FACILITIES

SELF-SUPPORTED WHARF

WHARF INTEGRATED W/RETAINING STRUCTURE

WHARF SUPPORTED BY RETAINING STRUCTURE
OR INDEPENDENT TIES TO SHORE

Figure 7.11. Wharf lateral stability configurations.

of wharf design is the analysis of backland soil pressures and overall slope stability both behind and below the wharf, as described in Sections 8.4 and 8.5.

Lateral loads may be considered to be distributed uniformly along the pier face as a presumptive uniform load (see Sections 4.6 and 6.1) or as discrete point loads at mooring hardware or fender locations. Figure 7.12 illustrates traditional assumptions regarding lateral load distribution according to

DESIGN OF FIXED STRUCTURES 255

Figure 7.12. Lateral distribution of berthing and mooring forces. (After NAVFAC (7).)

NAVFAC standards (7). In reality, pier deck systems typically are very rigid, and even point loads concentrated directly on a single bent will be well distributed throughout the structure. The pier deck can be treated as a beam on an elastic foundation, as described in the previous discussion of pile caps, with the pile bent lateral stiffness acting as elastic spring supports. In certain situations, better distribution can be obtained through the use of more flexible pile bents. Padron and White (38) and Padron and Elzoghby (39) have developed computer methods and presented findings for example piers.

The percentage of the total vessel impact or other concentrated load taken by a given bent will depend upon the number and spacing of batter piles and batter pile bents, the pile batter and stiffness, the deck stiffness and the ratio of bent spacing to deck width, and the location of the impact along the pier length. In general, end bents are subject to relatively large shares of impact and mooring point loads. The simplest, conservative, approach is to assume that a given pile bent takes 100% of the point load. The assumption of a 45° distribution about an affected bent or contact area may result in an approximately 50% to 67% reduction in this load, whereas the elastic beam analysis will often result in a 67% to 90% reduction of the local impact load. The elastic beam analysis may often lead the designer to the use of more flexible bents in order to obtain a favorable load distribution and reduction in batter pile requirements. In this case it is important also to consider the pile and overall pier deflections and dynamic response and possible vibration problems. Gouda (40) has presented a simplified approach to the pile load distribution problem suitable for manual calculations.

The placement of batter piles or other means to resist longitudinal forces should be given careful consideration. Such piles may be subject to large additional forces due to overall torque moments generated by lateral forces. Added loads in longitudinal piles due to overall pier torque moments can be greatly diminished by the judicious placement of lateral pile bents near the ends of the pier. In general, if longitudinal forces can be taken out at a single location or at a minimal number of locations, the deck can be constructed without expansion joints over long distances because of the great flexibility of the lateral bents normal to their plane. For example, the ship repair pier shown in Figure 1.5 is over 600 feet long with no expansion joints and is fixed against longitudinal forces only at its inshore end. Ultimately the optimum overall layout will represent a compromise with the deck system design, with the designer's goal being to balance all the external forces with an efficient well-proportioned structure.

At offshore locations and at those where water depths exceed approximately 50 to 75 feet and/or those where large cyclic wave loadings govern the design, the equivalent pile depth to fixity method is generally inadequate for design. At offshore locations, the pile/soil interaction must be calculated much more accurately than this method allows. Large-diameter steel pipe pile towers often are used, similar to offshore oil platform construction. The towers usually are braced and the piles slightly battered, usually on the order of $1:12$ to $1:8$. The design of such structures is treated in textbook manner in references (33), (34), (41), and (42). For deepwater steel pipe pile construction and for structures with tubular joints, the design guidelines of the American Petroleum Institute (API) (18) are most useful. Offshore terminal structures also may be constructed of precast/prestressed concrete piles, with large-diameter cylinder piles being favored in deep water.

Open pier and wharf construction also may consist of columns or pillars that bear directly on bedrock at sites where there is little or no soil overburden. The column bases must be socketed or doweled into the bedrock, and lateral stability is provided by batter piles or rigid bracing similar to driven pile piers or, alternatively for wharf construction, by horizontal ties to shore anchorages. Lamella cross walls may also be placed at strategic locations to provide lateral stability in some instances. The columns often consist of concrete-filled steel pipe sections or precast concrete. Alternatively, the columns can be of poured-in-place concrete using underwater tremie concrete techniques. Rigid quality control of the tremie concrete mix and placement is critical to the success of this method. The use of tremie concrete in column type construction has been covered in some detail by Thoresen (6).

Pile load tests for both lateral load and bearing capacity often are called for, especially in areas of complex subsurface conditions and where soil exploration data are minimal.

Deck Systems

Pier and wharf deck systems generally fall into one of the following broad categories: all-timber, concrete, and composite systems. Composite systems usually consist of a concrete deck slab supported on either timber or steel pile caps and stringers. All-steel deck systems are rare except perhaps on isolated offshore platform structures and on floating structures. Traditional all-timber pier construction has been used for waterfront structures of all sizes, including deck loadings of up to 800 psf or more and railroad and heavy crane rail loadings.

Today, all-timber pier construction usually is relegated to small craft harbors, public facilities, and other more lightly loaded sites where it often is the least expensive alternative. Typical timber pier construction usually consists of pile caps on cross-braced bents supporting longitudinal stringers to which the decking is secured (see Figure 7.3). Pile bent spacing in timber piers usually is within the range of 5 to 15 feet, depending upon the nature of the loadings. Pile caps usually consist of a single large timber placed across the pile butts and "drifted" into the piles with steel dowels, called drift pins; or of split cap construction, where two timbers are dapped into opposite sides of the pile head and through-bolted. Drift pins should be recessed and the holes filled with tar pitch.

Figure 7.13 illustrates some typical timber pier construction details. Single solid caps sometimes are additionally secured with uplift straps to resist uplift forces associated with wave and ice action and to resist lateral loads. Stringers likewise are drifted into the pile caps and also may need uplift connections and sometimes lateral bridging as required by timber design specifications (11). Deck planks should be of quarter sawn lumber and be spiked into the stringers. Deck-

Figure 7.13. Timber pier construction details.

ing intended to support truck loads should be a minimum of three inches thick. Adequate gaps should be left between planks (usually $\frac{3}{8}$ to $\frac{1}{2}$ inch) to allow for drainage and some reduction of peak wave uplift pressures. Fastening hardware is generally of mild steel and should be hot-dipped galvanized in accordance with the appropriate ASTM specifications. Within and below the splash zone, bolts should be a minimum of one inch in diameter with plates of $\frac{1}{2}$ inch mini-

mum thickness. Above the splash zone, minimum sizes should be $\frac{3}{4}$ inch diameter for bolts and $\frac{3}{8}$ inch thickness for plates. Ogee-type washers are preferred on all connections below the deck level for their ability to distribute load and their long-term corrosion allowance.

Allowable stresses in timber should be reduced for wet service conditions, although an increase in allowable stress applies to loads of short duration in accordance with the timber design specifications. Wood is excellent at absorbing impact loads, but there is evidence that the CCA pressure treatment process may make timber more brittle. The use of tropical hardwoods (see Section 3.5) avoids the treatment problem and generally results in smaller member dimensions that help to offset the wood's typically higher cost. Allowable stresses for exotic timbers usually are available through the timber suppliers.

In the design of deck systems in general, no uniform live load reduction factor is applied in designing caps, stringers, and decks. The distribution of wheel loads is normally in accordance with highway bridge design standards (12), to which a 15% impact factor is applied. Impact factors for vehicle loads on piers generally are lower than those for bridges because of the limited vehicle speeds and maneuvering room. In the design of pile caps, it sometimes is assumed that one pile is defective and the cap must span the gap. In the design of concrete pile caps, it may be necessary to treat the cap as a beam on an elastic foundation if the piles are very long and of high capacity. This is especially warranted where pile lengths vary greatly across long caps with many piles, where elastic shortening under load may result in large bending moments in rigid caps.

Concrete pile caps almost always are cast in place. Although precast caps have been used successfully, the means of assuring an adequate pile head connection often is problematical. Pile bent spacing in concrete decked structures usually is within a range of 10 to 20 feet, and slab thicknesses typically range from 8 to 16 inches. Pile caps usually are run transverse to the long dimension of a pier, but they may alternatively run longitudinally. Longitudinal pile caps may be an attractive alternative where there are several railroad and/or crane rails running the length of the pier. Section 7.6 deals with crane rail girder design. The design requirements for railroad trackage are given by the American Railway Engineering Association (AREA) (43).

Figure 7.14 illustrates common types of concrete deck framing, the simplest being a one-way continuous cast-in-place flat slab spanning between pile caps. Where longitudinal girders are used, as for rail girders, two-way slab action may be attained. Another simple system involves the use of precast/prestressed planks designed as simple spans over which a cast-in-place topping is poured (44). The topping may be thick enough to contribute to structural action, or it may serve only as a wearing surface. Unless the concrete topping can be reinforced to resist the maximum negative moment condition over the pile cap, a continuous nick joint with filler material should be made along the cap center-

Figure 7.14. Concrete deck systems.

line. Another alternative is the use of precast/prestressed planks, generally much thinner than the topping and used as leave-in-place formwork designed to provide composite action serving as positive moment reinforcing in the completed deck (45). Another method of deck construction, more common in trestles and offshore terminals, is the use of large precast elements such as double tees, which may be used without any topping.

Where precast planks or elements are used, it is important that adequate lateral ties be provided to prevent differential deflections between members acted upon by concentrated loads. This is sometimes done by using post-tensioned

rods between adjacent members. Where precast composite construction is employed, this is usually accomplished by the thickness and lateral distribution of reinforcing steel within the cast-in-place concrete. Reference (46) provides additional descriptions of typical bridge deck framing systems and design methods that may also be used in pier construction. Edge beams along the exposed pier faces help to distribute berthing and mooring point loads, provide a flat surface for mounting fenders, and contribute to the overall deck system rigidity. Edge beams should be of a depth equal to or greater than that of the pile caps.

Deck slabs nearly always must be designed to support moving wheel and other concentrated loads; so they should be reinforced top and bottom in both directions, in order to distribute the loads and accommodate the changing sign of the moments with load locations. It also is preferable to use a greater number of smaller well-distributed reinforcing bars than a smaller number of larger bars of equal area in order to reduce surface cracking and increase impact resistance. Slabs usually must be checked for punching shear resistance under highly concentrated loads such as crane outrigger floats. It usually is not economical to design slabs for crane float loads; therefore, floats must be spotted over caps or girders, and/or timber mats must be used to help distribute the load. References (47) and (48) are helpful in analyzing and designing slabs for heavy concentrated loads. For slabs on grade such as may be used at closed fill structures, reference (49) provides useful design information. In the preliminary design of deck slab systems that must accommodate both uniform and concentrated loads, it may be useful to develop equivalency relationships by equating the maximum uniform and concentrated load shears and moments, as described by Junius (50). Slab thicknesses and unit costs then can be readily plotted against load levels for various cap or beam spacings.

NAVFAC (7) requires that 0.1 square inch of reinforcing steel per inch depth of slab be provided in addition to the main reinforcement around all unframed deck openings. Additional steel usually is also required around mooring hardware and at fender attachments. Two inches minimum of clear concrete cover usually is called for in slabs and beams and $2\frac{1}{2}$ to 3 inches of cover for more massive elements (20). All exposed concrete edges should be chamfered from $\frac{3}{4}$ to $1\frac{1}{2}$ inches to prevent sharp edges from being chipped off and initiating further spalling. As discussed in Section 3.5, it is very important to specify a dense and durable concrete mix.

Composite structures usually consist of concrete slabs supported on steel or timber caps or stringers. These methods are not generally desirable in waterfront construction, as steel is placed in its worse environment, the splash zone, and where it is least accessible, and timber may be subject to decay or compression under heavy loads which pose severe replacement or repair problems. However, in some applications, such construction may be expedient. Guidance for calculating composite action can be found in reference (51). Note, however, that in some instances where different materials are used, composite action may

not be reliable (i.e., where the caps or stringers and the deck system are structurally independent).

Dynamic Response

Structure deflections may be enhanced by dynamic effects, especially when the frequency of the applied loading is near to the natural frequency of the structure. In particular, marine structures often are subjected to relatively large lateral loads of a cyclic or an impulsive nature, which give rise to large lateral movements compared to those of land-based structures. The dynamic analysis of marine structures follows the same principles as that of any other structure, but special consideration must be given to the nature of the loadings, buoyancy and the virtual mass of entrained water, damping, and the effect of marine growth. In general, more rigid structures such as batter-pile-supported piers in shallower water will have higher natural frequencies and hence shorter natural periods (T_n is usually on the order of 0.25 to 0.75 second), thus making them less likely to experience large deflections but more likely to develop high reaction forces from impact. By contrast, more flexible platform structures in deep water, often with $T_n > 1.0$ second, may absorb impacts well but are more susceptible to cyclic wave loading. Note that the structure or its elements also may respond at some harmonic or subharmonic of the natural frequency.

Loadings on marine structures can be broadly classified as follows, in terms of producing dynamically enhanced movements (14):

- *Static and long-term cyclic loads,* such as dead loads, stationary live loads, earth pressures, and time-averaged wind and current loads. Such loads are treated by conventional statical methods.
- *Impulsive loads,* such as berthing impact, sudden release of mooring lines, wave slam, crane snatch loads, and impact, traction, and braking loads from moving equipment. Such loads normally are accounted for by applying impact factors to the static live load. For more flexible structures such as offshore berthing dolphins, a dynamic analysis whereby the structure is idealized as a simple single degree of freedom system (SDOF) may be required.
- *Cyclic loads,* such as from regular waves, vortex shedding phenomena, rotating machinery, or traffic vibrations. Dynamic analysis again is carried out by modeling the structure as an SDOF system.
- *Random loads,* such as normal (irregular) wave loading acting directly on the structure or via the moored vessel, turbulent wind gust effects, and seismic loadings. For near-shore structures random loads may be approximated by assuming a combination of cyclic and impulsive loads. In offshore structures with water depths approaching and exceeding approximately 100 feet, a more complex spectral analysis may be required. The

spectral analysis usually is performed in the frequency domain and requires the development of transfer functions (RAOs) as introduced in Section 6.6.

For an idealized simple SDOF system, the natural period of the structure, T_n, is given by:

$$T_n = 2\pi \sqrt{\frac{M'}{K_s}} \tag{7-5}$$

where M' is the effective or equivalent mass of the structure, and K_s is the effective spring constant or ratio of applied load to deflection (F/δ) of the structure. The quantity M' often is calculated by lumping the distributed masses of the structural elements into a single mass with equivalent static inertia properties. Allowance must be made for marine growth and entrained water on the underwater portions of the structure.

The maximum displacement, according to reference (14), will be:

$$\delta_{max} = F/K_s \sqrt{\frac{1}{(1 - T_n/T)^2 + (2qT_n/T)^2}} \tag{7-6}$$

where T is the period of the applied force, F, and q is the proportion of critical damping, usually in the range of .01 to .05 for marine structures of interest herein. The square root term in equation (7-6) is known as the dynamic amplification factor (DLF); if it exceeds 1.2, more exact analytical methods must be used. Added mass and damping factors for marine structures and an excellent overview of dynamic analysis principles can be found in reference (52).

In calculating the value of K_s it is important that consideration be given to piles with high slenderness ratios ($K_c l_u / r$) and high axial loads, which may be subject to horizontal displacements at their top connections. For example, consider the case of a cantilevered pile bent where the pile is fixed in the soil at some depth below the surface and encast in concrete at the pier deck such that it is fixed against rotation but allowed to translate under a horizontal force at the deck level. For this case, the theoretical column length factor k_c equals 1.0, and the pile lateral stiffness is given by:

$$K_s = F/\delta = \frac{12EI}{l_u^3} \tag{7-7}$$

If the pile is simultaneously under a heavy axial force (P) such as a crane wheel load which is significant in proportion to its critical buckling load (P_{CR}), then

the initial deflection, δ, will be magnified, and the spring constant (K'_s) effectively made softer, in accordance with:

$$K'_s = F/\delta(1 - P/P_{CR}) \tag{7-8}$$

The critical buckling load is given by the Euler equation, which for this case is:

$$P_{CR} = \frac{\pi^2 EI}{l_u^2}$$

An interesting case study of the stability and dynamic response of a large pier without batter piles is presented in references (53) through (55). A more rigorous analysis involves calculation of the various possible natural frequencies and mode shapes of the pier. For a simple planar bent, the overall sway motion of the bent often is coupled with the bowstring action of the individual piles (see Figure 7.15). A more rigorous dynamic analysis involves solving the general equations of motion using a lumped mass mathematical model with viscous damping, linearized soil spring, and hydrodynamic forcing function. This can be expressed in matrix form as:

$$(M)\{\ddot{x}\} + (C)\{\dot{x}\} + (K)\{x\} = \{F\} \tag{7-9}$$

where

(M) = Total mass matrix of the system
(C) = Damping matrix
(K) = Structural stiffness matrix
$\{x\}, \{\dot{x}\},$ and $\{\ddot{x}\}$ = Nodal displacements, velocities, and accelerations
$\{F\}$ = Forcing function

The equation is solved either in the time domain by direct integration over a given time interval or more often in the frequency domain using response spectrum techniques. Computers are mandatory for this type of analysis. The reader is referred to the general literature on structural dynamics for further treatment.

Seismic loadings on near-shore waterfront structures usually are carried out in accordance with AASHTO design standards (12). The deck often is assumed to be loaded with from 10% to 50% of its design live load, as well as being checked for the dead-load-only condition. Twenty-five percent of the pile mass should be added to the pier deck or capping mass (52). Special attention must be given to the stability of rail-mounted cranes and increased pile loads as a result of crane tipping moments. Seismic loads on equipment often are calcu-

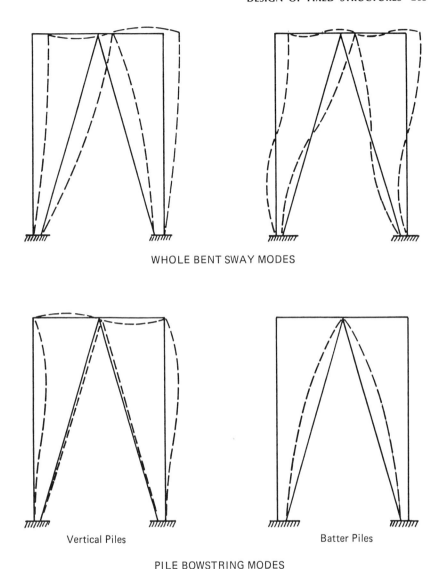

Figure 7.15. Pier bent mode shapes.

lated by applying a horizontal force coefficient to the equipment dead load. Japanese standards (17), for example, consider the seismic force on deck equipment as 20% of its dead weight. Up to a 35% increase in allowable stresses often is allowed under design earthquake conditions. (Refer also to the discussion in Section 4.5.) Often, however, earthquake loads for near-shore structures

are less significant than for land-based structures because marine structures generally are designed to accommodate large lateral forces.

An interesting seismic design problem arises in the design of piers and dolphins to which floating dry docks are permanently moored. Because the mooring connection usually is relatively rigid, and the dry dock is very massive, extremely large forces may develop if the pier structure is subjected to strong ground motions. The mass inertia of the dry dock is so large that it is essentially immovable under short-duration ground shaking. In order to prevent damage to the pier structure or mooring connections, flexible connections or release mechanisms must be employed (56, 57). In deeper water, platform structures may be more susceptible to seismic loads, and time history or spectral analysis methods may be required. The effect of earthquakes on fill stuctures and slope stability and the possibility of soil liquefaction are addressed in Chapter 8.

Structures subjected to repeated cyclic loading may suffer a reduction in strength known as fatigue, which manifests itself in a process of crack initiation and subsequent growth. As a general rule, near-shore/shallow-water structures and, in particular, concrete waterfront structures do not need to be designed to fatigue criteria. Tubular steel structures in deep water and exposed to wave loading are particularly susceptible, although it is important to check any structure with significant dynamic amplification of loads for fatigue effects. In practical design, it is customary to compare a structure's intended design life with its predicted fatigue life as limited by "hot spot" stresses (i.e., areas of high local stress reversals). One practical problem of fatigue design is uncertainty in determining the structure's life cyclic load history.

Guideline fatigue design criteria can be found in reference (18), from which Figure 7.16 is reproduced. The figure, called an S-N diagram, gives the maximum hot spot cyclic stress versus permissible load cycles for offshore tubular steel structures. The solid and dashed lines in the figure represent welding that merges smoothly with the adjoining base metal and welds without such profile control, respectively. The reader should consult the original source for further discussion of the application of these curves.

An in-depth case study of steel pipe pile fatigue failures during the construction of an offshore tanker terminal in the Arabian Gulf was provided in a series of papers by Weidler, Dailey, et al. (58). Free-standing vertical piles failed because of fatigue damage, the principal mechanism of which was a dynamic response to vortex shedding under storm wave conditions. These studies show the importance of considering that fatigue damage might occur during the construction phase. In this case, all free-standing piles that had been exposed like the failed piles had to be cut off at the mudline and were overdriven by larger-diameter piles as part of the repair program.

Fatigue design as applied to offshore structures is well covered in references (33), (34), (41), and (42). Highway bridges are designed to fatigue criteria by

Figure 7.16. Example fatigue S-N curves for offshore structures. (After American Petroleum Institute (18).)

reducing allowable stresses in accordance with the number of cycles of stress at a given stress level (12). In general, fatigue effects can by minimized by sound connection details that avoid stress concentrations. Within the splash zone in particular, corrosion-fatigue can result in an especially rapid propagation of fatigue cracks; so the use of corrosion-resistant materials and details will help reduce cumulative fatigue effects.

Caissons and Precast Elements

Large-scale precast concrete units, including box and pneumatic caissons, plain and buttressed L-walls, and cellular and solid blocks, often are employed in

massive waterfront construction. All of the above types of precast elements require a carefully prepared and leveled foundation base. Joints between abutting units require careful design, both to allow for some differential settlement and to form a tight soil-retaining seal. Application of L-wall and caisson elements, in particular, usually is economical only on large-scale projects where many units are required and where sufficient land staging areas and/or dry dock building basins are located nearby. L-wall elements may be constructed as simple L-shaped sections for walls up to around 20 feet in height, or they may be buttressed with ribs or webs to stiffen the base and stem, in which case wall heights of around 60 feet are practical. Individual units typically range from about 10 to 40 feet in width, depending upon the capacities of handling and placing equipment. Box caissons can be built to heights of 60 feet or more, with lengths of from around 100 to 150 feet being common and widths of up to 80 feet.

Box caissons usually are built in dry docks or building basins formed by temporary earth berms or sheet pile cell walls, and are floated into position where they are ballasted down and filled with earth or stone fill. As shown in Figure 7.17, box caissons usually are compartmented internally for strength and

Figure 7.17. Elements of precast concrete box caisson design.

ballast control. Also, internal compartments can be filled differentially when in position to help reduce peak overturning toe pressures due to lateral earth pressures. Caissons also may be of the pneumatic type with recessed bottoms, or they may be constructed open without a bottom slab. These types must be slid into position from the land side.

All types of precast concrete element walls usually have a continuous cast-in-place concrete capping beam to help tie adjacent units together and form a smooth continuous face into which mooring and fendering hardware can be cast. The capping beam should be placed after any expected initial settlements of the precast units have occurred. Capping beams, in general, should be designed as elastic beams on continuous elastic foundations and must be strong enough to distribute local berthing and mooring forces and to accommodate any long-term differential settlements between precast elements. Capping beams also may serve as crane rail trackage. The capping beam typically performs a soil-retaining function as well.

In the structural design of precast elements, the construction, handling, and/or towing and placement stresses may be more severe than the in-service design load stresses. Aspects of the structural design of precast elements are presented in references (16) and (17), and several case history design examples can be found in reference (59). Geotechnical aspects of precast element design include resistance to sliding and overturning, and the evaluation of bottom bearing pressures and lateral earth forces, as discussed in Chapter 8.

7.4 DOLPHINS AND PLATFORMS

Dolphins are discrete structures designed primarily to accommodate the lateral forces associated with vessel berthing and mooring. Some dolphins, referred to as *mooring dolphins*, are designed strictly to secure a vessel's mooring lines. *Berthing dolphins* are designed primarily to absorb the impact of berthing vessels, and usually are fitted with a fender system and sometimes also with mooring bollards. Dolphin structural types vary largely with water depth, bottom conditions, and vessel size. Figure 7.18 illustrates a progression of dolphin types, showing their approximate ranges of water depth and vessel displacement.

Timber pile clusters are a common feature of a working waterfront, usually constructed of groups of 3, 7, 19, or sometimes 30 piles secured with wire rope wrappings. The center pile of the group typically projects approximately three feet above the other piles to accommodate mooring lines (see Figure 7.19). Timber pile clusters provide relatively low energy absorption, and their lateral load capacity usually is limited to 10 to 15 tons maximum. They generally are suitable and certainly are economical in water depths of up to about 35 feet and for vessels of up to 2000 to 5000 DWT. The horizontal load capacity of timber

270 DESIGN OF MARINE FACILITIES

Figure 7.18. Ranges of application of dolphin types.

pile clusters is carried mainly by axial loads in the piles, and their ultimate capacity thus is limited by individual pile pullout capacities. Energy absorption is primarily via bending. Elms and Schmid (60) have shown that both load capacity and energy absorption can be increased by providing the proper shear flexibility at the pile heads. They present graphical data and a semi-empirical formula for calculating the dolphin energy and maximum horizontal load capacities for 7-, 19-, and 30-pile clusters. With repeated use over time, many timber cluster dolphins suffer from soil shakedown and do not spring back to their original upright positions.

Timber racks and platform structures with both battered and vertical piles and intermediate framing have been used for vessels up to around 17,000 DWT and may have a lateral load capacity on the order of 40 to 50 tons. As with pile clusters, however, their ultimate capacity is limited by pile pullout capacity, and their energy absorption is relatively low. Cantilevered pipe piles and occasionally prestressed concrete piles driven singly or in groups offer greater

DESIGN OF FIXED STRUCTURES 271

Figure 7.19. Typical seven-pile timber cluster dolphin.

energy absorption than these structures and may be the most economical solution where soil conditions permit.

Cantilevered steel pipe piles may be used in water depths of up to approximately 50 feet and sometimes more. Because they exhibit linear elastic behavior, the berthing energy to be absorbed can be related directly to the pile reaction

force (F_R) and the deflection (δ). Equating the energy of impact absorbed by the dolphin (E_F) to the work done by the pile deflection gives $E_F = F_R\delta/2$; and noting that the spring constant $K_s = F_R/\delta$, it can be shown that:

$$F_R = \sqrt{2K_s E_F} \tag{7-10}$$

For a single fixed end cantilever pile:

$$\delta = \frac{F_R l_u^3}{3EI}$$

where E and I are the pile modulus of elasticity and moment of inertia, respectively, and l_u is the unsupported length from the point of fixity to the point of load application; and the spring constant thus is given by:

$$K_s = \frac{3EI}{l_u^3} \tag{7-11}$$

It can be further shown that the pile deflection is related to the bending stress (f_b) by:

$$\delta = \frac{2f_b l_u^2}{3DE} \tag{7-12}$$

where D is the pile diameter. The allowable bending stress in a steel pipe usually is taken as two-thirds of the yield stress (F_y), implying a factor of safety of 1.5 for the material's yielding. Some engineers may apply the F.S. of 1.5 to the ultimate berthing energy, which varies inversely as the square of the deflection, resulting in an allowable stress in the steel of $\sqrt{.67}F_y = .82F_y$ (see reference 61). The pipe pile wall must be sufficiently thick that local buckling does not occur. Fatigue also should be considered for cantilevered steel pipe dolphins. Strong currents may result in vortex shedding oscillations. At sites with high corrosion rates and frequent use, especially under wave action, allowable stresses may be decreased about 50% to allow for fatigue effects. The use of high-strength steel, at least in the zone of high bending stress, increases the energy absorption for a given size pipe and water depth.

Suitable subsurface soil conditions are essential to the application of cantilevered piles (61). Kray (62) has presented some novel cantilever pipe pile applications. Cantilever dolphins that can be loaded only from one direction can be driven at a slight batter toward the direction of impact, resulting in an effective prestressing that increases the energy absorption potential of the pile.

Figure 7.20 Concrete cap breasting dolphins with resilient rubber fender units and face panels.

Where more energy absorption is required and lateral forces are higher, groups of battered and vertical piles may be capped with a massive concrete cap and fitted with resilient fender units. An example of such dolphins is shown in Figure 7.20. In this case, the c.g. of the concrete cap is designed to be aligned vertically with the center of action of the batter piles so that its weight reduces the uplift on the tension piles. In water depths approaching and exceeding approximately 50 feet and where mooring hardware is to be installed, concrete decked platforms may be constructed, such as the dry dock mooring platform shown in Figure 7.21. Figure 7.22 summarizes the loads acting on a typical berthing and mooring dolphin. Note that, in addition to the horizontal forces due to berthing and mooring line loads, vertical and longitudinal rubbing forces and vertical mooring line components also must be considered. The vertical lead of mooring lines should be considered in setting dolphin deck elevations. Loads applied to the dolphin that do not act through its c.g. will result in torsional moments that must be resisted by the pile foundation.

In water depths of up to 40 to 50 feet where rock or firm bearing strata are near the surface, a circular sheet pile cell with a concrete cap may be the most economical solution. Cells make excellent warping or turning dolphins and are capable of resisting large lateral loads. Although cells possess excellent impact

Figure 7.21. Mooring platform-type dolphin with concrete deck and steel pipe pile foundation.

resistance, they must be fitted with fendering to obtain suitable energy absorption, and their geometry makes the use of high-energy resilient fenders a little more problematic than for platform types. Caisson construction also may be used for berthing and mooring dolphins, but its use generally is feasible only in large-scale offshore terminal facilities where more than one or two dolphins are required. At offshore terminals where water depths approach or exceed approximately 100 feet, prefabricated tubular steel structures like the ones evident in Figure 1.1 normally are employed. Prestressed concrete cylinder piles also may be used in deepwater applications (27).

Oil and bulk cargo terminals, in particular, often have platform structures that support hose towers or conveyor equipment. Such platforms are designed like piers and dolphins except that lateral loads typically are less significant with the platforms. The designer must be careful, however, not to allow such structures to become too flexible, especially at exposed locations.

Two or more dolphins may be used to rigidly moor floating dry docks or floating pier structures. Floating dock systems for small craft commonly are

DESIGN OF FIXED STRUCTURES 275

Figure 7.22. Berthing and mooring dolphin load summary sketch.

moored by a series of cantilevered guide piles. Where the floating structure to be moored can be considered as a rigid body, the distribution of loads to the individual dolphins or guide piles can be calculated from basic statical principles. Where only two dolphins are used, it usually is considered that at least two-thirds of the total lateral force is taken by either dolphin for preliminary design purposes. For the general case of multiple dolphins of varying lateral stiffness, the total lateral force (F_t) will be divided among the dolphins in proportion to their stiffness, as found from:

$$F_n = F_t \frac{n(EI)}{\Sigma EI} \qquad (7\text{-}13)$$

where F_n is the load taken by the nth dolphin with stiffness (EI) and $\Sigma\ EI$ is the sum of the stiffnesses of all of the group. It is recommended that the force so found be increased by a factor of 1.3 for design purposes to account for possible misalignment of the dolphins (63).

7.5 ANCILLARY AND ACCESS STRUCTURES

Dolphins and platform structures normally are accessed by trestles, catwalks, or bridge structures. Trestles typically are of similar construction to that of the piers or platforms that they connect. Catwalks and utility bridges usually are of open truss construction with open grating deck surfaces. Access bridges and catwalks may be of open, through-truss construction cross-connected only at the deck level, or they may be rectangular or triangular in cross section such that the main carrying trusses are laterally supported at the top and the bottom. Tubular members often are employed on longer spans and to carry heavier loads such as utility systems. Long-span bridges should be constructed with an appropriate pre-camber to compensate for the structure and other permanent dead load. A critical aspect of the design of all access structures is their means of end connection and ability to tolerate the maximum expected differential movements between the structures being bridged. Sliding bearings nearly always will be required at one end of the catwalk or bridge. Elastomeric bearings frequently are used and must be properly designed and installed to accommodate the expected loads, displacements, and rotations at the bridge ends (16, 64). In addition, access structures should be set back from the berthing face to avoid being struck by vessels, and the bottoms should be high enough above water to avoid wave action or their being struck by waterborne ice or flotsam. Bridges also may support product lines and utilities subject to large thermal movements.

Catwalks intended solely for personnel access usually are designed for uniform live loads of from 40 psf for widths less than 3.5 feet to 100 psf for widths up to 6 feet. Catwalk spans normally are limited to around 100 feet for practical reasons, although much longer spans have been built. For narrow walkway

structures, spans may be limited by the lateral stiffness of the deck. Deflections must be checked to avoid springiness under foot and vibration problems. Lateral wind loads on catwalks can be significant and must be accommodated in the end connection design.

Trestles must be designed for all vehicular traffic or product lines to be accommodated. Trestles usually consist of relatively short spans over independent pile bents and lend themselves readily to precast concrete highway plank construction. Moles are essentially solid-fill roadways; they usually are most economical in shallow water less than 10 to 15 feet deep and typically are used at the shore end of access trestles.

Other ancillary structures include vessel boarding platforms and gangway structures called brows or pier stands. NAVFAC (7) requires that brows and brow platforms be designed for a uniform live load of 75 psf and be load-tested with a 150 psf live load. Brow platforms may be stationary or mounted on caster wheels and may have several platform levels to accommodate large tide changes and vessel boarding locations. Boarding platform and gangway structures are available in a variety of sizes and configurations of prefabricated aluminum or steel. (See Section 9.6 for additional discussion of gangway-type structures.) Handrails should be provided as described in Section 7.7.

7.6 CRANE TRACKAGE

Crane trackage and railroad trackage are common features on many piers and wharves. The design requirements of railroad trackage are covered in reference (43), and their layout on waterfront structures is discussed in reference (65). This section is oriented toward the general design of dockside cargo handling and ship repair type cranes, for which no complete design standard exists. Crane trackage generally consists of the following major elements:

- Rail and its mounting.
- Beam, cross-ties, and foundation system.
- Ancillary features, including: end bumpers, tie downs, power feed details, and drainage.

Crane rails are made of heat-treated carbon steel in a variety of cross sections. Rails normally are designated by their weight in pounds per yard, followed by CR. In the United States, crane rails are available in sizes ranging from 40-CR to 175-CR, although the 104-CR is the lightest rail usually acceptable for waterfront applications. European rails typically have broader, flatter heads and a lower overall height, and are available in sizes to just over 200 lb./yd. The maximum load a rail can carry is a function of the effective width of the rail head. Steel wheels generally range from 12 inches to 36 inches in diameter with maximum load capacities on the order of 35,000 to 176,000 pounds. Of U.S.

rail sections, the 171-CR is often preferred for heavy-duty crane service because of its relatively broad flat head. Depending upon the class of service, this rail can carry wheel loads up to and exceeding 176,000 pounds. References (66) through (70) contain additional information on crane rail selection and installation requirements. Standard U.S. rail lengths are 33 and 39 feet, and larger rails are available in 60-foot lengths.

Rails normally are provided with joint bars and predrilled holes at their ends. For dockside cranes, however, the use of continuous electric welded rails is strongly recommended. Rails can be supplied with prebeveled ends for welding on request. Crane rails usually are mounted on a continous steel plate, generally between $\frac{3}{4}$ inch and $1\frac{1}{4}$ inches thick, with bolted rail clips that allow the rail to slide longitudinally. The rail plate, in turn, is secured to the foundation with anchor bolts usually 1 to $1\frac{1}{2}$ inches in diameter spaced at 2.5 to 3.0 feet. The rail plate is leveled and set in a bed of nonshrink epoxy grout. An alternative to the plate and rail clip arrangement is the use of closely spaced cast steel chairs, also set into an epoxy grout. Because of very high local bearing pressures that can be developed and difficulty in achieving a perfectly uniform bearing surface, fabric-impregnated elastomeric bearing pads, usually ranging from $\frac{1}{4}$ inch to $\frac{3}{8}$ inch thick, often are placed below the rail. It has been noted that a heavy moving concentrated load causes the crane rail to bend, forming a "wave" ahead of the moving rail (70). This wave action may result in very high tension loads in rail clips and anchor bolts. The use of fabric pad washers below the rail clip nuts may help to reduce peak uplift loads. Figure 7.23 shows the installation of a crane rail for a large shipyard revolving gantry crane. Note the full-width fabric/elastomeric bearing pad below the base of the rail. Various crane rail mounting details are illustrated in references (7) and (16).

Crane beams and foundation piles must be analyzed as a beam on elastic supports. Although there is some distribution of peak wheel loads through the rail, wheel loads should be applied as concentrated loads for the beam-on-elastic-foundation analysis. It generally will be found that the beam bends over a long distance over the length of the crane bogie or series or wheels rather than between the pile supports (see Figure 7.24). Crane beams must be designed for adequate stiffness, and deflections should be as near to zero as possible. Operational wheel loads usually are given by the crane manufacturer for maximum traveling, and for over the side and over the corner lifting conditions. There may be as many as eight wheels per crane leg with spacings usually between 2.5 and 5 feet. A 20% impact factor should be applied to all crane lifting loads.

The beam and foundation should be analyzed for the following conditions:

1. Maximum crane operating loads.
2. Maximum operating load plus moderate wind (usually on the order of 25 mph).
3. Maximum wind—crane secured.

DESIGN OF FIXED STRUCTURES 279

Figure 7.23. Installation of crane rail for shipyard gantry crane prior to grouting below rail plate. Note anchor bolts, rail clips, and bearing pad.

Stresses for condition 1 should be within normal allowable working stresses. Increases in allowable stresses for conditions 2 and 3 sometimes are taken as 15% and 33%, respectively. Lateral forces arise from wind, earthquake, crane slewing, and crabbing. Longitudinal forces arise from wind, earthquake, and braking and traction forces. According to NAVFAC (7), braking loads should

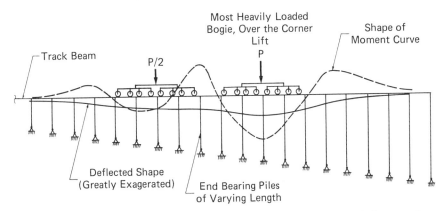

Figure 7.24. Schematic of crane trackage moment and deflection variation below traveling bogie.

be taken as 15% of the live load without impact and traction loads as 25% of the weight on the driving wheels on any one track without impact. Wind areas and vertical center of gravity and equipment weights required for wind and earthquake analysis must be obtained from the crane manufacturer, as should travel speeds and accelerations. End bumpers must be designed to stop the crane when traveling at its full speed. Reference (14) recommends a design velocity of 3.3 fps. Although the ends of the crane bogies often are equipped with cushioning devices, some resiliency should be built into the end bumpers as well.

It is generally desirable for crane rails to be recessed to allow passage of vehicles on deck. The crane rail pocket detail must allow for access to the rail clip bolts and have some provision for drainage. Cranes are secured when not in use or during high winds, usually by means of clamping devices that grip the rail or sometimes by special tie-down arrangements. Pockets may be provided at desired tie-down locations to allow use of the crane's clamping device. Contemporary dockside cranes are powered almost exclusively by electricity. Most traveling gantry cranes are equipped with a self-spooling power cable near one leg, which emerges from the pier deck near the crane's center of travel. Sometimes a recessed culvert may be provided for the power cable to rest in. Container and bulk handling cranes sometimes are fed power from elevated continuous trolley bus bars.

Crane rails should be located so that there is a minimum of four to five feet of clearance from the edge of the bogies to the pier face to allow safe passage of personnel for line handling. NAVFAC recommends 7.5 feet minimum from the centerline of the rail to the pier face. Additional description of crane rail design and installation requirements can be found in references (7), (16), and (69).

7.7 MISCELLANEOUS DESIGN FEATURES

This section addresses important design features common to almost all pier and wharf construction. Fendering and mooring hardware and their attachment are covered in Chapters 5 and 6, respectively.

Expansion Joints

Piers and wharves are subject to thermal movements and stresses associated with the extreme rise and fall of the structure's temperature about its normal mean. Because of the flexibility of pile bents normal to their plane, piers may be constructed over long distances without expansion joints, provided that longitudinal restraint is provided at only one location. Piers have been constructed up to 1000 feet or more without expansion joints, and NAVFAC (7) allows distances of up to 600 feet. Even when expansion joints are not provided, ther-

mal effects and structure restraint must be investigated for concrete and steel structures. The coefficients of thermal expansion for concrete and steel are very nearly equal and usually taken as 0.0000065 unit per unit length per degree Fahrenheit. The amount of temperature rise and fall about the normal mean temperature to be assumed for design will vary with the local climate and with structural materials. For steel structures, the amount of rise or fall typically ranges from around 40°F for moderate climates to around 60°F in extreme climates. For concrete structures, the amount of rise or fall typically ranges from around 30°F in moderate climates to 40° or 45°F in extreme climates. Thermal expansion effects normally are not considered for timber structures. In general, thermal extremes in waterfront structures are somewhat more moderate than for most land-based structures, although thermal gradients from deck top surfaces in direct sunlight to deck undersides can be significant. For further guidance, see AASHTO (12).

Expansion joint gap widths must be properly set for the ambient temperature at the time of construction. The possibility of relative lateral displacements between adjoining pier sections should be investigated and lateral shear keys provided as required. Care must be taken in detailing joints to prevent them from becoming clogged with debris. Preformed rubber units permanently bonded to galvanized steel angles cast into the concrete deck are a common solution. Another simple solution is the use of a sliding steel plate secured to a cast-in-place angle along one edge and allowed to slide atop a cast-in-place angle along the other edge. Figure 7.25 illustrates typical expansion joint details. Sufficient longitudinal restraint must be provided within each pier section between joints; otherwise, adjacent units may take a permanent set, effectively closing one joint and leaving too wide of a gap at the opposite end. Utility or product lines must be provided with suitable means, such as sliding sleeve arrangements, to accommodate relative movements at the joint. Continuous welded crane rails should be allowed to remain continuous across expansion joints, but the anchored rail plate must not span the joint. Sufficient space at the ends of continuous rails must be allowed for their overall expansion.

Deck Drainage

Decks must be adequately drained with regard to local rainfall intensities or, in some cases, wave overwash. Deck drain openings may also provide an important venting function to wave uplift or air compression pressure, especially below open pile wharves. In general, decks should be pitched a minimum of $\frac{1}{16}$ inch per foot toward drains and/or scuppers (curb openings) along the pier face. The down pipe should be at least four inches in diameter or an equivalent size, and a grate inlet should be provided to prevent the pipe from clogging with debris. Drains normally should be provided between every pile bent, or other-

282 DESIGN OF MARINE FACILITIES

PREFORMED COMPRESSION SEAL

SLIDING PLATE

ELASTOMERIC BEARING PAD

Figure 7.25. Typical pier expansion joint details.

wise sized and spaced to suit local runoff conditions. Crane rail pockets and slots should be adequately drained, usually with their own smaller, $1\frac{1}{2}$-inch-diameter, pipe drains. Deck drain hardware is usually of cast iron or cast steel construction, consisting of a main frame, grate, and down pipe. Frames should be adequately proportioned to reinforce the opening in the concrete deck, and grates and frames should be heavy duty load-rated to suit traffic conditions. At petroleum oil loading (POL) facilities and at certain other facilities, interceptor/collection systems may be required to collect deck drainage runoff and conduct it to an oily water waste treatment facility.

Fire Holes and Stops

Piers and wharves with combustible substructures, such as all timber and concrete decks on timber piles, should be fitted with framed fire hole openings to allow insertion of fire fighting "fog" nozzles from above. Fire holes are usually on the order of 6- to 12-inch-diameter clear opening and may be fitted with grated covers to double as deck drains. Their spacing and location will vary with the arrangement of the substructure and the local fire marshal's requirements. All timber structures also should have fire stop cutoff walls extending from below deck to MLW. The spacing, location, and details of fire stops and deck openings should conform to the requirements of the National Fire Protection Association (NFiPA) (71).

Curbs, Handrails, and Ladders

As a general rule, all piers and wharves where vehicular traffic is allowed should be fitted with continuous curbs along their waterside face and edges. One exception may be Ro/Ro facilities that must accommodate vessels using their own transfer ramps. In reinforced concrete piers, a typical curb is on the order of 12 inches high by a minimum of 10 inches wide. Curbs, sometimes referred to as bull-rails, may be made wider at bollard locations so that mooring hardware can be mounted directly atop them. Sometimes the curb is discontinuous, and the mooring hardware is mounted directly to the deck slab. At some facilities where it is not possible to prevent mooring lines from leading down across the curb, chafing plates may be installed along the curb's outboard edge. Chafe plates often consist of a quarter round of galvanized steel pipe with anchor bars or studs cast into the concrete curb. Timber curbs may be employed on concrete decks as well as on all-timber piers. They usually consist of 10 × 12 or 8 × 12 treated timbers, which are elevated slightly above the deck by chocks to facilitate drainage and prevent the onset of timber decay. Curbs may be omitted entirely at passenger handling facilities where handrails are required. Handrails should be a minimum of 3 feet 6 inches high from the deck to the top rail, and preferably have three rails. At passenger terminals, webbing or wire mesh should be installed between rails and stanchion posts for passenger safety. Steel pipe or tube rails should be hot-dipped galvanized, and the top rail should be of a larger dimension than the lower ones to provide a comfortable hand grip. Removable sections of rail often must be provided. Vertical stanchions with hooks for chains in place of rails may be used at such locations. Rails and their anchorage to the pier deck should be designed to withstand a continuous horizontal and vertical force along each rail of 50 plf, or as otherwise required by highway bridge design standards (12) for combined traffic loading.

Ladders often are provided to facilitate access from small craft and for safety reasons. The ladder should extend downward to approximately one to one and

one-half feet above MLW for boarding from small craft. Ladders that are provided primarily for "man-overboard" safety reasons preferably should extend two to three feet below MLW to facilitate climbing out of the water at low tide. However, the lower portions are then subject to fouling and corrosion. At some locations, the use of ladders is discouraged for security reasons; thus the lengths and locations of ladders will vary with facility types and owner/operator requirements. Ladders should have hand rails at the deck level to help users to pull themselves on deck, and they should be recessed flush with the dock face or be sufficiently set back from the fender face to preclude damage. Also, ladders must be adequately secured against sway in all directions, and the design of rungs should consider the impact of the concentrated weight of a large person carrying gear. Rungs are usually from 1.5 to 2.0 feet wide, spaced at approximately 12 inches. Rungs of wooden ladders must be let into the verticals, and in no case should metal fastenings be relied upon as the sole vertical support.

Anchorage Details

Various hardware and equipment must be anchored to pier and wharf structures by anchor bolts, including fender systems, mooring hardware, pad eyes, davit/hoists, product lines, services, and so on. Anchor bolts must be amply sized to secure the base of the fitting being anchored under all expected horizontal and vertical loads and overturning moments, and to develop the full capacity of any load-rated hardware. The anchor bolts normally are made of mild steel, although high-strength steel bolts may be used, and should be hot-dipped galvanized. It is desirable wherever possible, especially for critical hardware, that bolts be installed in pipe sleeves to allow for their eventual inspection and replacement. Critical mooring hardware should be fitted with backing bars between adjacent bolts. Anchor bolt connections on timber pier construction should be solidly chocked. An ample corrosion allowance should be provided for anchor bolts, and an additional protective coating such as tar pitch should be applied to the nuts and threads after tightening. Nuts should be properly torqued in accordance with AISC specifications (10). Additional information on minimum hardware sizes and general design requirements can be found in reference (72).

Markings and Navigation Lighting

Pier decks often are painted with yellow safety lines to delineate the outline of crane bogie travel or the swing radius of rotating equipment. Berth identification numbers also may be painted on the dock face or inside face of the curb. Such

marks are designated by the owner/operator of the facility. Navigation marks such as day marks, radar reflectors, fog signals, and lighting may be required at exposed berths, especially at offshore locations; these marks fall under the jurisdiction of the Coast Guard in the United States and usually must conform to accepted international standards. In general, a fixed or flashing amber or yellow light, which has no lateral significance, is used to define the extremities of a fixed or special-purpose structure. Navigation lights often have their own power supply and may have solar cells to keep their batteries charged. Navigation marks for a specific project must be approved by the Coast Guard and local harbor authorities. Range lights may be employed, usually at head-on berth approaches, to assist vessels docking at night and under reduced visibility.

7.8 SHIP SERVICES AND UTILITY SYSTEMS

Marine facilities may be equipped with a wide variety of service and utility systems, depending upon the nature of the operations. This section is intended to review the various kinds of systems and their general requirements; the mechanical and electrical design of such systems is beyond the scope of this book. The most common dockside services usually provided are water, electric power, sanitary waste, and communications systems. Water is supplied to the vessel as potable water for drinking and washing, and the pier also may have a nonpotable water supply for washing down and flushing and for fire protection.

Fire protection systems may be salt water supplied from the pier's own pumping station. Fire fighting requirements will vary and be subject to the approval of local fire officials. The U.S. Navy's requirements (73), for example, range from 1500 gallons per minute (gpm) for a single berth up to 2000 feet in length with a minimum residual pressure of 40 psi to a maximum of 3500 gpm with a minimum of 125 psi residual pressure at the most remote outlet for an unlimited number of berths. Hose connections are typically $2\frac{1}{2}$-inch, spaced so that any one location can be reached from more than one outlet with a maximum hose length of 300 feet. In addition, the fire protection system should have $4\frac{1}{2}$-inch pumper connections. International shore connections should be provided for direct supply to the ship's fire mains.

Potable water requirements range from 1000 gpm at 40 psi minimum for a single berth up to 2000 gpm at 40 psi for multiple berths over 2000 feet in length. The ship service system should be equipped with double check backflow preventers; $2\frac{1}{2}$-inch connections taken from a four-inch main usually are spaced at approximately 120-foot intervals.

In freezing climates, water lines must be protected by heat tracing or being run inside utility trenches or enclosures with live steam lines. Water supply lines that are infrequently used should be self-draining to prevent freeze-ups. Salt

water systems are often kept "live" with small inductor pumps that continally circulate a small volume of water, which is discharged through a small orifice or valve at the end of the line.

Sewage usually is handled with an eight-inch line with flexible hose connections from the ship. Sewage connections may be spaced up to 200 feet apart or in some instances at a single location for single berths.

Electrical power supply requirements range from a minimum 440 volts/3 phase to 10 kilovolts (kV) to operate container cranes and illuminate large marshaling yards. As many oceangoing ships utilize direct current, transformers are required for the ship's shoreside power supply. It is essential that all electrical services be properly grounded in accordance with the National Electrical Code requirements. Electric conduits are often cast in place inside the beams of concrete piers. Telephones or voice-activated sound-powered communications often are provided at locations where electrical and other utilities are grouped together for ship connections. Sometimes small shelters or covered boxes are provided at utility connection locations. Pier deck lighting normally is provided by light standards ranging from 20 to 50 feet in height. For security only, a minimum illumination of 0.2 to 0.5 foot-candle should be provided. Nighttime cargo loading operations require a minimum of 5 foot-candles and passenger waiting areas in excess of 20 foot-candles.

Steam lines may be provided for space heating and for warming utility trenches, and are essential in shipyard operations. Compressed air may be supplied for operating equipment, especially in hazardous areas where electrical sparks present a safety hazard. In shipyards, various gases, such as oxygen, natural gas, argon, acetylene, and others, frequently are piped from the pier, and they have significant spatial requirements. Many shipyards also provide connections for the removal of oily ballast water from the vessels.

Bunkering or fueling of vessels usually is carried out from barges or oilers supplying bunker C or diesel fuel; so such services generally are not provided from dockside.

Ship services and utility systems may be carried directly on the pier deck, although they are most commonly carried below the deck on pipe hangers or within cast-in-place utility trenches. Utility trenches may run along the inshore margin of a wharf with periodic crossovers to utility stations or bases; or there may be one crossover, with the trench running along the outshore face with risers at box locations. On piers with two-side berthing, the trench often runs down the middle of the pier with branches to both sides. Utility trenches must be designed for ready access to piping and must have adequate drainage; so the bottom of the trench must be kept above the highest tide levels. Where this is not possible, the trench structure must be watertight, and other provisions made for keeping it dry inside. Figure 7.26 illustrates some typical utility trench de-

Figure 7.26. Typical utility trench details.

tails. Exposed pipe risers, manifolds, and utility connections should be protected by parapet walls, bumping posts, enclosures, or other suitable means. Further description of dockside utility systems can be found in references (5) and (73).

REFERENCES

1. AAPA (1973). *Port Planning, Design and Construction.* American Association of Port Authorities, Washington, D.C.
2. Agerschou, H. et al. (1983). *Planning and Design of Ports and Marine Terminals.* Wiley, Chichester, England.
3. Brunn, P. (1981). *Port Engineering,* 3rd ed. Gulf Publishing, Houston, Tex.
4. Cornick, H. F. (1962). *Dock and Harbour Engineering,* 4 vols. Griffen, London.
5. Quinn, A. D. F. (1972). *Design and Construction of Ports and Marine Structures.* McGraw-Hill, New York.
6. Thoresen, C. A. (1988). *Port Design—Guidelines and Recommendations.* Tapir Publishers, Trondheim, Norway.
7. NAVFAC (Nov. 1980). *Piers and Wharves,* Waterfront Design Manual, DM-25.1. U.S. Navy, Alexandria, Va.
8. UNCTAD (1985). *Port Development Handbook,* 2nd ed. Geneva, Switzerland.
9. ACI (1983). *Building Code Requirements for Reinforced Concrete,* ACI 318-83, American Concrete Institute, Detroit, Mich.

10. AISC (1978). *Specification for the Design, Fabrication and Erection of Structural Steel for Buildings.* American Institute of Steel Construction, New York.
11. NFoPA (1986). *National Design Specification for Wood Construction.* National Forest Products Association, Washington, D.C.
12. AASHTO (1983). *Standard Specifications for Highway Bridges.* American Assoc. of State Highway and Transportation Officials, Washington, D.C.
13. NAVFAC (1981). *Seawalls, Bulkheads and Quay Walls*, DM-25.4. Dept. of the Navy, Arlington, Va.
14. BSI (1984). *British Standard Code of Practice for Maritime Structures*, BS 6349, Part I: "General Criteria." British Standards Institution, London.
15. BSI (1988). *British Standard Code of Practice for Maritime Structures*, BS 6349, Part 2: "Design of Quay Walls, Jetties and Dolphins." British Standards Institution, London.
16. EAU (1985). *Recommendations of the Committee for Waterfront Structures*, Society for Harbor Engineering and the German Society for Soil Mechanics and Foundation Engineering, 5th English ed. Wilhelm Ernst & Sohn, Berlin–Munich (1986).
17. PHRI (1980). *Technical Standards for Port and Harbor Facilities in Japan.* The Overseas Coastal Area Development Inst. of Japan, Bureau of Ports and Harbors, Ministry of Transport, Tokyo.
18. API (Apr. 1987). *Recommended Practice for Planning, Designing and Constructing Fixed Offshore Platforms*, RP-2A, American Petroleum Institute, Washington, D.C.
19. DNV (1977). *Rules for the Design, Construction and Inspection of Fixed Offshore Structures*, Det Norske Veritas, Oslo, Norway.
20. ACI (1984). "Guide for the Design & Construction of Fixed Offshore Concrete Structures," 357R-84, American Concrete Institute, Detroit, Mich.
21. FIP (1985). *Recommendations for the Design and Construction of Concrete Sea Structures.* Federation Internationale de la Précontrainte, Thomas Telford, London.
22. NAVFAC (May 1982). *Foundations and Earth Structures*, DM-7.2. U.S. Navy, Alexandria, Va.
23. ASCE (1984). "Practical Guidelines for the Selection, Design and Installation of Piles", ASCE Comm. On Pile Foundations, Geotechnical Div.
24. AWPI (1976). "Technical Guidelines for Pressure Treated Wood, AWPI, PI, Timber Piling," American Wood Preservers Inst.
25. ACI (1974). "Recommendations for Design, Manufacture and Installation of Concrete Piles," ACI 543-74, American Concrete Institute, *Manual of Concrete Practice*, Part 3, Detroit, Mich.
26. PCI (Mar./Apr. 1977). "Recommended Practice for Design, Manufacture, and Installation of Prestressed Concrete Piling," PCI. Comm. Report, *PCI Journal*, Vol. 22, No. 2.
27. ———. "The Raymond Cylinder Pile & Cylinder Pile Structures," Brochure of Raymond International, Inc., Houston, Tex.
28. ——— (1979). "Bethlehem Steel H-Piles," Handbook by Bethlehem Steel Corporation, Bethlehem, PA.
29. Chellis, R. D. (1961). *Pile Foundations.* McGraw-Hill, New York.
30. U.S. Army (1963). "Pile Construction," TM-5-258, Dept. of the Army.
31. AWPI (Oct. 1969). "Pile Foundations: Know-How," American Wood Preservers Institute, Washington, D.C.
32. AISI (1982). "The Steel Pile Cap Connection," American Iron and Steel Institute, Washington, D.C.
33. McClelland, B. and Reifel, M. D., eds. (1986). *Planning and Design of Fixed Offshore Platforms.* Van Nostrand Reinhold, New York.
34. Dawson, T. H. (1983). *Offshore Structural Engineering.* Prentice-Hall, Englewood Cliffs, N.J.

35. AWPI (1968). "Laterally Unsupported Timber Pile Study," Design report prepared by Cornell, Howland, Hayes and Merryfield for AWPI, McLean, Va.
36. Hopkins, D. A. (1955). "The Design of Piers, Jetties, and Dolphins," *Trans. ASCE*, Paper No. 2846, New York.
37. Seely, F. B. and Smith, J. O. (1952). *Advanced Mechanics of Materials*. Wiley, New York.
38. Padron, D. V. and White, S. M. (May 1983). "Optimizing Pier Design by Utilizing Deck Stiffness," *Proc. ASCE*, Ports '83, Spec. Conf., New Orleans, La.
39. Padron, D. V. and Elzoghby, H. M. (May 1986). "Berthing Impact Force Distribution on Pile Supported Pier Structures," *Proc. ASCE*, Ports '86, Spec. Conf., Oakland, Calif.
40. Gouda, M. A. (June 1960). "Analysis of Load Distribution in the Piles of Piers," *Proc. ASCE*, Vol. 86, No. WW-2.
41. Graff, W. J. (1981). *Introduction to Offshore Structures—Design, Fabrication, Installation*. Gulf Publishing, Houston, Tex.
42. Hsu, T. H. (1984). *Applied Offshore Structural Engineering*. Gulf Publishing, Houston, Tex.
43. AREA (1981). *Manual for Railway Engineering*. The American Railway Engineering Association.
44. Gustafson, D. P. and Long, M. F. (Dec. 1981). "Precast Concrete Deck System for Two New Orleans Wharves," *Concrete International*, ACI, Detroit, Mich.
45. PCI (Mar./Apr. 1983). "Recommended Practice for Precast/Prestressed Concrete Composite Bridge Deck Panels," *Precast Concrete Institute Journal*.
46. ACI (1988). "Analysis and Design of Reinforced Concrete Bridge Structures," ACI 343R-88, American Concrete Institute, Detroit, Mich.
47. Appleton, J. H. (Aug. 1980). "Design of Concrete Slabs for Concentrated Loads," *Concrete International*, Vol. 2, No. 8, ACI, Detroit.
48. ——— (June 1980). "Shear Strength of Flat Slab Decks Subjected to Heavy Loads," *The Dock and Harbour Authority*, London.
49. PCA (1966). "Thickness Design for Concrete Pavements," *Concrete Information*, Portland Cement Association, Skokie, Ill.
50. Junius, R. W., Jr. (1983). "Wharf Decks—Design Load vs. Cost," *Proc. ASCE*, Ports '83, Spec. Conf., New Orleans, La.
51. NAVFAC (1980). *Structural Engineering: Aluminum, Masonry, and Composite Structures and Other Structural Materials*, DM-2.6. Dept. of the Navy, Alexandria, Va.
52. CIRIA (1978). *Dynamics of Marine Structures: Methods of Calculating the Dynamic Response of Fixed Structures Subject to Wave and Current Action*, Report UR-8, CIRIA Underwater Engineering Group, London.
53. Michalos, J. and Billiington, D. P. (May 1961). "Design and Stability Considerations for Unique Pier," *Proc. ASCE*, Vol. 87, No. WW-2.
54. Michalos, J. (Aug. 1962). "Dynamic Response & Stability of Piers on Piles," *Proc. ASCE*, Vol. 88, No. WW-3.
55. Reese, L. C. and Haliburton, T. A. (May 1963). "Dynamic Response and Stability of Piers on Piles—Discussion," *Proc. ASCE*, Vol. 89, No. WW-2.
56. Keith, J. M. et al. (1986). "Seismic Isolation System for a Floating Dry Dock," *Proc. ASCE*, Ports '86, Spec. Conf.
57. Mazurkievicz, B. et al. (1977). "Loads Acting on Mooring Facilities of Floating Docks and Ship Hulls," PIANC, 34th Int. Nav. Congress, Leningrad.
58. Weidler, J. B., Dailey, J. E., et al. (May 1987). "Pile Fatigue Failures: Damage Appraisal, Mechanism Studies, Motions in Seas, Fatigue Description" (published in 4 parts), *Proc. ASCE*, Vol. 113, No. 3.
59. PIANC (1986). "Bulletin No. 54," Brussels, Belgium.

60. Elms, D. G. and Schmid, W. E. (Nov. 1965). "Structural Action of Pile Cluster Dolphins," *Proc. ASCE*, Vol. 94, No. WW-4.
61. PIANC (1984). "Report of the International Commission for Improving the Design of Fender Systems," Suppl. to Bull. No. 45, Brussels, Belgium.
62. Kray, C. J. (1976). "Concepts in Design of Coastal Structures," *Proc.* of 15th Coastal Engineering Conf., ASCE.
63. Tsinker, G. P. (1986). *Floating Ports: Design and Construction Practices*. Gulf Publishing, Houston, Tex.
64. Husain, S. I. (1983). "Design and Full Scale Testing of Elastomeric Bridge Bearings for Trestle and Jetty Trestle of Super Tanker Offshore Terminal," *Proc. ASCE*, Ports '83, Spec. Conf., New Orleans, La.
65. NAVFAC (Apr. 1974). *Civil Engineering*, DM-5. U.S. Navy, Alexandria, Va.
66. —— (n.d.). "Crane Rails," Brochure of the Bethlehem Steel Co.
67. —— (1971). "CMAA Specification No. 70," Crane Manufacturers Associates.
68. Molyneux, G. and Molyneux, G. M. O. (1978). "Crane Rails and Their Installation," *Iron and Steel Engineer*, Assoc. of Iron and Steel Engineers.
69. Mazurkiewicz, B. K. (1980). *Design and Construction of Dry Docks*, Trans. Tech Publications, Rockport, Mass.
70. Jenkins, I. D. H. (1980). "Crane Rail Support and Fixity," *The Dock and Harbour Authority*, London.
71. NFiPA (1975). *Standards for the Construction and Protection of Piers and Wharves*, National Fire Protection Association No. 87.
72. NAVFAC (1981). *General Criteria for Waterfront Construction*, Waterfront Design Manual, DM-25.6. U.S. Navy, Alexandria, Va.
73. NAVFAC (1981). *Dockside Utilities for Ship Service*, Waterfront Design Manual, DM-25.2. U.S. Navy, Alexandria, Va.

8

Geotechnical Design Considerations

*by David R. Carchedi, P.E., Associate and
Russell J. Morgan, P.E., Senior Project Manager,
Goldberg-Zoino & Associates, Inc.*

The design of waterfront and near-shore structures is greatly dependent on site history, local geology, and geotechnical characteristics of site soils and bedrock. As in many foundation designs, structural members and natural materials combine to form a soil–structure interaction design challenge. This is particularly evident in the design of marine structures such as flexible bulkheads, solid fill piers and wharves, and basin-type dry docks. In these facilities, soil and rock not only provide a foundation for the structure, but they may also furnish resistance to overturning, uplift pressures, and stability failure. Ironically, while providing these benefits, the soil or rock mass also may be the source of the greatest structural loads.

The characteristics of the soil and rock at near-shore sites can vary widely in areal extent as well as with depth because of the dynamic environment in which they were deposited or reworked. These materials may be found in a natural undisturbed state or may have been altered by previous human activities. Consequently, the marine designer not only must be proficient at structural engineering and the determination of environmental loads, but also must be acutely aware of the many variables associated with geotechnical design in the coastal zone. An intimate knowledge of soil mechanics and foundation engineering principles is essential also, along with a firm understanding of site geotechnical characteristics, geology, and history.

This chapter first outlines site investigation and subsurface exploration techniques, material characteristics, and testing methods. A discussion of geotechnical design considerations for various types of waterfront and near-shore structures follows. Particular attention is given to the details that make geotechnical design in the coastal zone different from more traditional upland work. Finally, a brief discussion of site improvement techniques is presented to illustrate possible impacts on waterfront structures.

8.1 MARINE SITE INVESTIGATIONS

Development history, local and regional geologic and geotechnical characteristics, and coastal processes can have a profound effect on the design of waterfront and near-shore facilities. The structural type and construction costs may be greatly impacted by these factors. Therefore, adequate site investigations are essential to the design of marine facilities.

A well-planned investigation program is fundamental to attaining sufficient data for design. Site investigations should be performed in three phases, including: (1) collection of existing data and historical information, (2) completion of a detailed site reconnaissance, and (3) implementation of a subsurface exploration and field testing program.

Existing Data Collection

Existing site data and historic information can be obtained from several sources. Geologic, topographic, and coastal feature information is often available in the form of surficial geology, soil science, agricultural, and topographic maps, navigation charts, or summary reports compiled by local or national geologic and oceanographic organizations. Hydrologic data and well logs also may be available through water resource departments. Additionally, many colleges and universities have amassed large amounts of detailed local geologic and oceanographic information.

Other potential sources of information are historical data, maps, and photographs, which are often obtainable at libraries, historical societies, registries of deeds, or local building departments. Aerial photographs are also available in many locations and can be used to document the existence and location of previous structures, site filling history, and changes to the shoreline. Old surveys, site plans, construction documents for existing structures and utilities, and subsurface exploration information may be available through physical plant or facilities personnel at currently developed industrial or military facilities. Communications with long-term residents or employees of the facilities in question can lead to valuable information pertaining to site history, filling, former construction, and previous industrial practices.

When existing or former industrial sites are being studied, it is essential that the investigator determine any previous site use, activities, and products manufactured, processed, or handled. Many cases warrant a full environmental site assessment and internal audit of buildings. The cost of removal and disposal of contaminated soils and hazardous materials may far outweigh the benefit of the project, and the selection of another site may become appropriate.

Site Reconnaissance

A site reconnaissance should be performed following a thorough review of available information. The reconnaissance usually consists of a detailed walkover and visual observation. General site characteristics, surface soils, rock outcrops, subsided or eroded areas, existing structures and foundations, or any other unusual topographic feature should be noted and located. The effect of coastal processes, littoral currents, and natural deposition and erosion also may be inferred. Existing data should be compared with previous information in order to determine changes to the site. Updated aerial, topographic, and bathymetric surveys may be warranted if existing information is not current or adequate.

Existing upland and waterfront structures should be investigated to evaluate their general condition and foundation type. If it is anticipated that these structures will be rehabilitated or reused, a much more extensive structural and materials condition survey should be performed (1) (see Chapter 11). A determination of how existing bulkheads are anchored also is important and may have serious implications on future near-shore construction. Special attention should be given to the possibility that former seawalls and other marine structures may have been buried by subsequent land reclamation.

Subsurface Explorations

Subsurface explorations on land and in the near-shore environment will be required to evaluate soil, rock, and groundwater conditions for the proposed development. Subsurface explorations generally consist of drilled borings and test pits, and may include geophysical surveys. Borings and test pits allow for detailed observation of subsurface soils and rock while providing investigators the opportunity to obtain samples for laboratory testing, perform in situ tests, and install groundwater monitoring wells and equipment. Drilling and test pitting operations tend to be costly, however, and only supply information pertaining to a limited portion of the site. Geophysical methods allow for broad interpretation of large areas, but lack the detailed information required for design. A combination of two or more methods may be the most efficient and cost-effective way to map site subsurface conditions. In addition to these methods, aerial and satellite photography may be useful in locating large-scale geologic features.

Drilled Explorations. The most widely used method of subsurface exploration is the boring of holes into the ground using rotary drilling, augering, percussion drilling, cable-tool methods, or casing drive and wash techniques.

294 DESIGN OF MARINE FACILITIES

Figure 8.1. Typical float-mounted drill rig. (Photo courtesy of GZA Drilling, Inc.)

Powered rotary equipment can be mounted on trucks, tracks, all-terrain vehicles, ships, barges, and small floats (see Figure 8.1). Shallow, nonrotary drilling can be accomplished in areas inaccessible to other rigs by using a small tripod and a gas-operated cathead and pump. The methods and equipment employed will depend on the size and depth of the explorations, the type of sampling and field testing required, and the physical character and accessibility of the project site. Table 8.1 lists several types of drilling equipment with their common uses and capabilities.

A typical borehole is advanced by alternately drilling and sampling. Samples generally are taken at intervals of five feet, or less if changes in soil type are encountered, using a split-spoon sampler and standard penetration test (SPT) methods (2). SPT data can be empirically related to soil engineering characteristics pertaining to foundation design (3–10). The recovered soil sample is used

for visual and laboratory classification of soil types. Larger-diameter undisturbed samples of cohesive soils can be obtained for laboratory testing using one of several types of thin-walled sampling devices (11). Where the quality and the character of bedrock are of concern, rotary-driven rock coring equipment will be required to obtain samples (11).

In Situ Testing. In addition to the SPT, several types of downhole tests are available (12). In situ testing is used to determine design parameters that may be incorporated into a direct design approach. Certain types of in situ equipment can provide a continuous record of the subsurface profile. This gives a good indication of site variability, and it can provide characterization of soils that normally are difficult or impossible to sample and test in an undisturbed state.

Typical methods of testing in situ strength or properties of soils include vane shear, flat dilatometer, electric piezocone penetrometer, and the pressuremeter. Installation of piezometers or monitoring wells can be used to determine piezometric levels at specific locations or the phreatic groundwater surface. Field testing equipment and instrumentation also may be required during construction to monitor subsurface soil and structure response, and may include earth pressure cells, strain gages, inclinometers, settlement gages, and piezometers (13). Table 8.2 lists several types of in situ testing and field monitoring equipment commonly used during the exploration and construction phases of waterfront projects.

Test Pit Explorations. Test pit explorations are used extensively on land, and less commonly below the water, to supplement drilled explorations. They are useful in determining existing foundation types, locating buried seawalls and other waterfront structures, accessing subsurface structures for materials condition and structural surveys, characterizing the nature of fill or shallow natural soils, and exploring for environmental contamination. Test pit excavations are relatively inexpensive and can reach a practical depth of approximately 20 feet, depending on soil and water conditions.

Geophysical Methods. Several geophysical methods are available to the marine engineer to facilitate the characterization of subsurface conditions. Seismic reflection methods, depending on the operating frequency, can be used to determine the bathymetry, the location of the bedrock surface, the depth to sound rock, and the location of other distinctive soil horizons (14). Seismic reflection, in conjunction with limited borings, is of particular use in determining types and qualities of spoil material for channel and berth dredging. Side-scan sonar can be used to locate and characterize bottom features, outcrops, and obstructions. When coupled with a magnetic survey, side-scan sonar also can be useful in locating metallic obstructions, such as sunken vessels, that may hamper port operations. Table 8.3 lists several types of geophysical equipment that can be used for site characterization below water. Several types of land-operated geophysical equipment also are available to aid in the assessment of the site.

Table 8.1. Subsurface exploration equipment.

Exploration	Use	Capabilities	Comments
Hand excavated test pit or auger hole	Sampling for visual identification and laboratory testings.	Shallow explorations in soils above the water table.	Provides data in inaccessible areas, limited depth of exploration.
Backhoe (test pit excavation)	Bulk sampling and visual identification of soils, in situ testing, determination of groundwater and bedrock elevations.	Possible depths range from 20' to 30', but generally shallower below water in cohensionless soils.	Access may be difficult, limited undisturbed sampling, difficult below water.
Tripod boring rig	SPT sampling using casing drive and wash techniques.	Limiting depths range between 25' and 40'.	Provides boring data in otherwise inaccessible areas.
Truck-mounted drill rig	Deep drilling with SPT, cone penetration, undisturbed sampling, in situ testing and well installation. Boring may be advanced using augering, rotary drilling, percussion drilling, cable tool methods, and casing drive and wash techniques.	Possible depths range to greater than several hundred feet.	Limited access in some situations, efficient and cost-effective method of data collection.
All-terrain vehicle or track-mounted drill rig	Similar to truck-mounted rig.	Similar to truck-mounted rig.	Provides drilling access to undeveloped sites, production somewhat slower than truck-mounted equipment.

GEOTECHNICAL DESIGN CONSIDERATIONS 297

Float mounted drill rig (anchored)	Water borings with SPT and undisturbed sampling, using casing drive and wash techniques.	Operation in 40' to 60' of water, drilling to approximately 100' below mudline.	Limited to protected sites, sizes and power of rig limited, susceptible to sea condition, limited rock coring.
Barge-mounted drill rig (spud barge)	Water borings with SPT and undisturbed sampling, using rotary drilling or drive and wash techniques.	Operation in 30' to 40' of water, drilling to several hundred feet below mudline.	Limited to relatively shallow water, somewhat less susceptible to sea condition than float-mounted rig.
Barge-mounted drill rig (jack up)	Similar to spud barge	Operation in up to 80' of water, drilling to several hundred feet below mudline.	Provides stable platform that avoids problems with sea condition and tides, can be fitted with large drilling equipment.
Barge-mounted drill rig (anchored)	Similar to spud barge	Operation in over 100' of water, drilling to several hundred feet below mudline.	Can operate in relatively deep water, susceptible to sea condition, may require large anchoring equipment at exposed sites.

298 DESIGN OF MARINE FACILITIES

Table 8.2. In situ testing and monitoring equipment.

Device/test	Description	Use	Comments
Standard penetration test (SPT)	Split spoon sampler driven into soil at bottom of borehole in accordance with ASTM D-1586.	Determination of soil density and consistency and inferred soil properties based on sampler resistance (blow count).	A disturbed sample is retrieved for visual identification; test results vary depending on equipment and operator.
Cone penetration test (CPT)	A hardened steel, cone-tipped probe instrumented with load cells, pressure cells, and other site-specific devices. A continuous record of data is obtained as the probe is pushed into the ground.	Development of a detailed stratigraphy profile, determine ϕ angle, relative density, pore pressure, coefficient of consolidation, and other properties depending on instruments used.	Penetration into dense soil may be limited, depending on equipment used. The CPT has been extensively used, and many design procedures have been developed based directly on the results.
Pressuremeter	A cylindrical membrane is inserted into a borehole. The membrane is inflated, while corresponding pressure and strain measurements are recorded.	Determination of elastic moduli of soils and ultimate bearing capacity. Other properties may be inferred.	Test is sensitive to borehole preparation; test procedure is relatively simple; data reduction and interpretation are completed using a computer program.
Dilatometer	A thin blade fitted with an earth pressure cell on the face. The blade is pushed to some depth into the soil where pressure measurements are taken.	Determination of pore pressure, stress history, in-situ stress, and other properties.	A rugged device that may penetrate dense soils; test may be run rapidly at each depth with little operator variability.
Field vane	A four-bladed vane which is pushed into undisturbed soil to some depth and rotated. Resistance is measured until the in situ shear strength is exceeded.	Determination of peak and residual undrained shear strengths.	Test procedure is highly variable, may produce erratic results; it is impossible to determine direction of weak shear planes.
Borehole shear device	A device that applies a normal force to the borehole walls and is pulled upward at a constant strain rate while the resistance is measured.	Determine of shear strength, ϕ angle, and cohesion.	Simple device, tests may be run quickly, and results tend to be reproducible.

GEOTECHNICAL DESIGN CONSIDERATIONS 299

Table 8.3. Geophysical methods.

Method/equipment	Purpose	Capabilities	Comments
Fathometer (depth sounder)	Determines bathymetry.	Typical depths range between 0 and 500'.	Usually operates in the 200 kHz range; results in little subbottom penetration.
Seismic reflection profiler	Determination of subbottom profile.	Operates in a wide range of water depths with penetration depths depending on operating frequency.	Typical operating frequencies range from 3.5 to 12-kHz. Lower frequencies provide deeper penetration with less definition.
Side-scan sonar	Locating and defining bottom surface features.	Exploration range of 50' to a few hundred feet to either side of towed transducer.	Typical operating frequency of several hundred kHz. Transducer (transmitter and receiver) towed close to bottom; sound waves transmitted from either side of towed equipment.
Magnetometer	Determines size, location, and depth of large-scale geologic features or locates near-surface metallic objects buried in the sediment.	Operates in any water depth; sensitivity can be adjusted in order to resolve different size features.	Equipment towed near water surface; operates in any water depth. May be towed closer to mudline in search of small objects.

Subsurface Exploration Program Planning. There are no hard and fast rules as to how many explorations are required, where they should be located, or how deep they should penetrate. These decisions are different for each project and depend upon available information gathered during the site reconnaissance, as well as the project location and configuration, the structural type, the local geology and subsurface conditions, and the size and cost of the project (10, 15–18). Geophysical methods can be used to help locate preliminary borings and to reduce the overall drilling program, should uniform conditions be encountered. In general, preliminary borings usually are taken along the face of proposed bulkheads, quays, and marginal wharves at a spacing generally not exceeding 200 feet. For anchored bulkheads, another series of borings usually is required at the location of the anchor system. Additional boreholes may be drilled offshore and upland to obtain information on alignments perpendicular to the structure. This configuration may be desirable for the development of subsurface cross sections for stability analyses. The same scheme would be used for cut slopes and embankments running parallel to the shore.

Narrow pile-supported piers usually require preliminary explorations along the proposed centerline not exceeding a spacing of 200 feet, and isolated pile-supported structures and dolphins may require one or more borings at each location. Solid fill piers and breakwaters may require additional lines of borings perpendicular to the structure centerline if stability is an issue.

The depth of the explorations depends on the type of anticipated foundation support. In the case of pile foundations, explorations must extend into the bearing layer or to bedrock, or to soil depths sufficient to develop adequate frictional resistance. Explorations for solid fill structures, quays, bulkheads, and slopes must penetrate to below any possible failure planes. Where compressible soils exist, explorations should be advanced through the compressible layer or to a depth where the calculated increase in stress due to the new construction approaches zero.

The total number and the depth of borings required will depend on the preliminary exploration information and the type of structure to be constructed. Depending on the scope and results of the preliminary drilling program, additional borings at closer spacings may be required. Test pits may supplement initial explorations where detailed shallow subsurface information is required.

Because of the high cost of mobilizing certain drilling rigs, it may be more cost-effective to adjust the exploration program as drilling progresses in order to complete the program in one phase, rather than two or more phases. Table 8.4 presents guidelines for the spacing and depth of explorations generally required for different types of structures. It must be noted that these are only guidelines and that changes to the exploration program often result due to subsurface conditions encountered.

Table 8.4. Subsurface exploration program planning.

Structure/site	Exploration configuration and depths
Large undeveloped waterfront site	Preliminary borings 200' to 500' on center, advanced to competent bearing soils. Layout of final phase of borings should reflect proposed construction locations, foundation requirements, and construction procedures.
Filled sites	Exploration program should consist of borings supplemented with test pits. Proposed boring locations and depths should be based on the proposed development, and allowance should be made for slow drilling or refusal from obstructions. Test pits should be used to define vertical and lateral extent of fill and classification of the material.
Embankments, retaining structures, and marginal wharves	Preliminary borings spaced at approximately 200' on center along the length of the structure. Final borings at 50' to 100' on center, with additional borings inboard and outboard to develop cross sections for stability analyses. Borings to be used for stability analyses should penetrate all possible failure surfaces.
Pile foundations	Borings should be carried into dense soils or to bedrock for end bearing piles, or deep enough to develop adequate frictional resistance for friction piles.
Shallow foundations	For individual footing or mat foundations, borings should be completed to a depth not less than twice the width of the loaded area.

8.2 SOIL MECHANICS AND PROPERTIES

Subsurface soils found in the near-shore environment typically include sand and gravel, inorganic and organic silt, clay, peat, and fill. These materials possess specific engineering properties that can affect the design, construction, and performance of structures. The engineering properties of a soil may remain constant with time, or can vary with changes in stress, pore water pressure, and physicochemical reactions (19). The location, type, and characteristics of the bedrock also can play an important role in design.

The following sections briefly discuss general soil characteristics and describe some of the more important material properties and their implications in the design and construction of near-shore structures. A brief discussion of rock properties and their engineering implications also has been included. More complete discussions of soil and rock mechanics and properties can be found in various references (3–10, 16).

Soil Properties

Soil is by nature nonhomogeneous and anisotropic, displaying markedly variable physical and mechanical properties, even when samples are obtained from the same deposit in close proximity. This variability is a result of the soil's composition, formation, and geologic and stress history, as well as physical and chemical changes it has undergone, and its current location and state of stress. Fortunately, many soils of comparable makeup display similar behavior when subjected to changes in stress from structural loads.

Natural soils transported to and deposited in the near-shore environment generally are terrigenous, consisting of gravel, sand, silt, and clay (20). These materials are predominantly river-borne, and may be either well sorted and stratified or mixed and reworked because of currents and coastal processes. Coarser materials tend to settle out first, with finer particles carried on to estuaries, embayments, and eventually offshore, where they are finally deposited (21). Soil types, as well as the geologic and coastal history of a particular site, can seriously affect foundation design.

Soil is composed of mineral or biogenic particles, air, and water. The particle-to-particle contacts of the mineral or biogenic materials form a matrix called the soil skeleton. Within the skeleton are gaps or void spaces that can be filled with gas (air) and water. Figure 8.2 presents a simplified representation of a soil in a phase diagram, along with several definitions and equations used to describe the material.

The actual configuration of the soil matrix depends on many factors, among them the mineral or biogenic composition, the gradation and packing of the particles, the rate and way in which the soil was formed, physicochemical reactions, and the stress history. All of these factors affect not only the physical appearance of the soil, but also its engineering characteristics such as strength, compressibility, and permeability. When considering the behavior of a soil, one must understand these characteristics as well as basic soil mechanics principles.

Principle of Effective Stress

Though much if not all of the soil that the marine engineer analyzes is in a submerged state, basic soil mechanics principles still hold. An important concept in soil mechanics, the effective stress principle, states that:

$$\sigma_T = \sigma' + u \tag{8-1}$$

where σ_T is the total stress on a plane, u is the hydrostatic or pore water pressure, and σ' is the effective stess on the plane.

As shown in the example in Figure 8.3, if a soil is submerged for its entire

GEOTECHNICAL DESIGN CONSIDERATIONS

Figure 8.2. Idealized soil phase diagram and associated relationships.

Unit Weight

Total $\quad \gamma_t = \dfrac{W}{V}$

Water $\quad \gamma_w = \dfrac{W_w}{V_w}$

Solids $\quad \gamma_s = \dfrac{W_s}{V_s}$

Total (Dry) $\quad \gamma_d = \dfrac{W_s}{V}$

Specific Gravity

Total $\quad G_t = \dfrac{\gamma_t}{\gamma_w'}$

Water $\quad G_w = \dfrac{\gamma_w}{\gamma_w'}$

Solids $\quad G_s = \dfrac{\gamma_s}{\gamma_w'}$

Volume

Porosity $\quad n = \dfrac{V_v}{V}$

Void ratio $\quad e = \dfrac{V_v}{V_s}$

Degree of saturation $\quad S = \dfrac{V_w}{V_v}$

Weight

Water content $\quad wc = \dfrac{W_w}{W_s}$

Buoyant unit weight $\quad \gamma_b = \gamma_t - \gamma_w$

Note: $\gamma_w = \gamma_w' = 62.4$ pounds per cubic foot (PCF) for fresh water @ $4°C$.
$\gamma_w = 64$ PCF for saltwater.

depth, the effective stress at a particular location within the deposit remains unchanged by fluctuations in the water level. However, if the water level decreases to such a point that the entire depth of the soil no longer is submerged, the effective stress in the soil is increased. An increase in the effective stress can lead to consolidation of compressible soils and settlement of structures. Additional loading due to loss of soil buoyancy can lead to stability and slope failures.

The engineering properties of a soil depend on both the current effective stress

304 DESIGN OF MARINE FACILITIES

Calculation of effective stress on soil element located 10 feet below mudline with fluctuating water level.

CASE 1:
σ_t = 10' (64 PCF) + 10' (120 PCF) = 1840 PSF
μ = 20' (64 PCF) = 1280 PSF
σ' = 1840 − 1280 = 560 PSF
or
σ' = 10' (56 PCF) = 560 PSF

CASE 2:
σ_t = 1 (64) + 10 (120) = 1264 PSF
μ = 11 (64) = 704 PSF
σ' = 1264 − 704 = 560 PSF
or
σ' = 10 (56) = 560 PSF

CASE 3:
σ_t = 5 (110) + 5 (120) = 1150 PSF
μ = 5 (64) = 320 PSF
σ' = 1150 − 320 = 830 PSF
or
σ' = 5 (110) + 5 (56) = 830 PSF

Figure 8.3. Effect of water level on soil effective stress.

and the stress history. If equation (8-1) is rewritten in the form:

$$\sigma' = \sigma_T - u$$

it becomes evident that the effective stress in the soil, σ', is inversely related to the pore water pressure u if σ_T is held constant. Therefore, changes in the pore pressure at constant total stress will affect not only the effective stress within the soil mass, but also the mechanical properties of the material. As discussed in the following sections, this may become an important factor, affecting the strength of a soil when structural loads are applied.

Shear Strength

Of the major soil engineering concerns, shear strength frequently is the property of primary interest because the performance of nearly every structure founded in soil in some way depends on it. Slopes, gravity structures, pile foundations, bulkheads, retaining walls, and anchor systems all rely on soil shear strength to resist failure.

A complete discussion of the shear strength of soils is beyond the scope of this book. However, as outlined in reference (21), two major principles related to shear strength must be understood. First, strength is uniquely related to effective stress at failure, as shown by the Mohr-Coulomb equation:

$$s = c + \sigma' \tan \phi \tag{8-2}$$

where s is shear strength, c is cohesion, σ' is the effective stress, and ϕ is the soil friction angle. Second, soil strength at failure is also related to water content. As water content increases, soil strength decreases. This may not be the case, however, for highly sensitive soils, as will be discussed later.

There are many ways to determine the shearing resistance of a soil in the laboratory as well as in the field (12, 22–24). Unfortunately, different tests often produce conflicting and confusing results; so testing methods should be chosen carefully to best simulate actual field conditions for a particular foundation problem.

Consolidation

Soil consolidation, an important factor in determining the settlement potential of a structure or a fill, generally is of most concern with cohesive and organic soils, which tend to produce relatively large strains in response to an increase in stress. When a saturated soil is subjected to increased stress, all the load is taken initially by the water in the pore space. The elevated water pressure that results, known as excess pore pressure, dissipates with time as stress is transferred from the water to the soil skeleton. Consolidation then is the process by which the void ratio of a soil is decreased as the soil particles move into a denser packing. It should also be noted that the water content decreases during the consolidation process, leading to a corresponding increase in shear strength.

For loads that are of large extent relative to the depth of the compressible layer, it is reasonable to assume that the soil is subjected to one-dimensional consolidation:

$$\Delta H = \frac{C_c}{1 + e_0} H \log \frac{\sigma'_0 + \Delta \sigma}{\sigma'_0} \tag{8-3}$$

where ΔH is the consolidation settlement, C_c is the soil compression index, e_0 is the initial soil void ratio, H is the thickness of the compressible soil layer, σ'_0 is the average initial effective stress in the compressible layer, and $\Delta \sigma$ is the change in stress due to the applied load.

If loads of limited lateral extent were applied to the ground surface, a stress distribution analysis would have to be performed prior to the calculation of settlements.

Consolidation is a time-dependent process directly related to soil permeability, thickness of the compressible layer, and drainage conditions of the confining soils. Many clays have such low permeabilities that consolidation under constant stress may take place over decades. This can make site improvement techniques, such as preloading (see Section 8.9), impractical and may preclude the use of shallow foundations.

When the excess pore water pressure has completely dissipated, and all the load has been transferred to the soil skeleton, primary consolidation is complete. Some soils, however, experience secondary compression, known as creep, even after excess pore pressures are dissipated. Creep is common with highly organic soils and may result in substantial amounts of settlement over long periods of time (25).

The compression of cohesionless soils is nearly elastic in nature, takes place almost immediately upon loading, and tends to be much smaller in magnitude relative to that of cohesive soils. However, it can be of concern where heavy nonuniform loadings produce substantial differential settlements.

The consolidation characteristics of cohesive soils can be found by performing laboratory tests on undisturbed samples or can be calculated from large-scale field loading tests. The elastic properties of cohesionless soils also may be determined in the laboratory. As undisturbed samples of cohesionless soils are difficult to obtain, and the physicochemical effects are minor, test specimens usually are reconstituted to simulate field conditions (i.e., dense or loose state). Elastic properties also can be inferred from SPT or field cone test results, which are compared to existing published data (16).

The natural consolidation of soils is a slow ongoing process. As sediments build up, the additional soil load drives the consolidation process. Soils that have been consolidated under their own weight are termed normally consolidated. Once a soil has been loaded and subsequently unloaded, it is termed preloaded or overconsolidated. Overconsolidation can result from the removal of the overburden by erosion, desiccation, former loading of glacial ice masses, or human activities such as the placement and removal of surcharges. Overconsolidated soils can sustain loads up to their maximum past pressure without significant settlement. Once the maximum past pressure is reached, however, the soil again acts like a normally consolidated material, and settlements increase significantly with increasing stress.

Some marine clays exhibit overconsolidated behavior even though they have not been subjected to past loading. This phenomenon, known as apparent overconsolidation, is caused by physicochemical bonding that takes place at the particle contacts. Apparent overconsolidation generally is found in clays that have sedimented slowly, allowing sufficient time for particle bonding to occur before the soil structure is placed under significant stress (20, 26).

Conversely, clays that have sedimented quickly can show underconsolidated

behavior. Rapid sedimentation increases natural loading at such a rate that underlying soils cannot consolidate in equilibrium with the loading. These soils, typically found in river deltas, tend to have high water contents and low shear strengths, and are highly susceptible to stability failure (20).

Permeability

The permeability of a soil can have a profound impact on structural design. As previously mentioned, the rate of consolidation is directly dependent on permeability, which is also important in the design of retaining structures and cofferdams, where the drainage characteristics of the backfill or the potential for piping beneath the structure is of concern (see Section 8.5).

A number of field and laboratory test procedures are available to determine soil hydraulic characteristics (22–24, 27). Permeability also can be calculated from data obtained from consolidation tests and certain types of shear strength tests. It is important to realize that natural soil deposits tend to be nonhomogeneous and anisotropic and may not be adequately represented by a small soil sample. Additionally, the permeability of a natural deposit or a fill may be markedly different in different directions. For projects where massive dewatering or preloading is required, large-scale pumping or field surcharging tests should be performed.

It is evident that there is a strong interrelationship between permeability, shear strength, and consolidation characteristics of a particular soil, all of which can affect the design of a structure. Extensive testing of these properties can be performed, but it may not always be justified because of varying soil conditions, difficulty in obtaining undisturbed samples, and cost. Quite often these characteristics can be reasonably inferred from SPT or field cone data, or from relatively simple classification methods including the Atterberg limits, natural water content, and grain size distribution tests, all of which can be performed on relatively low-quality split-spoon samples. Additionally, an approximation of the undrained shear strength of a cohesive soil can be determined by using a hand-held penetrometer or a vane shear device on split-spoon samples or block samples obtained from test pits. It should be noted, however, that conservatism in the design must increase to compensate for the lack of substantive geotechnical data. Although this approach may be justified for small projects, larger more costly projects usually justify more extensive field and laboratory testing (10).

Bedrock

The bedrock characteristics that are usually of concern to the marine designer include the degree of weathering and the amount of fracturing of the bedrock

surface, which are important in the design of piles and sheeting that must be founded on the rock, in determining the capacity of rock anchors, or where rock excavation or dredging is required. Also, if slopes must be cut into the rock, the degree and the direction of fracturing are important in the evaluation of block stability and the need for rock bolting. The degree of weathering and fracturing can be determined from high-quality rock cores, but fracture orientation is more difficult to determine and may require directional drilling techniques if local bedrock exposures are insufficient or unavailable for inspection.

Soil Types

The following paragraphs briefly describe general soil types commonly found in the near-shore environment, as well as their implications in the design of marine structures. Many special subsurface conditions are indigenous to specific geographic areas, but their description is beyond the scope of this book.

Sand and Gravel. Sand and gravel deposits found in the near-shore environment may be stratified by glacial meltwater flow or river processes, or may be highly mixed from subsequent reworking. These soils also are found in various gradations and may contain silt, clay, boulders, and cobbles.

Sand and gravel generally provide good bearing resistance for either shallow or deep foundations if they are relatively dense. The presence of appreciable amounts of cobbles or boulders, however, can lead to difficulties in the construction of deep foundations or in the driving of sheeting.

If relatively free of silt and clay, sand and gravel can be used as free-draining backfill for retaining and solid fill structures. Piping can become a problem, however, in very clean sand where large differential head conditions exist (see Section 8.5).

If clean, uniform, fine sand is present in a loose state below the water table, whether as a natural deposit or from hydraulic fill operations, it may be susceptible to liquefaction. Earthquake events (28) or other vibrations may lead to liquefaction and a resultant loss of strength, thereby causing large settlements and increased loads against retaining structures. If saturated loose fine sand is present, deep densification by one of several ground modification techniques (discussed in Section 8.9) may be required.

Cyclic loading of granular soils due to wave action may cause a fatigue-type failure at shear stress levels below the ultimate static stress levels. Cyclic shear testing data can be used to determine the soil pore pressure response relative to stress levels and the number of loading cycles. This information then can be applied to stability analyses for structures subjected to multiple loadings (29). Research is ongoing on this subject, with the recent construction of many large offshore gravity-type structures (30).

Because it is difficult to obtain undisturbed samples of clean sand and gravel,

sophisticated laboratory tests normally are not performed on these materials. Values for their permeability, shearing angle, and in situ unit weights generally are inferred from grain size distribution, SPT values, or field cone results (3–10).

Inorganic Silt. Inorganic silt can occur in relatively homogeneous deposits, or varved with clay as a result of the cyclic depositional environment of glacial lakes. Silt deposits often are mixed or interbedded with appreciable amounts of clay and sand. Depending on its actual makeup and stress history, a silt deposit may behave predominantly as either a cohesive or a cohesionless material.

Deposits that resemble a sand or clay usually can be treated as such. However, particular attention must be given to the microcharacteristics of the deposit and the orientation of its planes of weakness. This is of particular importance when stability is an issue (31).

Uniform inorganic silt deposits can exhibit appreciable strength values even though natural water contents are near or above the liquid limit (31). These soils, if not heavily preconsolidated, tend to be somewhat sensitive, however, and may lose much of their strength when disturbed. For this reason it is difficult to obtain high-quality undisturbed samples for laboratory testing. Loose silt also may be susceptible to loss of strength due to cyclic loading.

Some denser or overconsolidated silt deposits are adequate for the support of shallow foundations provided that disturbance is minimized. A working mat of filter fabric and stone often is required to avoid disturbance by construction equipment. Preloading of silt often is successful in reducing post-construction settlements. Many silt deposits tend to be relatively free-draining, especially when appreciable amounts of sand or numerous sand strata are present.

Low-capacity pile foundations also can be successfully installed in silt deposits. These soils may lose much of their strength when disturbed during pile driving, but load tests show that they regain much if not all of their original strength with time.

Clay. Thick deposits of clay are typical in many near-shore environments. Strength, compressibility, sensitivity, and mineralogic makeup often vary widely from region to region, as well as with depth within a deposit. Clay can exhibit relatively high compressibility and low permeability, resulting in large settlements over long periods of time following loading; so it often is not practicable to preload a site underlain by thick clay deposits. Many clays, however, are overconsolidated and produce small settlements for loadings below their maximum past pressure.

Lightly loaded structures may be founded on friction piles, and, in some cases, caissons can be founded directly on stiff clay. Deep foundations may be used to transfer heavy structural loads to strata below the clay deposits. If the site grade is raised, however, downdrag forces resulting from consolidation of the clay may be transferred to deep foundations (32, 33) (see Section 8.7).

Stability is often an issue when clay deposits are present. The stability of slopes and near-shore structures depends upon the shear strength of the soil in question. Some marine deposited clays exhibit high sensitivity (ratio of undisturbed shear strength to remolded strength). These soils may have relatively high undisturbed strengths at high water contents, but may suddenly lose most if not all of their strength during loading. This brittle type of failure is likely due to the breakdown of chemical bonds between the soil particles.

Uplifted marine clays that have been leached with fresh water also display this phenomenon due to a breakdown of cementation and physicochemical bonding (34–36). Clays that have not been uplifted above sea level also may exhibit these characteristics because of leaching by fresh water from artesian conditions. Sensitivities of normally consolidated marine clays typically range from 4 to 8, but may be greater than 100 for highly sensitive deposits (21).

Cyclic loading also can be an issue in working with clays, causing stress reversals that can increase strain and lower ultimate shear strength (37).

Organic Silt and Peat. Many waterfront environments are frequently overlain by deposits of organic silt or peat. In some areas, these materials have been buried by subsequent natural deposits or filling. Organic soils tend to be soft and highly compressible, exhibiting low shear strengths. Water contents are usually near or above the liquid limit, so that the soil approaches the liquid state with loading or disturbance (25).

The presence of organic soils can have a major impact on development because of their inherent weakness and loss of strength when loaded, and their high degree of compressibility. Low shear strength can lead to the instability of slopes or extremely high pressures against earth retaining structures. High compressibility can result in excessive settlement of shallow foundations, utilities, and bulkhead anchor systems, or can cause sizable downdrag forces on deep foundations.

Although not generally used, ground modification techniques usually have little effect on organic soils. Even the removal of these materials can be problematic because of the high cost and difficulty of handling and disposal. Preloading is not generally viable because of relatively large amounts of secondary settlement. With the help of sand or wick drains and reinforcing fabrics, settlements may be reduced to the extent that these portions of the site may be suitable for staging, storage, or stockpile areas. Another alternative is the use of stone columns, which can aid in soil drainage while adding stability to the site (see Section 8.9).

Fill. Many waterfront sites are located on land created by filling, frequently in an uncontrolled manner. The fill may consist of a miscellaneous and random mixture of soft or loose soil containing large boulders, blasted rock, construction debris and trash, or hydraulically placed sand and silt. The amount and the depth of fill present likely reflect the original topography and required grade

elevations. In many instances, the fill was placed directly over soft organic soils. Existing fill at many sites is unsuitable for supporting shallow foundations in its current state and may require ground modification or deep foundations. If the fill contains large obstructions, even the deep foundation scheme may be a costly and difficult alternative.

8.3 MARINE FOUNDATIONS

The development of port and ship repair facilities generally requires some modification of the natural shoreline in order to handle deep draft vessels. This modification may be severe, requiring massive dredging and filling and the construction of bulkheads or marginal wharves. In contrast, the shoreline may remain relatively unchanged, with the construction of piers and dolphins for handling vessels offshore, in naturally deep water. Even in the latter case, however, some shoreline protection usually is required to prevent scour and erosion.

The design of near-shore facilities is a soil–structure interaction problem where often the structure is also the foundation. Waterfront structures are exposed to relatively severe berthing and mooring loads as well as dynamic environmental forces. These conditions can cause large lateral and uplift forces that in addition to the gravity loads from backfill, stockpiled materials, cranes, and other handling and transfer equipment, must be resisted by the structure. Such facilities may also require deep draft capacities, resulting in long unsupported pile lengths, or very high retaining walls. To further complicate the design, construction materials are subject to ongoing degradation, which decreases their structural integrity.

The remaining discussion in this chapter is devoted to design considerations for near-shore structures from a foundation engineering standpoint. Although geotechnical considerations may dictate the use of a particular structural type, cost and other factors may favor other alternatives.

Structural Type

There are many types and configurations of waterfront structures, each of which presents different geotechnical design issues. These structures include slopes, piers, wharves, bulkheads, quay walls, dolphins, and dry docks, and may be constructed of soil, stone, masonry, concrete, timber, or steel. Figure 7.1 depicts some of the many types of waterfront structures, and Chapter 10 presents a discussion of several types of dry docks.

Slopes differ from other structures in that soil and rock generally provide the building material as well as the foundation. Slopes can be either cut into existing soil or constructed over in situ soil utilizing other fill materials. A major concern with all slopes is stability, a situation that can be aggravated by erosion, scour,

or surface sloughing. Slope protection, therefore, becomes a major consideration in construction along the waterfront.

Retaining structures provide a vertical or near-vertical wall that retains backfill. They may be constructed to allow dredging for berthing and mooring of vessels, or simply to allow for land reclamation. Retaining structures include flexible bulkheads, mass masonry or concrete gravity walls, steel cofferdam bulkheads, and concrete caissons. The designer of retaining structures generally is concerned with issues of bearing capacity, sliding, overturning, and stability, with additional concerns such as bending stress, toe resistance, anchor design, and interlock tension for sheet pile structures. Scour and erosion often lead to the failure of bulkheads and retaining structures and must be considered in the design.

Solid fill piers and dolphins are similar in design to many of the solid fill retaining structures and share many of the same structural features. The major geotechnical difference is that, in most cases, the piers and dolphins are not used to provide retainage for backfill behind the structure itself. Alternatively, open pile-supported construction can be used for piers, wharves, or dolphins. Because of their relatively light weight, these types of structures generally rely on batter piles to resist lateral forces. Major geotechnical concerns for pile design are compression, uplift, and lateral loads.

Foundations for dry docks, such as marine railways, vertical lifts, and floating dry docks (see Chapter 10), can take on many different forms. Marine railways may be supported on either shallow or deep foundations, or a combination of the two; winches for vertical lift systems generally are supported by open, pile-supported piers; and floating dry docks may be moored to solid or open-type piers, wharves, or dolphins. The walls of basin-type dry docks generally take on the form of one of the above-referenced retaining structures. Buoyancy and uplift pressures are of great concern with this type of structure. Shallow, narrow docks can be designed to resist these forces with the use of deadweight and structural floors. However, most major basin docks require either tension piles or grouted anchors to reduce spans and counteract buoyancy forces.

Construction Considerations

Construction near or over the water tends to be more difficult and costly than upland construction, and the logistics of handling equipment and materials becomes an important consideration. Work that can be performed from land usually is more cost-effective than similar work performed using barge-mounted equipment. Environmental conditions such as tides, currents, and wave action must be constantly dealt with, and may require the construction of temporary dikes or cofferdams. Additionally, certain parts of the construction can be per-

formed only for limited periods of time in some cases, because of tidal conditions or ongoing shipping and handling operations.

Construction operations and equipment used in waterfront work often differ from those used on land. Pile driving operations may require the use of free instead of fixed leads, and elaborate driving templates often are needed to ensure alignment and proper mating to other structural members. Backfilling operations can be slow, requiring the use of clamshell or conveyor equipment to place materials. Compaction of the backfill may require the use of one or more of the ground modification techniques discussed in Section 8.9. Dewatering for deep structures also can be difficult and expensive. Often the construction of extensive cofferdams is the only effective way to perform the work.

The design of waterfront structures must take these factors into account in order to obtain the most efficient and economical construction. The geotechnical portion of many marine projects is large and requires careful planning. Facilities may be designed so that subsequent portions of the work can be constructed by using completed portions for staging of equipment and materials. This approach may dictate the type of construction or the sequence of filling and dredging operations.

8.4 SLOPES AND SLOPE PROTECTION

An important consideration in slope design is stability, which depends on soil and rock characteristics and controls the steepness and configuration of the slope. Generally calculated in terms of the factor of safety against failure, stability can change with time. Changes in the factor of safety can occur as a result of the construction sequence, the pore pressure response during or following construction, fluctuations in the groundwater table, surcharge loadings, changes in slope configuration due to scour, erosion, and dredging, or earthquake events.

Problems with scour and erosion due to current and wave action are common along the shorefront. For this reason, the exposed faces of slopes must be adequately protected. Additionally, many waterfront land reclamation projects are constructed over soft soils or use weak or hydraulically placed soils as backfill (38). These situations can lead to severe stability problems and must be carefully engineered to avoid failures.

Analysis of Slopes

There are many different modes of failure and associated methods of analysis for slopes constructed of earth or rock (39, 40). In general, each method compares the forces tending to cause failure (driving forces) to the forces tending to resist failure (resisting forces) in order to determine a factor of safety against

slope failure or sliding. All of these methods can be calculated either by hand or by utilizing the numerous computer programs that have been developed to assist in the process. Details of the calculations can be found in reference (39).

Natural erosion or scour at the toe of a slope or dredging can affect stability. Seepage pressures are also a concern for waterfront slopes, especially in areas subject to coastal flooding. Such flooding can substantially raise the groundwater level within the slope, producing increased driving forces and seepage pressures if the flood waters subside quickly.

Minimum recommended factors of safety range from 1.3 to 1.5, although lower values are acceptable under certain conditions (39, 40). As discussed below, the factor of safety of a slope can change with time because of the pore pressure response of the soil, which must be accounted for in the analysis. Recommended factors of safety assume that the most critical failure surface has been analyzed, that soil strength parameters are adequately known, and that good control has been maintained during construction (39). Where there is a high potential for loss of life or property, more conservative factors of safety may be warranted.

Cuts and Fills

Although the final configuration of two slopes may be identical, their factors of safety against sliding at any point in time may be quite different, depending on construction methods and the soil pore pressure response. The major contribution to the forces resisting failure comes from the shear strength of the soil along the failure surface. As discussed in Section 8.2, shear strength is related to the effective stress in the soil. Any change of the pore pressure within the soil results in a change of effective stress, which, in turn, affects shear strength and overall slope stability.

Highly permeable granular soils are not generally affected by pore pressure because dissipation of the pressure takes place very quickly compared to the construction time. Pore pressure dissipation in cohesive soils is slow, however, and can greatly affect slope stability. When filled embankments are constructed over normally consolidated to slightly overconsolidated cohesive soils, pore pressures increase, resulting in a decrease in the shear strength of the soil, as well as a reduction in the stability of the slope. Pore pressures dissipate as the soil consolidates with time, however, and the factor or safety against sliding increases correspondingly. When cuts are made into or above cohesive soils, pore pressures become negative, increasing the short-term factor of safety of the slope. As the negative pore pressure dissipates, the soil expands, the water content increases, and the shear strength and slope stability decrease. Depending on how the slope is constructed, either the short-term (end-of-construction) or the long-term factor of safety will determine overall slope stability.

When calculated factors of safety are low, several steps can be taken to decrease the possibility of failure. The rate of fill placement can be slowed to allow underlying soils to consolidate and gain strength under the new loads. Pore pressures also can be monitored to determine when factors of safety are adequate for filling operations to continue. The rate of consolidation of underlying soils also may be increased with the use of sand or wick drains (see Section 8.9). Where the long-term factor of safety is too low, preloading of the underlying soils can be used to increase shear strengths prior to the construction of cut slopes. Slopes also may be benched, counterweighted, or decreased in steepness to reduce the sliding potential.

The strength of loose fills and underlying soils may be increased in place by using one of the ground modification techniques discussed in Section 8.9. In addition to the densifying and strengthening of in situ soils, shearing resistance along the failure plane may be increased by installing piles or stone columns through the anticipated slip surface (41–43). Also, steeper and more stable embankments may be constructed by using reinforcing fabrics or grids within the fill (44).

Slope Protection

Numerous methods and materials are used to protect the exposed faces of waterfront slopes from scour and erosion due to wave action, currents, and propeller wash. A common method is to build up graded layers of stone with a protective layer of riprap on the slope face (45). The individual stone size and thickness of the protective layers will depend on the forces to which it will be subjected. Exposed slopes subjected to heavy wave action also may be protected with man-made armor units, which are carefully placed on the slope face (see below, Figure 8.30). These units are made of concrete and tend to be relatively expensive to make and place, but they may provide necessary protection in areas where large rock for riprap is not available or economical.

Several other proprietary systems of slope protection are available for use in areas where wave and current forces are limited. Figure 8.4 shows a number of these systems.

One slope protection system uses gabions, which consist of wire baskets laced together and tightly filled with stone. The baskets can be coated with PVC for marine use and are available in many shapes that can be connected to form surface mats or stacked to form retaining structures or revetments (46).

Concrete block revetments can be used in low-energy areas; the blocks are laid on the slope to provide continuous face protection (47). Grout-filled fabric mats are another form of slope protection that can be used where there is limited wave action. The fabric form is laid on a prepared slope, anchored, and filled with pumped concrete. The segmented mats have a quilted appearance when

Figure 8.4. Typical slope protection systems for areas of low wave and current energy.

completed and are usually designed to allow water to pass through the seams (48).

Slope protection design must include a filter to stop migration of fines from below. Graded stone and gravel filter layers generally are used with the large stone riprap or armor unit type of protection. In recent years, geosynthetic filter fabrics have been used extensively in coastal engineering (49, 50). They are

routinely used with most of the proprietary systems to reduce labor and materials costs while providing durable and effective filtration.

The portion of a slope below low water, as well as the bottom adjacent to the toe, must be protected from scour due to wave action, currents, and propeller wash (51) (see Sections 3.4 and 4.6). This has generally been accomplished by using stone mats, with the material sized such that an adequate factor of safety with respect to threshold eroding velocities is maintained (52). A layer of high-strength geotextile usually is placed between the stone bed and the existing bottom soils to avoid washout of fines. Concrete-filled geotextile "mattresses" also have been used under water for scour protection (48).

8.5 BULKHEADS AND RETAINING STRUCTURES

Retaining structures are utilized extensively in the development of waterfront facilities. These structures are used to stabilize the shorefront, allow dredging, and provide for land reclamation. Although these goals can be accomplished by creating slopes or embankments, retaining structures provide a vertical face for berthing and mooring of vessels without wasting valuable land area. Retaining walls generally are more costly than slopes, but may be more cost-effective than open pile-supported structures, depending on the site physical and subsurface conditions. Even if a pile-supported wharf is used, generally some sort of retaining structure is used in conjunction with it.

General Discussion

Retaining structures used along the shorefront include cantilevered and anchored flexible bulkheads, solid fill steel sheet pile structures, concrete caissons, and mass gravity type concrete or masonry quay walls (see Figure 7.1). Each type of structure has advantages, disadvantages, and particular design considerations, as discussed in the following sections, but some general design considerations are common to retaining-type structures. Most retaining structures require stable subsurface conditions to provide an adequate foundation. When backfilled, these structures tend to be quite heavy, thereby greatly increasing the stresses in subsurface soils. The increased loads can cause bearing failures, instability, or intolerable settlements. In addition, these structures can severely limit future dredging unless they are specifically designed for it.

Retaining structures also must resist lateral earth pressure loads. These loads increase with the square of the height of wall, thus requiring very heavy sheeting and anchoring or excessively large gravity structures for high walls. Wall heights of approximately 50 feet for anchored sheet pile bulkheads, 60 feet for concrete caissons, and 80 feet for sheet pile cofferdam type construction are

practical limits. Mass gravity concrete or masonry walls, rarely used today because of their relatively high cost, are generally less than 40 feet in height.

Lagging water levels behind a retaining structure can increase lateral forces significantly (see Figure 8.5). In addition, differential piezometric heads on either side of the wall will cause water to flow under the structure. If the situation becomes severe enough, piping—leading to a "quick" condition—can take place. This causes the soil to lose its shear strength and can lead to a failure (4, 53). To avoid this situation, retaining structures should provide for free drainage of water from behind the wall, usually accomplished by using free-draining backfill above extreme low water, and providing adequate weep holes as close as practicable to that level. In any case, where there is any question of drainage, structures should be designed for a minimum water lag of 25% to 50% of the tidal variation.

Reference (54) recommends that a tidal lag of 50% of the normal tidal variation be used for impermeable backfill, with a coefficient of permeability (K) less than or equal to 10^{-3} ft/min. or for walls of low permeability, such as sheet piling, with backfills having K values between 10^{-3} and 1.0 ft/min. This reference also provides a graph for determining the tidal lag for soils with various permeabilities when a permeable wall, such as masonry block or timber crib, is used. It should also be noted that even if the appropriate tidal lag is used, much of the lateral pressure above the groundwater level should be calculated using wet rather than dry unit weights.

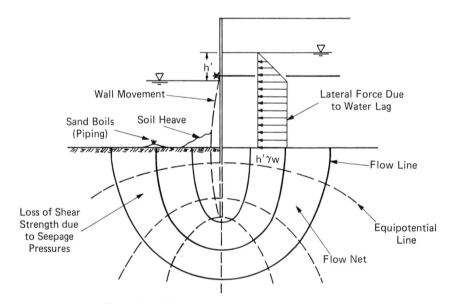

Figure 8.5. Effect of water lag on a retaining structure.

These procedures or recommendations do not account for inundation of the backfill from overtopping. Where overtopping from wave action or floods is possible, the structure should either provide positive surface drainage to avoid saturation of the backfill or be designed to handle a full hydrostatic load to the top of the wall. In such situations of temporary loading, lower factors of safety, overstressing, or reduced load factors can be allowed for many of the structural members. It is imperative, however, that adequate factors of safety be maintained for critical elements of the design, such as anchor systems, resistance to piping, and overall stability.

Backfilling of the structure must be considered during design. Not only must a free-draining granular backfill material be used, but the method of placement and densification also must be determined. Organic silt and soft sediments should be removed from immediately behind a retaining structure to avoid large lateral stresses. If existing bottom contours behind the structure are not steep, soft sediments may be displaced by progressively placing a granular fill dike immediately behind the structure. Placing fill from the landward side can aggravate the situation by causing a mud wave against the structure.

Hydraulic filling can result in high lateral loads on retaining structures during construction and generally controls the design. Hydraulically placed sand tends to settle into a very loose state and may be susceptible to liquefaction during earthquake events or from other vibrations. To ameliorate the situation, densification can be performed using one of the ground modification techniques discussed in Section 8.9.

In general, waterfront walls fail more often due to scour or undermining at the toe or from buildup of hydrostatic pressure behind the wall than from direct environmental, mooring or berthing, or surcharge loads. Cantilevered and anchored bulkheads also can suffer serious damage or failure due to scour, overdredging, or shallow embedment at the toe. It is important to provide protection from scour, as discussed in Section 8.4, or to design the structure for such. Corroded or undersized tie rods and connections also lead to many bulkhead failures. Figure 8.6 depicts general modes of failure for several types of retaining structures.

Flexible Bulkheads

Flexible bulkheads can be designed with steel, timber, or concrete and can be constructed in either cantilevered or anchored fashion (see Figure 7.1). Cantilevered construction usually is limited to wall heights of 10 to 15 feet because of excessive bending moments in the sheeting and deflection at the top of the wall. These structures depend solely on passive earth pressures for support and must be driven relatively deep to attain adequate support.

Flexible bulkheads are severely affected by scour and erosion, so they nor-

320 DESIGN OF MARINE FACILITIES

Figure 8.6. General modes of failure of waterfront retaining structures.

mally are used only for very low walls or as end return walls in shallow water for other major structures. The recent development of piling systems with very large bending resistance, however, has made the use of much larger cantilevered structures possible where soil conditions are favorable and scour is not a problem. Two of these systems include H-Z type steel sheet piling (55) and large-diameter, prestressed concrete cylinder piles (see Figures 8.7 and 8.8).

Conversely, anchored bulkheads derive their support from a combination of passive earth support and anchoring and are somewhat less susceptible than cantilevered bulkheads to damage due to scour. The wall height usually is lim-

Figure 8.7. Installation of H-Z steel sheeting for a cantilevered retaining wall.

ited by anchoring requirements rather than bending of the piling, because of the availability of sheeting systems with very large section moduli (55, 56). The use of multilevel tiebacks is also possible, to decrease the size of sheeting and individual anchor components. This is usually impractical in waterfront structures, however, because the lower anchor generally would have to be located well below low water to have any appreciable effect.

References (57) through (59) present design considerations for several types of bulkheads. Analytical methods used for the design of anchored bulkheads range from the liberal Danish Rules and free earth support methods to the more conservative fixed earth method. Tschebotarioff (8) proposed a simple equivalent beam method that assumes a hinge at the dredge line and sets the driving depth of the sheeting at 0.43 times the exposed height of the wall. This method tends to give reasonable values for the required section modulus, depth of penetration, and anchor force, which generally fall somewhere between the more conservative and liberal methods.

As piling is driven deeper, rotation of the sheeting and the required section modulus decrease. In certain cases, simply driving longer sheets may result in the use of lighter sections and overall cost savings. No matter which method of analysis is used, there is a tradeoff between the anchor force and the required sheeting section modulus, depending on where the anchor is located. As the

322 DESIGN OF MARINE FACILITIES

Figure 8.8. Prestressed concrete cylinder piles used as a cantilevered retaining wall, City of Newport News, Virginia, Seafood Industrial Park. (Photo courtesy of The Maguire Group.)

level of the anchor is lowered, the anchor force increases, and the required sheeting section modulus decreases. Anchors, then, normally are placed as close to low water as possible in order to decrease the required section modulus, which, in turn, increases the anchor force.

In most marine structures, however, the distance between the mudline and low water is much greater than the distance between low water and the top of the wall. This usually makes it more cost-effective to decrease the sheeting size by lowering the anchor as far as possible. As the highest moments in the sheeting occur over a relatively short distance, the section modulus can be increased

over this distance by welding plates to the flanges of the steel sheeting. Because of the labor costs involved, however, this may or may not provide an economical solution.

Anchoring of a bulkhead generally is accomplished by using an anchor rod that spans between the wall and the anchor system. Anchor rods are spaced some distance apart, generally 5 to 12 feet, and are attached to the wall through a wale. A typical wale consists of a pair of channels spaced back to back, through which the anchor is bolted. Two possible positions for the wale are on the face of the wall and behind the wall (see Figure 8.9). From a mechanical standpoint, the exterior wale is the best alternative because the piling bears directly against it, and the whole system is available for periodic inspection. From a practical point of view, however, most wales are placed behind the wall. This avoids conflict with vessels and fendering systems while affording some protection from corrosion and mechanical damage. Care must be taken in the design of this type of installation because the wall is in essence supported by a series of bolts rather than the wale system itself.

Many types of systems can be used to anchor a wall, as shown in Figure 8.10. Generally, the most economical system utilizes a deadman, which uses passive earth resistance to provide anchorage. The deadman can consist of concrete blocks, a continuous concrete wall, or driven sheet piling with another wale system. Use of this type of system may be difficult when the proposed bulkhead is located a good distance from the existing shorefront. Other anchor alternatives consist of battered pile anchors for the tie rods, batter piles (either tension or compression) attached directly to the wale, or drilled and grouted anchors.

The location of the anchor system is also of particular importance, as shown in Figure 8.11. An anchor placed within the active wedge of soil behind the

Figure 8.9. Typical wale details for anchored sheet pile bulkhead: (a) wale located outside of wall; (b) wale located inside of wall.

324 DESIGN OF MARINE FACILITIES

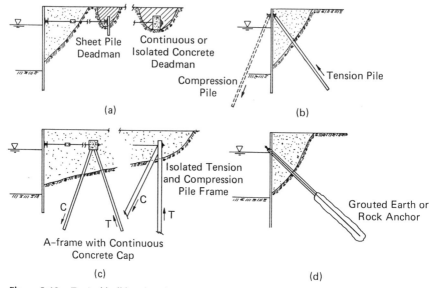

Figure 8.10. Typical bulkhead anchor systems: (a) tie rods and deadman; (b) H-pile tension or compression anchor; (c) tie rods and pile anchors; and (d) drilled grouted anchor. (After Pile Buck (57).)

wall provides no resistance. A deadman system must be placed behind line OB in order to provide full resistance without placing additional load on the wall (57, 60). Pile-supported systems may be placed between lines OA and OB be-

Figure 8.11. Effect of location of anchor system. (After U.S. Navy (85).)

cause resistance is derived from compression and tension in the piles rather than from passive earth pressure. When tension pile anchors and grouted anchors are used, only the zone outside the active wedge can provide resistance. Several other possible modes of failure and design considerations must be addressed when grouted anchors are used, as discussed in the references (61–65).

All anchor systems should be designed for a minimum factor of safety of 2.0. This provides redundancy in the system so that if one anchor fails, progressive failure of the whole structure will not take place.

Solid Fill Gravity Walls

Two major types of solid fill gravity structures are in common use as waterfront retaining walls: sheet pile cofferdam type structures and concrete caissons (see Figure 7.1). Both systems resist overturning and sliding by virtue of their weight and dimensions, and must be analyzed for bearing capacity failure and stability when founded in soil. These structures provide a particular advantage over anchored bulkheads in that they can be designed to resist very large lateral loading and are suited for use where shallow rock exists.

Sheet pile cofferdams are constructed with straight web or shallow arch-type sheeting and, as shown in Figures 7.1 and 8.12, are generally driven in the form of interconnecting circles or diaphragms. The circular-type construction has the advantage that each individual circular cell can act as an independent structure, thus greatly facilitating construction. Diaphragm bulkheads must be constructed in stages because excessive filling within an individual cell will cause collapse of the common diaphragm. Soft organic sediments must be removed prior to the filling of the cells in order to avoid excessively high lateral stresses.

Circular cells act in hoop tension, which is dependent on cell diameter and lateral stress; so they are limited in size and height by the allowable interlock tension in the piling. Diaphragm cells, however, can limit hoop tension by using a smaller arc radius; thus this configuration allows greater width of the structure, which translates into greater lateral resistance.

Cellular structures must be analyzed for several modes of failure not common to other gravity-type structures, such as slippage between the sheeting and cell fill, shear failure within the fill, and interlock tension. Design details and results of laboratory and field testing of cellular structures are provided in the references (57, 66–78).

Concrete caissons are prefabricated structures built on land or in dry dock and transported for placement at the site (see Figure 8.13). Caissons can be either open- or closed-bottom. Open-bottom caissons with cutting edges are lowered into place, usually in sections, and are sunk into the sediments to firm bearing. Sinking may be facilitated by predredging or jetting. Closed-bottom caissons are usually floated to the site, then ballasted to a prepared bed of scour-

326 DESIGN OF MARINE FACILITIES

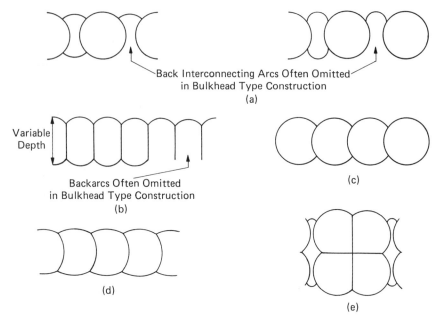

Figure 8.12. Steel sheet pile cellular structures used for cofferdams, bulkheads, or dolphins: (a) circular cells with interconnecting arcs; (b) diaphragm cells; (c) modified circular cells; (d) modified diaphragm cells; (e) cloverleaf cells.

resistant crushed stone or rock. Both types of structures then are filled with rock or granular fill. Because of difficulties in attaining precise alignment during placement, filters or grout must be placed between adjacent sections to avoid washout of backfill. The top portion of these structures is generally cast in place to adjust further for misalignment.

Quay Walls

Mass masonry or concrete quay walls also are designed as gravity structures. This type of wall, while long-lasting, is very costly to construct and thus is rarely used today. These designs may be competitive for smaller walls, however, if precast concrete sections are used or if construction can take place in the dry. Precast cantilevered or L-type walls also may have merit for smaller wall applications. At locations where bearing capacity or settlement is a problem, quay walls may be supported on pile foundations.

8.6 SOLID FILL STRUCTURES

Solid fill structures are often used to construct piers or breasting and mooring dolphins. A solid fill pier can be a massive structure providing several acres of

GEOTECHNICAL DESIGN CONSIDERATIONS 327

Figure 8.13. Concrete caissons: (a) open-bottom; (b) closed-bottom.

land reclamation for upland activities, or it can be a relatively narrow structure for handling vessels and product. The mass-fill-type structure incorporates one or more of the retaining systems discussed in the previous section to form the exterior of the pier. Typical solid fill pier configurations are shown in Figure 8.14.

Isolated dolphins often are required for the breasting and mooring of vessels. Circular steel sheet pile cells commonly are used for this type of structure (see Figure 8.15).

Design Considerations

The design concepts and considerations applicable to solid fill piers and dolphins are similar and in many cases identical to those outlined in Section 8.5. Mass fill structures of sufficient width simply are designed as normal backfilled retaining structures. The design of narrow piers, however, may be controlled by exterior loadings rather than internal forces. These structures, generally constructed as a series of interconnecting circular steel sheet pile cells, concrete caissons, or anchored parallel sheet pile walls, are sized by the overturning resistance required for berthing, mooring, and wave forces. Because of the narrow width and heavy gravity load, bearing capacity or stability may become the controlling issue.

Solid fill dolphins normally are constructed as isolated circular sheet pile cells. Where the water is deep and lateral loadings are high, a cloverleaf-type configuration may be used to reduce hoop tension while increasing stability (see Figure 8.12). Dolphins used for breasting and turning of vessels are subjected to large torsional as well as lateral forces that can result in racking and twisting of the cell and rupture of the interlocks. An exact analysis of interlock stress for these conditions is difficult because of the nature of the structure. These effects, however, can be reduced by stiffening the top of the cell with a deep cast-in-place concrete cap, by welding interlocks together, or by outfitting the dolphin with a low-friction fendering system (see Chapter 5). Isolated cells also are used to support loading and cargo handling facilities, catwalks, trestles, and conveyor systems.

8.7 PILE FOUNDATIONS

Piles, which are used extensively in the marine environment to carry structural loads through the water column and soft marine deposits to suitable foundation soils or rock, may be designed as either end bearing or friction-type foundations. End bearing piles derive their main resistance from the bearing capacity of the hard or dense soils or rock to which the piles are driven. Conversely,

GEOTECHNICAL DESIGN CONSIDERATIONS 329

Figure 8.14. Solid fill pier configurations: (a) mass fill pier; (b) narrow solid fill pier.

Figure 8.15. Solid fill steel sheet pile dolphin.

friction piles mainly support their load through frictional forces between the piles and the surrounding soils.

Primary design considerations include: (a) the magnitude, direction, and nature of the loading; (b) allowable stresses of the pile material; (c) the limiting supporting capacity of the soil or rock for single piles or pile groups; and (d) the total and differential deflections in the pile–soil and structural systems.

As a result of the unique environment in which they are used, marine and near-shore pile foundations are subjected to different loadings and conditions from those of upland pile foundations. Marine piles often are only partially embedded, leaving much of the pile material exposed to severe environmental conditions (52) (see Chapter 3). The design of piles, then, must consider corrosion, marine organism attack, rot, abrasion, impact, and ice damage, as well as cyclic and dynamic loading. Often the presence of soft surface sediments or scour further increases the unsupported pile length, thereby increasing applied moments and decreasing axial capacity.

Marine piles are commonly subjected to substantial uplift forces, lateral loadings, and downdrag forces. The ability of the piles to resist these forces is dependent on the pile material, pile section properties, soil–structure interaction, and soil stress–strain characteristics. Evaluation of ultimate pile capacity and deflection is based on a combination of theory, load test data, and empirical relationships.

The following sections briefly discuss several general pile types that are used in the marine environment, along with some of their advantages and disadvan-

tages (see also Figure 7.5). Also, a discussion of pile analysis and design considerations for several situations is presented. Section 7.3 covers structural design considerations of piles.

Timber Piles

Timber piles, the oldest type of piling, still are commonly used in marine applications. In the past, piles were driven with the bark on for protection against marine borer attack. Today, timber piles are generally pressure-treated with either creosote or chromated copper arsenate (CCA) for protection (refer to Sections 3.5 and 7.3). Some tropical woods, such as greenheart, are naturally resistant to marine borer attack. All wood piles, however, must be protected from fresh water, which can cause severe damage through dry rot and decay (79).

Typical pile capacities range from 15 to 20 tons, with some of the tropical woods extending capacities to 30 tons or more (80, 81). Standard timber pile lengths range from 50 to 65 feet, although lengths of up to 125 feet are possible as special orders. The natural taper of the pile and a reasonably high pile–soil friction angle make them well suited for friction-type foundations. The same taper, however, detracts from the pile's ability to resist uplift forces.

The end bearing resistance of the pile may be considerable (81), especially in situations where larger-tip diameters are possible. In instances where timber piles are used in end bearing, extreme care must be taken to avoid crushing, cracking, or brooming of the pile tip during driving. A protective tip may be attached to piles where hard driving or obstructions are anticipated.

Prestressed Concrete Piles and Cylinders

Prestressed concrete piles, which are among the most commonly used high-capacity marine piles, can be cast in a variety of shapes, lengths, and diameters. The prestressing strands act as tension elements to resist bending and the large tension stresses induced into the pile during driving (82).

Although they are displacement-type piles, which perform well in friction, prestressed concrete piles often are driven as high-capacity end bearing foundations. Capacities range up to approximately 120 tons, with lengths up to 120 feet. A steel plate generally is cast at the tip for protection of the concrete during driving. If hard driving or obstructions are anticipated, an H-pile "stinger" can be welded to the tip plate (see Figure 8.16).

Large-diameter prestressed concrete cylinders are being used in applications where high vertical as well as lateral capacities are required. Use of the cylinder shape provides moment capacities several times greater than those obtained with solid piles of the same weight (81). Three-hundred-ton, 54-inch-diameter cyl-

Figure 8.16. Steel H-pile "stinger" attached to prestressed concrete pile for hard driving conditions.

inders, 224 feet in length, were driven for the construction of a naval pier at Staten Island, New York (83). These large-diameter cylinders also can be used for cantilevered-type retaining walls where satisfactory subsurface conditions exist (see Section 8.5).

The engineer must be careful when determining the length of concrete piles. Although several types of mechanical splices are available (82, 84), unanticipated field splices are difficult and expensive to make. Splices should be designed to be located below the mudline whenever possible, and in no case should they end up in the splash zone. If splicing is anticipated for batter piles, an analysis of how driving stresses will be affected by the splice should be performed. Splices may have to be specially designed to take into account eccentric loading due to the dead weight of the spliced section.

Steel Pipe Piles

Concrete-filled pipe piles are used in both friction and end bearing. Typical load capacities range between 40 and 150 tons (81), with lengths in excess of 200 feet possible because of the relative ease of field splicing. Pipe piles also provide relatively high section modulus-to-weight and capacity-to-weight ratios, with section properties being the same in all directions. Because of the

thin wall thickness used in many pipes, drivability should be checked with a wave equation analysis.

Pipes may be driven either open- or closed-ended. Either case results in a displacement-type pile because of the formation of a soil plug in the open end of the pipe. Water and mud generally have to be removed from the piles before concrete is cast. For this reason, pile tips or end plates normally are welded to the pipe before driving. This also allows for internal inspection of the pipe after driving. Use of an open-ended pile, however, permits a grouted anchor to be driven through the pile in cases where large uplift loads must be resisted.

Steel H-Piles

The low-displacement, high-capacity H-pile generally is used where hard or dense soils or bedrock must be penetrated (81). H-piles typically are designed for loads in the range of 40 to 120 tons (85), with lengths in excess of 200 feet possible due to the ease of splicing. These piles are relatively light, are easy to handle, and can withstand high driving stresses. With their low-displacement cross section, H-piles can be driven close to existing structures without causing severe damage from heave. H-piles may become damaged during driving if obstructions are encountered and the tip is not protected with a driving shoe.

Analysis and Evaluation

Pile performance is a function of the pile–structure interaction, pile material and configuration, soil–pile interaction, and soil properties. Evaluation of pile performance generally is based on a combination of theory and empirical relationships. Pile performance may be separated into axial capacity, lateral capacity, and buckling. A discussion of structural design considerations is presented in Section 7.3.

Axial vertical load capacity is dependent on either pile buckling above some point of fixity in the soil or the pile–soil interaction below the mudline. Vertical pile capacity may be evaluated by using static formulas based on (a) the ultimate capacity of the supporting soil or (b) deflections determined from rheological modeling of the pile–soil system using nonlinear springs. Lateral load capacities also may be found by using limit equilibrium methods or deflections based on subgrade reaction theory.

Analysis of the structure can follow two routes. First, the pile can be idealized as a cantilevered column that is fixed at some depth below the mudline (see Figures 7.7 and 8.17). This simplified model is used for analysis of the pile–structure system as well as for column buckling, but the method results in

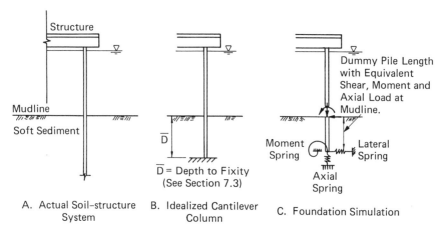

Figure 8.17. Methods of analysis for waterfront piles.

conservative moments and required pile cross sections below the mudline. Although this analysis may be justified for smaller coastal structures, it is rarely adequate for large structures in deep water. Alternately, the pile–soil system can be modeled along with the structural system to effect a more economical solution. The following sections discuss different methods of analysis for axial and lateral loading conditions.

Axial Load Capacity

The ultimate axial load capacity of the pile may be developed through skin friction, end bearing, or both. The theoretical static load capacity method of determination of axial capacity is outlined in Figure 8.18 (82, 85–87). In applying the static analysis as indicated in Figure 8.18 to marine piles, particular attention should be given to friction factors, especially for tension piles.

This classical static analysis typically is used to approximate ultimate capacities based on pile material, configuration, and length. For some marine structures this type of design approach suffices, but larger, more complicated structural designs often require an accurate description of pile behavior, including capacities with resulting deflections and rotations, for pile–structure compatibility.

Vertical pile deformation may be approximated by using a nonlinear pile–soil system model, referred to as the subgrade reaction or t–z method (88, 89). Figure 8.19 illustrates this pile–soil model. The pile–soil response along the shaft and tip is represented by t–z and q–z curves, which describe the relationship between shear stress (t), tip resistance (q), and displacement (z). The analysis of skin friction using this approach is separated into two modes: prefailure,

Figure 8.18. Static axial pile analysis.

which is based on elasticity, and post-failure, which is based on residual stress-deformation behavior.

At any depth along the pile, the prefailure t–z curve is a function of displacement, shear stress, radius of influence, and soil modulus. Post-failure t–z analysis is dependent on maximum skin friction, the strain at which maximum skin friction is mobilized, residual skin friction, and strain behavior between maximum and residual stress (89) (see Figure 8.19). Pre- and post-failure t–z curves are developed using displacement equations, laboratory soil testing, modeling, and engineering judgment.

Similarly, a q–z curve may be developed for the pile tip from the results of instrumented pile load tests or laboratory model tests, or theoretically from the following equation (89):

$$z = q \frac{D(1 - \mu)^2}{E_s} I \qquad (8\text{-}4)$$

where z is displacement, q is stress at the pile tip, D is the tip diameter, E_s is the modulus of elasticity of the soil, μ is Poisson's ratio, and I is the influence coefficient.

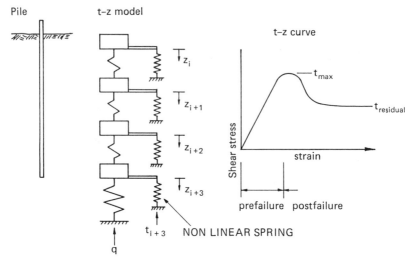

Figure 8.19. Rheological pile–soil model for t–z analysis of vertical pile deformation. (After Focht and Kraft (88).)

Illustration and explanation of the t–z method may be found in references (88) and (89). Data from the t–z curves for each stratum of soil and the tip q–z curve are input into a finite difference computer program along with certain boundary conditions to model the pile–soil system. The model uses the load displacement curves to simulate foundation response.

Lateral Load Capacity

Lateral load capacity analysis of piles may be separated into two methods, based on (a) the ultimate lateral resistance of soils and (b) the allowable deflection of a loaded pile. In most cases, the lateral load capacity is limited by the tolerable deflection of the structure. The two methods are addressed separately below.

Classical methods of earth pressure analysis have been applied to ultimate lateral pile capacity analysis. This method now is considered an invalid representation because mobilization of active and passive earth pressures requires excessive movement, which is generally intolerable to most structures. However, a simple static method for determining lateral load capacity was developed by Brinch-Hansen and later modified by Broms (86, 90–92). Broms assumed that short piles fail as a result of rotation or translation, depending upon whether the pilehead is free or restrained. Long piles were assumed to fail by the development of one or two plastic hinges along the pile length, again depending upon whether the pile head is free or restrained. The ultimate lateral resistance

is based upon the lesser of the lateral capacity of the soil or the maximum pile bending stress (see Figure 8.20).

This empirical method assumes a maximum soil resistance for cohesive soil of nine times the product of the undrained shear strength and pile diameter (see Figure 8.20). For cohesionless soils, the resistance along the pile shaft is taken as three times the Rankine passive pressure, where active pressure along the back of the pile is neglected. A family of curves was developed by Broms to solve the problem, based on the assumptions discussed above but with different boundary conditions. A detailed description of this method and the associated curves is presented in the references (86, 90–92).

There are presently two methods of analyzing lateral pile deflection and associated moments for a given lateral load and pile–structure system: the subgrade reaction theory method and an elastic approach (86, 93, 94).

The subgrade reaction theory is based on the Winkler soil model. The theory assumes that the pile acts as a thin beam with horizontal springs representing discrete soil layers penetrated by the pile (see Figure 8.21). The pile reaction (p) at a point is related to the corresponding deflection (y) at the same point by a spring constant. The spring constant (k) of each spring is the coefficient of subgrade reaction.

A widely used approach to the subgrade reaction model is the $p-y$ method

Where:
 C is the cohesion
 D is the pile diameter
 γ is the unit weight of soil
 k_p is the passive earth pressure coefficient
 f is the distance from the ground to the location of the maximum moment
 P is the applied lateral load
 Z is the depth below ground surface

Figure 8.20. Failure modes for short and long unrestrained piles. (After Broms (90–92).)

338 DESIGN OF MARINE FACILITIES

A. Actual Pile-soil system B. Rheological Model Using Linear Springs

Figure 8.21. Subgrade reaction theory, p–y method of analysis for laterally loaded piles. (After Cox and McCann (94).)

(94). A solution for the elastic deflection along the length of the pile is obtained by using the differential equation for the deflected shape:

$$E_p I_p \frac{d^4 y}{\partial x^4} + p = 0 \tag{8-5}$$

where E_p is the pile's modulus of elasticity, I_p is the pile's moment of inertia, x is the depth along the pile, y is the lateral deflection at depth x, and p is the soil reaction. The soil reaction is given by the equation:

$$p = k_s y \tag{8-6}$$

where k_s is the horizontal modulus of subgrade reaction.

A computer solution for this approach was developed to account for variability of the subgrade reaction with deflection and depth. Actual values of the subgrade reaction usually are based on empirical correlations with other soil properties, but they also may be obtained by using pile load test results and in situ tests such as the pressuremeter. From these data, representative stress–strain plots (p–y curves) are developed for each soil layer. Consequently, a relative stiffness factor is developed for each element along the beam based on the p–y curves. The deflection of the beam is modeled through trial and error until the deflection and resistance patterns of the beam agree as closely as possible with the resistance–deflection relations of the soils (94).

A noncomputerized solution to this problem was developed (for homogene-

ous nonlayered soils) (94) by using a nondimensional approach and estimated coefficients of subgrade reaction. A detailed explanation of this solution is contained in the references (85, 94).

Unlike subgrade reaction theory, the elastic analysis considers the soil to be an elastic continuum (86, 93) with an associated constant modulus of elasticity. The pile is considered as a thin beam on a semi-infinite elastic material and is divided into discrete elements that are acted upon by a uniform soil pressure. Displacements are determined at the center of each element. To determine these displacements, differential equations for bending of a thin beam are derived for a fixed- or free-head floating pile. From the equations, load versus displacement curves for the pile may be developed. A full discussion with representative examples is contained in the references (86, 93).

A comparison between the two methods indicates that the subgrade reaction method is slightly more conservative than the elastic method when estimating displacements and rotation, and it is also easier to develop than the elastic method (86). Both methods require considerable experience in estimating elastic soil properties.

Determination of Depth to Fixity of Partially Embedded Piles

When partially embedded piles penetrate a surface stratum of weak marine deposits, only moderate lateral pile support can be expected from these soils. Intuitive reasoning suggests that the amount of lateral resistance should increase with depth to full lateral restraint in competent soils. As discussed previously, an idealized model of this condition was developed to simplify the analysis of the pile–structure system and to determine an effective unsupported pile length for column buckling analysis. The idealized pile–soil reaction is represented by a cantilevered pile of equivalent stiffness, fixed at some depth below the mudline. The depth between the mudline and the point of fixity is called the depth to fixity (\overline{D}) (see Figure 7.7), a term that was introduced in Section 7.3, along with associated equations (7-1) and (7-2). Depth to fixity, as calculated by these equations, is equal to one-half of the critical length as determined in lateral pile deflection analysis using subgrade reaction theory. The critical length is defined as the length at which the pile behaves as an infinitely long beam (86). Therefore, equations (7-1) and (7-2) actually are modifications of equations developed from subgrade reaction theory as applied to laterally loaded piles.

For equations (7-1) and (7-2), soil resistance to lateral movement is represented by a subgrade reaction modulus (k_s) when k_s is constant with depth, or by a coefficient of subgrade modulus (n_h) when k_s varies with depth. Determination of these soil properties is done by one of the following methods: full-scale lateral pile load tests, plate load tests, or empirical correlations with other

soil properties. Several empirical values are presented in Section 7.3 and in references (85) and (87).

In summation, static analysis methods for determining pile–soil capacity and the depth of fixity method for analyzing pile buckling generally are used in the design of most near-shore structures. The analysis then is verified with full-scale pile load tests. For larger, more complicated structures, more sophisticated analysis is warranted for accurate simulation of the foundation response. These methods generally are performed using computer solutions with a combination of empirical information and test data.

Uplift Resistance

Because of the existence of relatively large uplift and lateral loads common to marine structures, piles often must be designed for tension as well as compression. Uplift capacities typically are calculated in the same way as shaft friction resistance for compression piles. Conservative values for the coefficient of lateral earth pressure for cohesionless soils and adhesion for cohesive soils should be used (85, 86).

When insufficient tension resistance is available through shaft friction, several options are available. The simplest method of increasing uplift resistance is to drive the pile deeper than normally required for compression, thereby increasing the shaft length. This is not always possible because of shallow rock or hard driving conditions. Alternatively, additional piles can be driven in order to increase uplift resistance. If several piles are driven close together, the uplift may be limited by group action. The addition of lagging may be useful for piles that are driven through soft soil (81), where the lagging consists of bolting or welding additional sections to the pile shaft to effectively grab surrounding soil and mobilize its dead weight for uplift resistance. If large tension forces must be resisted, grouted anchors may be drilled through open-ended pipe piles.

Batter Piles

Batter piles, sometimes referred to as raker piles, often are used to resist lateral loads applied to the structure. These piles usually are driven on batters ranging from 1 to 5 horizontal to 12 vertical. Batters shallower than 5 on 12 become difficult to obtain because of pile driving equipment limitations.

The pile heads must be attached to the structure or other piles in order to adequately transfer loads. If the attachment is assumed to be a pinned connection, the lateral load results in an axial load in the sloped pile; so these piles generally are analyzed in the same way as axially loaded vertical piles.

Because a batter pile transmits load axially, any horizontal load will have a resulting vertical component that must be resisted in order to satisfy equilib-

rium. This vertical component of load may be resisted by the dead weight of the pile cap and structure or by tension resistance in other piles.

Batter piles often are driven in opposite directions and coupled, so that when one pile acts in compression, the other acts in tension. Compression loads are resisted by a combination of shaft friction and end bearing. The tension load, however, can be resisted only by friction. When insufficient tension capacity is available through shaft friction of a single pile, additional piles may be required. Also, other methods of increasing tension resistance, as discussed in the previous section, may be used.

Batter piles are susceptible to downdrag forces (see below) where compressible soils exist. Also, because of the sloped configuration of the pile, large bending stresses may be introduced.

Downdrag

Downdrag, or negative skin friction, is developed when all or a portion of the overburden soil settles relative to the pile. The force mechanism is the same as normally calculated for shaft resistance, but acts in the opposite direction. The downdrag forces are developed not only in the layer that experiences settlement but also from all the natural soil or fill above this layer.

The effect of these often substantial loads may be pile settlement if the remaining shaft friction and bearing capacity is exceeded, or pile crushing if yield stress is reached before soil failure.

Negative skin friction may be decreased by preaugering before pile driving, thereby reducing the friction coefficient or horizontal stress along the shaft. In some cases, the preaugered hole can be filled with a bentonite slurry to eliminate negative skin friction in the areas subject to downdrag forces. The bentonite might be used only temporarily until settlement is complete, at which point it is displaced by backfilling with coarse gravel or crushed stone. Still another method used to reduce negative skin friction is to cover the pile with a bituminous coating (32, 33, 95). The coating tends to lubricate the pile, thereby reducing the detrimental effects of settlement.

8.8 DRY DOCKS

There are several types of dry dock and shiplift facilities, including marine railways, vertical lifts, floating docks, mobile straddle lifts, and basin or graving type docks (refer to Chapter 10). These facilities must handle heavy vertical loads as well as appreciable lateral loads due to wind and, in some cases, wave and current forces. Foundation types are varied and will depend on the type of dock, facility configuration, and subsurface conditions.

342 DESIGN OF MARINE FACILITIES

Geotechnical Considerations

Marine railways, vertical shiplifts, straddle lift piers, and floating dry dock mooring dolphins typically are supported by pile foundations. In certain situations, many of these dry dock systems may be used in conjunction with solid fill structures. Shallow foundations also are commonly used for transfer systems where competent soils exist, and in some cases are used for marine railways.

Basin dry docks provide some of the most interesting and challenging geotechnical design problems. Basin docks can be constructed within a cofferdam built out into the water, or they can be constructed inland to be finally opened to the water by excavating and dredging in front of the gate.

Basin walls must be designed for lateral earth pressures as well as hydrostatic forces when the dock is dewatered. The walls generally are constructed of reinforced concrete. However, if a cofferdam is used, it may be incorporated into the final structure (73, 74). Anchored slurry or diaphragm-type walls (96), steel sheet piling, or concrete cassions also can be used in conjunction with a concrete floor in certain situations (see Section 10.2).

The basin floor must be designed to resist hydrostatic pressures, and the whole dock must resist overall buoyancy. This can be accomplished by constructing a heavily reinforced concrete floor to resist the induced bending stresses. The floor dead weight along with the weight of the walls and soil can be used to resist uplift (see Figure 8.22). An approach that usually is more cost-effective than this, however, is to use grouted anchors to reduce spans, resist buoyancy forces, and decrease floor thickness (63). Vibrated, jetted, or augered plate anchors have also been used for the same reason (15).

Grouted anchors, embedded plates, and augered anchors derive their resistance from the dead weight of soil or rock within the cone of influence (see Figure 8.23). The size or half angle of the cone is dependent on the friction

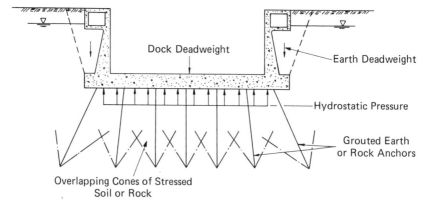

Figure 8.22. Basin dry dock with grouted tiedown anchors.

GEOTECHNICAL DESIGN CONSIDERATIONS 343

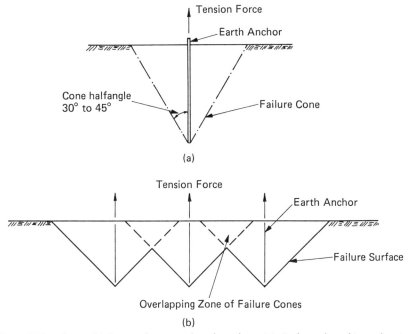

Figure 8.23. Cone of influence for grouted earth anchors: (a) single anchor; (b) overlapping cones for multiple anchors.

angle of the soil, or the fracturing, bedding, and degree of weathering of the rock. Half angles of the cone typically range from 30° to 45°, but may be less depending on bedding angles (15) (see Figure 8.24). In any case, where anchors are closely spaced, an allowance must be made for overlapping of the cones.

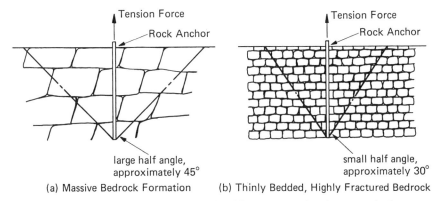

Figure 8.24. Influence of rock fracturing and bedding on grouted anchor cone of influence.

Because of the cyclic nature of a dry dock operation, care must be taken in the design of the anchors. Prestressing forces must be greater than the change in loading from dewatering in order to maintain contact pressure between the ground and the dock. Although the mechanisms are not well understood, repeated loading reduces anchor capacity (63). Bember and Kupferman (97) suggest a factor of safety of 2.5 for ultimate static load compared to the peak cyclic load. More extensive information on the design of tension anchors is provided in the references (63, 64).

Tension piles can provide resistance to uplift and are advantageous when a pile foundation already must transfer loads across a weak or compressible layer. The calculation of anchor pile resistance is discussed in the previous section. Where closely spaced piles are used, single pile as well as group action must be checked to determine limiting values.

As an alternative to designing the dock to resist hydrostatic pressures and buoyancy forces, a pumped pressure-relieving subdrain system may be installed. Hydrostatic pressures on both the floors and walls may be relieved by such a system. A pumped system located above cohesive soils, however, can lead to consolidation of the layer and settlement of the dock.

8.9 SITE IMPROVEMENT, METHODS, AND MATERIALS

Site improvement or ground modification can be an important phase in the development of waterfront sites. Several methods and materials can be used to densify loose soils, reduce post-construction settlements, and increase stability. Some of these methods also can be employed to densify backfill materials, to increase soil strength and friction angle, and to reduce the possibility of liquefaction during earthquakes (98–101). In addition, there are several relatively new products, as well as products whose applications are new to geotechnical and marine engineering, that can assist the engineer in the design of waterfront structures.

Vibro-compaction and Vibro-replacement

Vibro-compaction and vibro-replacement are two commonly used methods for the improvement of deep soft or loose soils (41, 102, 103). Vibro-compaction is performed by using a probe that vibrates as well as water-jets its way into loose granular soils as it is lowered by crane.

When the appropriate depth is reached, the forward water jets are turned off, and the equipment is extracted. Granular backfill material is continually fed into the void left by the probe, and the vibration compacts the backfill as well as the surrounding material (see Figures 8.25 and 8.26). The operation leaves a col-

GEOTECHNICAL DESIGN CONSIDERATIONS 345

| Borehole is Formed by Sinking Vibrator into the Ground | Backfill Material is Placed into the Borehole in Stages and Compacted | Alternating Backfilling and Compacting Forms Granular or Stone Column |

Figure 8.25. Schematic of vibro-compaction and vibro-replacement technique. (Courtesy of GKN Hayward-Baker.)

umn of densely compacted material approximately seven to nine feet in diameter.

Vibro-replacement, or the use of stone columns, is a variation on the vibro-compaction technique used for in situ stabilization of cohesive soils or densification of loose cohesionless materials (41–43, 104–106). Stone columns are

Figure 8.26. Stone column vibro-replacement equipment. (Photo courtesy of GKN Hayward-Baker.)

composed of coarse gravel or crushed stone that is compacted as it is placed. A vibrating probe is used to open the hole for the column, and the stone is either dumped in from the surface or placed directly at the probe tip through a chute. Each load of stone then is compacted by the weight and vibration of the probe as the backfill material is pushed laterally into the surrounding soils. The column diameter depends on existing subsurface conditions.

Stone columns typically are placed on a grid with a center-to-center spacing ranging from 3 to 10 feet. A series of columns can be placed under a new structure to improve bearing capacity and allow for the use of a shallow foundation, as opposed to a pile-type foundation. The use of stone columns can be less expensive than a pile foundation, and can aid in the drainage and consolidation of subsurface layers. This can be important in placing fill at sites that have underlying deposits of cohesive soils.

As mentioned in Section 8.4, vibro-replacement can be used to increase the stability of slopes (41). The strength and the friction angle of the subsurface soils are improved by the placement of stone columns. Also, the columns can penetrate through possible slip surfaces, thereby increasing factors of safety against sliding. Figure 8.27 presents a schematic of the use of stone columns to aid slope stability.

Vibrating Pile Compaction

The vibrating pile compaction technique uses a specially fabricated, open-end, large-diameter pipe pile that is vibrated into loose sand and gravel (107-108). Vibrations are delivered to the pile by a vibratory pile driving hammer, and cause a densification of loose materials as the pile is placed and extracted. Raising and lowering of the vibrating pile can take place several times at each location until the desired density is obtained (see Figure 8.28). Reference (109) describes a comparison of the vibro-compaction and vibrating pile techniques

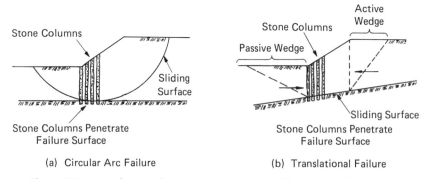

Figure 8.27. Use of stone columns to increase slope stability. (After Dobson (41).)

Figure 8.28. Vibrating pile compaction.

based on full-scale field tests. The project consisted of the densification of hydraulically placed backfill for the construction of a basin dry dock.

All of the above methods are carried out in grid-type fashion over the entire area to be compacted. The grid spacing usually is determined by field testing prior to development of the compaction program. Densification normally is verified by performing SPT or cone penetration tests at the center of grid locations during construction, supplemented with pressuremeter or dilatometer tests after construction (110).

Deep Dynamic Compaction

Deep dynamic compaction is a relatively new method that is being used to compact loose granular materials (111–113). The method consists of raising and dropping a large mass, usually a concrete or steel block, onto the ground surface, with the induced vibrations causing densification of the subsurface material. Different energies are easily obtained by increasing or decreasing either the mass of the falling block or the free fall height (see Figure 8.29).

348 DESIGN OF MARINE FACILITIES

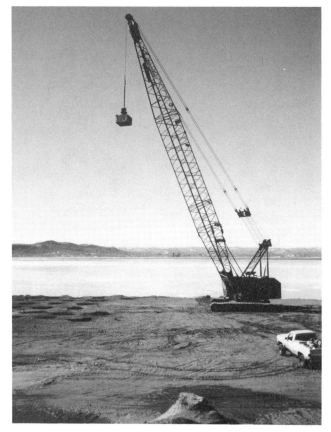

Figure 8.29. Deep dynamic compaction.

As in the vibratory methods described above, an in situ test program is performed in the field and checked with the SPT or cone penetration test to determine the grid spacing, mass of the block, free fall height, and number of blows required at each location. A particular advantage of deep dynamic compaction is that it can be used in areas that have been filled with heavy bouldery or rock fill, a situation that precludes the use of most other methods. Unfortunately, testing of compaction performance in this type of material is difficult to impossible, although measurement of the average total settlement will give an indication of the void ratio reduction.

Preloading

An increase in site grade or additional loads due to proposed structures can result in unacceptable settlements where underlying compressible layers are

present. However, this problem can be ameliorated by preloading, a method of consolidating compressible soils to a stress level greater than the stresses due to new construction loadings (114). The method generally is performed by stockpiling fill on the site until the underlying soils have consolidated. Other methods, for example, temporarily storing bulk products such as coal, scrap steel, or iron ore on site, or constructing dikes in order to flood large areas, also have been successful.

Preloading can be a time-consuming process, depending on the permeability of the soil and the thickness of the compressible stratum. A thick deposit of low-permeability soil can require an excessive time span to modify the site. The use of closely spaced vertical sand drains or wick drains can speed the process considerably, however, by providing a relatively short drainage path for pore pressure dissipation.

Vertical Wick Drains

Vertical wick drains, corrugated plastic strips covered with filter fabric, have been used extensively to increase drainage of pore water from deep deposits of cohesive materials (115–119). The vertical drains are used to accelerate consolidation of fine-grained materials and avoid a buildup of excess pore water pressure that can lead to slope stability problems. Closely spaced wick drains provide a short drainage path for pore water to travel to the surface where it can easily be carried away.

In the few years that wick drains have been available, they have received wide acceptance and have almost totally replaced the use of sand drains. They are cleaner, easier, and faster to install, and are more reliable, efficient, and cost-effective than sand drains (118). It should be noted, however, that in certain situations the filter fabric can become smeared with fine-grained soils that reduce the effectiveness of the method.

Lightweight Fill

Lightweight aggregate, commonly used in the production of lightweight concrete, currently is being used as a geotechnical backfill material. Its high strength-to-weight ratio can be used to solve stability, settlement, and high lateral earth pressure problems associated with many waterfront structures.

Rotary-kiln-produced expanded shale aggregate has a number of important properties that make its use as a lightweight backfill material viable. The angle of internal friction is in the range of 40° to 45°, and compacted dry unit weights below 65 pounds per cubic foot are possible (120). Individual particles have fairly high abrasion resistance and do not exhibit appreciable breakdown in the field. The material also can be supplied in a number of standard gradations.

Though lightweight fill is more expensive than granular backfill material, it

can lead to an essentially economical solution by allowing the use of lighter structural members for new construction. Lightweight backfill also may be used to relieve lateral earth pressures on existing distressed structures. Not only has lightweight aggregate been used successfully for numerous small rehabilitation projects; its use on large-scale projects also has proved economical (55).

Low-density cellular concrete recently has been used to reduce loads on existing waterfront structures (121). The material is aerated prior to being pumped in a slurry consistency, and resultant unit weights are on the order of 36 pounds per cubic foot.

Reinforced Earth

Reinforced earth is another relatively new concept in foundation engineering. Originally developed for a vertical retaining wall system (44), it has proved useful in many other foundation applications. Reinforced earth retaining walls are built with individual interlocking members attached to a series of rows of reinforcing strips or geosynthetic grids. The strips or grids reinforce the soil by acting as a series of tension members, reducing lateral earth pressure to such an extent that only lightweight face panels or blocks are required to prevent the loss of fill.

The use of reinforced earth in the marine environment has been reasonably successful (44, 122, 123) (see Figure 8.30). Construction can be somewhat difficult, however, because of the nature of the environment, especially where deep water or large tidal fluctuations exist. The use of filter fabrics or face unit sealants usually is required to prevent migration of fines from the backfill. Concerns about the corrosion of metal reinforcing strips have been addressed with coatings and the recent use of geosynthetic grids.

The concept of soil reinforcement also can be applied to the design of embankments and fills over soft ground. The reinforcing material allows the use of steeper slopes and reduces the possibility of stability failure.

Fabrics and Filter Materials

The use of fabrics in geotechnical engineering has boomed in recent years. Numerous types of fabrics are being used for filters, impermeable barriers, reinforced earth, wick drains, drainage mats, and erosion control.

In waterfront engineering, the use of fabrics has been mainly for filters and erosion control. Normally, filters for granular soils used in coastal structures are made up of graded layers of gravel and stone (45). Materials in the proper gradation are often of limited availability or costly, and proper placement is often time-consuming and difficult to control. Woven plastic fabrics have been used successfully in a number of coastal structures to alleviate filter problems

GEOTECHNICAL DESIGN CONSIDERATIONS 351

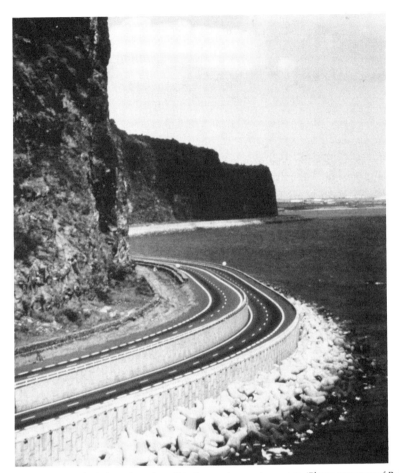

Figure 8.30. Use of reinforced earth wall in the marine environment. (Photo courtesy of Reinforced Earth Company.)

and preclude the possibility of structural failures due to leaching and erosion of construction materials (49, 50).

Fabrics have been widely used to distribute loads from equipment, fills, and stockpiled materials. When placed over soft or organic soils, they also can help to control mud waves from subsequent filling or capping operations.

H-Z Sheet Piling

The H-Z type of steel sheet piling has opened up a whole new realm of possibilities for bulkhead construction and rehabilitation (55, 56). The name H-Z

Figure 8.31. H-Z steel sheet piling.

actually describes the type of pile system, in that it is composed of a series of specially designed H-piles integrated with intermediate pairs of Z-sheets. The H-piles are sized to resist anticipated bending stresses, and the Z-sheets act as a diaphragm to transmit load to the H-piles. Because the Z-sheets are not used to develop appreciable resistance to bending, they need only be driven to a safe scour depth below the mudline.

The H-Z system is set up so that the section modulus of the wall can easily be changed by selecting from a number of different H-pile sizes. All combinations of H-piles and Z-sheets fit together with a common interlock, resulting in a wide range of available section moduli (Figure 8.31).

REFERENCES

1. Gaythwaite, John (1984). "Rehabilitation of Waterfront Structures: Design Criteria and General Considerations," Waterways Seminar on Rehabilitation of Waterfront Structures, Boston Society of Civil Engineers, Boston, Mass.
2. ASTM (1989). *Annual Book of ASTM Standards*, Section 4, Vol. 04.08, "Standards Relating to Natural Building Stones, Soil and Rock, and Geotextiles." Philadelphia, Pa.
3. Bowles, Joseph (1988). *Foundation Analysis and Design*, 4th ed. McGraw-Hill, New York.
4. Holtz, R. D. and Kovacs, W. D. (1981). *An Introduction to Geotechnical Engineering*. Prentice-Hall, Englewood Cliffs, N.J.
5. Lambe, T. W. and Whitman, R. V. (1969). *Soil Mechanics*. Wiley, New York.
6. Peck, R. B., Hanson, W. E., and Thornburn, T. H. (1974). *Foundation Engineering*, 2nd ed. Wiley, New York.
7. Teng, W. C. (1962). *Foundation Engineering*. Prentice-Hall, Englewood Cliffs, N.J.
8. Tschebotarioff, G. P. (1979). *Foundations Retaining and Earth Structures: The Art of Design and Construction and Its Scientific Basis in Soil Mechanics*, 2nd ed. McGraw-Hill, New York.
9. Winterkorn, H. F. and Fang, H. Y., eds. (1975). *Foundation Engineering Handbook*. Van Nostrand Reinhold, New York.
10. Terzaghi, K. and Peck, R. B. (1968). *Soil Mechanics in Engineering Practices*. Wiley, New York.
11. Hvorslev, M. J. (1949). "Subsurface Exploration and Sampling of Soil for Civil Engineering Purposes," Soil Mechanics and Foundation Division, ASCE, Committee on Sampling and Testing, Vicksburg, Miss.
12. Robertson, P. K. (1986). "In-Situ Testing and Its Application to Foundation Engineering," *Canadian Geotechnical Colloquium, Canadian Geotechnical Journal*, Vol. 23, No. 4, pp. 573-594.

13. Dunnicliff, J. (1988). *Geotechnical Instrumentation for Monitoring Field Performance*. Wiley, New York.
14. ——— (Spring 1977). "Sound in the Sea," *Oceanus*, Vol. 20, No. 2, Woods Hole Oceanographic Institution, Woods Hole, Mass.
15. BSI (1984). *British Standard Code of Practice for Maritime Structures*, BS 6349, Part I, "General Criteria." British Standards Institution.
16. NAVFAC (1982). *Soil Mechanics*, DM-7.1. Dept. of the Navy, Naval Facilities Engineering Command, Alexandria, Va.
17. Fletcher, G. A. (1969). "Marine-Site Investigations," in *Handbook of Ocean and Underwater Engineering*, John J. Myers, Carl H. Holm, and R. F. McAllister, eds. Section 8, pp. 18-31. McGraw-Hill, New York.
18. National Research Council of Canada (1975). *Canadian Manual on Foundation Engineering*. Ottawa.
19. Mitchell, J. K. (1976). *Fundamentals of Soil Behavior*. Wiley, New York.
20. Rocker, J. (1985). *Handbook for Marine Geotechnical Engineering*. Naval Civil Engineering Laboratory, Port Hueneme, Calif.
21. Nacci, V. (1975). "Submarine Soil Mechanics," in *Introduction to Ocean Engineering*, Hilbert Schenck, Jr., ed., pp. 43-75. McGraw-Hill, New York.
22. Bowles, Joseph (1970). *Engineering Properties of Soils and Their Measurement*, 3rd ed. McGraw-Hill, New York.
23. Lambre, T. W. (1951). *Soil Testing for Engineers*. Wiley, New York.
24. USACE (Nov. 1970). "Laboratory Soils Testing," EM 110-2-1906, Dept. of the Army, Washington, D.C.
25. Pierce, F. C. and Calabretta, V. (1978). "Unique Challenges Associated with Marine Structures in New England," *Journal of the Boston Society of Civil Engineers Section, ASCE*, Vol. 65, No. 2, pp. 43-53.
26. Carchedi, David (1979). "Apparent Overconsolidation and Intrinsic Strength of Deep-Sea Sediments," Masters of Science thesis submitted to the University of Rhode Island.
27. USACE (Apr. 1951). "Time Lag and Soil Permeability in Ground-Water Observations," Bull. No. 36, Department of the Army, Vicksburg, Miss.
28. ASCE (1976). "Liquifaction Problems in Geotechnical Engineering," ASCE Annual Convention and Exposition, 27 Sept.-1 Oct. 1976, Philadelphia, Pa.
29. Frankel, E. G. and Pollalis, S.N. (1979). "Considerations in the Design of Gravity Structures," *PIANC*, Bull. No. 34, pp. 14-24.
30. Watt, Brian J. (1978). "Foundation Design of Offshore Structures for Dynamic and Cyclic Effects from Wave Loading," Geotechnical Lecture Series on Foundation Design for Dynamic and Repeated Loading, BSCES/ASCE, 14, 21, 28 Mar. and 11, 18, 25 Apr. 1978, Boston, Mass.
31. Nacci, Vito (1969). "Shear Strength Properties of Varved Silt," Division of Engineering Research and Development, Dept. of Civil Engineering, Engineering Bull. No. 11, University of Rhode Island, Kingston, R.I.
32. ——— (Mar. 1973). "Downdrag on Piles," Professional Symposium, M.I.T., Cambridge, Mass.
33. ——— (June 1971). "Downdrag of Piles by Negative Skin Friction," Research Report R 72-12, Soils Publication 296, M.I.T., Cambridge, Mass.
34. Bjerrum, L. (1967). "Engineering Geology of Norwegian Normally-Consolidated Marine Clay as Related to Settlement of Buildings," *Geotechnique*, Vol. 17, No. 2, pp. 81-118.
35. Kenney, T. C. (1966). "Shearing Resistenace of National Quick Clays," Ph.D. thesis submitted to University of London.
36. Kenney, T. C., Moum, J., and Berre, T. (1967). "An Experimental Study of Bonds in a Natural Clay," *Proc. of Geotechnical Conf.*, Oslo, pp. 65-69.

37. Sangrey, S. A., Henkel, D. J., and Esrig, M. I. (1969). "The Effective Stress Response of a Saturated Clay Soil to Repeated Loading," *Canadian Geotechnical Journal*, Vol. 6, No. 3, pp. 241-252.
38. Enkeboll, W. and Smoots, V. A. (1980). "Case Histories of Harbor Developments in Weak Soils," *Journal of the Waterways, Port, Coastal and Ocean Division, ASCE*, Vol. 106, No. WW-4, pp. 429-447.
39. Huang, Yang (1983). *Stability Analysis of Earth Slopes*. Van Nostrand Reinhold, New York.
40. Mitchell, R. J. (1983). *Earth Structures Engineering*. Allen & Unwin, Boston, Mass.
41. Dobson, T. (1986). "Vibro Techniques for Geotechnical Engineering Practice," *Proc.* Pennsylvania Department of Transportation and Central Pennsylvania ASCE, 17-18 Apr., Pa.
42. ——— (1978). "Stone Columns—A Foundation Treatment (In-Situ Stabilization of Cohesive Soils)," Demonstration Project No. 46. U.S. Department of Transportation, Federal Highway Administration, Region 15, Demonstration Projects Division, Arlington, Va.
43. Rowe, R. K. and Poulos, H. G. (1979). "A Method for Predicting the Effect of Piles on Slope Behavior," *Proc.* 3rd Int. Conf. Numerical Methods in Geomechanics, Aachen, Balkema.
44. Ingold, T. S. (1982). *Reinforced Earth*. Telford, London.
45. USACE (1984). *Shore Protection Manual*, Vols. 1 and 2. Dept. of the Army, Washington, D.C.
46. Burroughs, M. A. (Jan. 1979). "Gabions: Economical, Environmentally Compatible Bank Control," *Civil Engineering*, pp. 58-61.
47. USACE (n.d.). *Low Cost Shore Protection—A Guide for Engineers and Contractors*. Department of the Army, Washington, D.C.
48. Welsh, J. P. and Koerner, R. M. (1979). "Innovative Uses of Synthetic Fabrics in Coastal Construction," *Proc.* Speciality Conf. on Coastal Structures 79, ASCE, 14-16 Mar., Alexandria, Va.
49. Barrett, R. J. (Sept. 1966). "Use of Plastic Filters in Coastal Structures," *Proc.* Tenth Int. Conf. on Coastal Engineering, Tokyo, pp. 1048-1067.
50. Dunhan, J. W. and Barrett, R. J. (1974). "Woven Plastic Cloth Filters for Stone Seawalls," *Journal of the Waterways, Harbors and Coastal Engineering Division, ASCE*, Vol. 100, No. WW-1, pp. 13-221.
51. Herbich, J. B. et al. (1983). *Sea Floor Scour—Design Guide for Ocean Founded Structures*, Marcel Dekker, New York.
52. Gaythwaite, John (1981). *The Marine Environment and Structural Design*. Van Nostrand Reinhold, New York.
53. Cedergren, H. R. (1989). *Seepage, Drainage and Flow Nets*, 3rd ed. Wiley, New York.
54. NAVFAC (1981). *Seawalls, Bulkheads and Quaywalls*, DM-25.4. Department of the Navy, Naval Facilities Engineering Command, Alexandria, Va.
55. Carchedi, D. R. and Porter, D. L. (Mar. 1983). "Marine Terminal Modifications-Port of Albany, New York," *Proc.* of the Spec. Conf. on Port Modernization, Upgrading and Repairs, ASCE, New Orleans, pp. 584-598.
56. Nolan, A. A. (May 1976). "Innovative Design at Port Everglades Bulkhead Reduces Material and Construction Cost," *Civil Engineering*, pp. 81-84.
57. ——— (1987). *Pile Buck Steel Sheet Piling Design Manual*. Pile Buck, Inc., Jupiter, Florida.
58. Terzaghi, K. (1953). "Anchored Bulkheads," *Trans. ASCE*, Paper No. 2720.
59. Casagrande, L. (1973). "Comments on Conventional Design of Retaining Structures," *Journal of the Soil Mechanics and Foundations Division, ASCE*, Vol. 99, No. SM-2, pp. 181-198.
60. Terzaghi, Karl (1943). *Theoretical Soil Mechanics*. Wiley, New York.
61. Brandl, E. D., Davis, J. S., and Houghton, R. C. (1983). "Tiebacks for Waterfront Bulk-

heads," *Proc. of the Speciality Conference on Port Modernization Upgrading and Repairs,* ASCE, 21-23 Mar. 1983, New Orleans, La.
62. Goldberg, D. T., Jaworski, W. E., and Gordon, M. D. (1976). "Lateral Support Systems and Underpinning," Vols. 1-3, Report No. RD-75/128-130. FHWA, Washington, D.C.
63. Hanna, Thomas (1982). *Foundations in Tension—Ground Anchors.* McGraw-Hill, New York.
64. Schnabel, H., Jr. (1982). *Tiebacks in Foundation Engineering and Construction.* McGraw-Hill, New York.
65. Weatherby, D. E. (1982). "Tiebacks," Report No. RD-82-046. FHWA, Washington, D.C.
66. Cummings, E. M. (1957). "Cellular Cofferdams and Docks," *Proc. ASCE,* Vol. 83, WW-3, pp. 13-45.
67. Fang, H. Y. and Dismuke, T. D., ed. (1970). *Proceedings of the Conference on Design and Installation of Pile Foundations and Cellular Structures.* Envo Publishing Co.
68. Khuayjarernpanishk, T. (1975). "Behavior of a Circular Cell Bulkhead during Construction," Ph.D. thesis submitted to Oregon State University at Corvallis.
69. Krynin, D. P. (1945). "Discussion of Stability and Stiffness of Cellular Cofferdams by K. Terzaghi," *Trans. ASCE,* Vol. 110 (2253), pp. 1175-1178.
70. Maitland, J. K. and Schroeder, W. L. (1979). "Model Study of Circular Sheetpile Cells," *Journal of the Geotechnical Engineering Division, ASCE,* Vol. 105, No. GT-7, pp. 805-821.
71. Schroeder, W. L., Marker, D. K., and Khuayjarernpanishk, T. (1977). "Performance of a Cellular Wharf," *Journal of the Geotechnical Engineering Division, ASCE,* Vol. 103, No. GT-3, pp. 153-168.
72. Schroeder, W. L. and Maitland, J. K. (1979). "Cellular Bulkheads and Cofferdams," *Journal of the Geotechnical Engineering Division, ASCE,* Vol. 105, No. GT-7, pp. 823-837.
73. Sorota, M. D. and Kinner, E. B. (1981). "Cellular Cofferdam for Trident Drydock: Design," *Journal of the Geotechnical Engineering Division, ASCE,* Vol. 107, No. GT-12, pp. 1643-1655.
74. Sorota, M. D. and Kinner, E. B. (1981). "Cellular Cofferdam for Trident Drydock: Performance," *Journal of the Geotechnical Engineering Division, ASCE,* Vol. 107, No. GT-12, pp. 1657-1676.
75. Swatek, E. P. (1967). "Cellular Cofferdam Design and Practice," *Journal of the Waterways and Harbors Division, ASCE,* Vol. 93, pp. 109-132.
76. Tennessee Valley Authority (Nov. 1966). *Steel Sheet Pile Cellular Cofferdams on Rock,* Technical Monograph No. 75, Vol. 1, TVA, Knoxville, Tenn.
77. Teraghi, K. (1945). "Stability and Stiffness of Cellular Cofferdams," *Trans. ASCE,* Vol. 110 (2253), pp. 1083-1119.
78. White, A., Cheney, J. A., and Duke, M. (1963). "Field Study of a Cellular Bulkhead," *Trans. ASCE,* Vol. 128, pp. 463-508.
79. U.S. Navy (1965). *Marine Biology Operational Handbook,* NAVDOCKS, MO-311, Washington, D.C.
80. DefQuinn, A. (1972). *Design and Construction of Ports and Marine Structures,* 2nd ed. McGraw-Hill, New York.
81. Chellis, R. D. (1961). *Pile Foundations,* 2nd ed. McGraw-Hill, New York.
82. Flemming, W. G., Weltman, A. S., Randolph, M. F., and Elson, W. K. (1985). *Pile Engineering.* Wiley, New York.
83. Buslov, V. B., Gould, J. P., and Koch, R. C. (1988). "The Precast Pier," *Civil Engineering,* Vol. 58, No. 12, pp. 46-49.
84. Bruce, R. N. and Hebert, D. C. (1974). "Splicing of Precast, Prestressed Concrete Piles: Part 1—Review and Performance of Splices," *PCI Journal,* Vol. 19, No. 5, pp. 70-97.
85. NAVFAC (1982). *Foundations and Earth Structures,* DM-7.2. Dept. of the Navy, Naval Facilities Engineering Command, Alexandria, Va.

86. Poulos, H. G. and Davis, E. E. (1980). *Pile Foundation Analysis and Design*. Wiley, New York.
87. Matlock, H. and Reese, L. C. (1961). "Foundation Analysis of Offshore Pile Supported Structures," *Proc.* Fifth Int. Conf. on Soil Mechanics and Foundation Engineering, Vol. 2, Paris.
88. Focht, J. A., Jr. and Kraft, L. M. (1986). "Axial Performance and Capacity of Piles," in *Planning and Design of Fixed Offshore Platforms*, B. McClelland and M. D. Reifel, eds., pp. 763-799. Van Nostrand Reinhold, New York.
89. Kraft, L. M., Ray, R. P., and Kagawa, T. (1981). "Theoretical t–z Curves," *Journal of the Geotechnical Engineering Division, ASCE*, Vol. 107, No. GT-11, pp. 1543-1561.
90. Broms, B. B. (1965). "Design of Laterally Loaded Piles," *Journal of the Soil Mechanics and Foundation Division, ASCE*, Vol. 91, No. SM3, pp. 79-99.
91. Broms, B. B. (1964). "Lateral Resistance of Piles in Cohesive Soils," *Journal of the Soil Mechanics and Foundations Division, ASCE*, Vol. 90, No. SM2, pp. 27-63.
92. Broms, B. B. (1964). "Lateral Resistance of Piles in Cohesionless Soils," *Journal of the Soil Mechanics and Foundation Division, ASCE*, Vol. 90, No. SM3, pp. 123-156.
93. Poulos, H. G. (1971). "Behavior of Laterally Loaded Piles: I-Single Piles," *Journal of the Soil Mechanics Division, ASCE*, Vol. 97, No. SM5, pp. 711-731.
94. Cox, W. R. and McCann, J. W. (1986). "Analysis of Laterally Loaded Piles," in *Planning and Design of Fixed Offshore Platforms*, B. McClelland and M. D. Reifel, eds., pp. 801-832. Van Nostrand Reinhold, New York.
95. Balign, M. M. and Vivatrat, V. (Nov. 1978). "Downdrag on Bitumen-Coated Piles," *Journal of the Geotechnical Engineering Division, ASCE*, No. GT-11, pp. 1355.
96. Xanthakos, Petros (1979). *Slurry Walls*. McGraw-Hill, New York.
97. Bember, S. M. and Kupferman, M. (1975). "The Vertical Holding Capacity of Marine Anchor Flukes Subjected to Static and Cyclic Loading," *Proc.* 7th Annual Offshore Technology Conf., Houston, Tex., Paper No. OTC 2185, pp. 363-369.
98. "Soil Improvement—History, Capabilities, and Outlook." (1978). Report by the Committee on Placement and Improvement of Soils of the Geotechnical Engineering Division of the American Society of Civil Engineers, ASCE, New York.
99. Welsh, J. P., ed. (1987). *Soil Improvement—A Ten Year Update*, Proceedings of a symposium sponsored by the Committee on Placement and Improvement of Soils of the Geotechnical Engineering Division, Special Publication No. 12. ASCE, New York.
100. Welsh, J. P. (1986). "Construction Considerations for Ground Modification Projects," *Proc.* of the Int. Conf. on Deep Foundations, 1-5 Sept., Beijing, China.
101. Mitchell, J. K. (1970). "In-Place Treatment of Foundation Soils," *Journal of Soil Mechanics and Foundations Division, ASCE*, Vol. 96, No. SM1, pp. 73-110.
102. Brown, R. E. (1977). "Vibroflotation Compaction of Cohesionless Soils," *Journal of the Geotechnical Engineering Division, ASCE*, Vol. 103, No. GT-12, pp. 1437-1451.
103. Harder, L. F. Jr., Hammond, W. D., and Ross, P. S. (1984). "Vibroflotation Compaction at Thermalito Afterbay," *Journal of Geotechnical Engineering, ASCE*, Vol. 110, No. 1, pp. 57-70.
104. Hughes, J. M. O. and Withers, N. J. (1974). "Reinforcing of Soft Cohesive Soils with Stone Columns," *Ground Engineering*, Vol. 7, No. 3, pp. 42-44, 47-49.
105. Hughes, J. M. O., Withers, N. J., and Greenwood, D. A. (1975). "A Field Trial of the Reinforcing Effect of a Stone Column in Soil," *Geotechnique*, Vol. 25, No. 1, pp. 31-44.
106. McKenna, J. M., Eyre, W. A., and Wolstenholm, D. R. (1975). "Performance of an Embankment Supported by Stone Columns in Soft Ground," *Geotechnique*, Vol. 25, No. 1, pp. 51-60.
107. Anderson, R. D. (1974). "New Method for Deep Sand Vibratory Compaction," *Journal of the Construction Division, ASCE*, Vol. 100, No. CO-1, pp. 79-95.

108. James, H. W. (1973). "Densification of Sand for Drydock by Terra-Probe," *Journal of Soil Mechanics and Foundation Division, ASCE*, Vol. 99, No. SM6, pp. 451–470.
109. Brown, R. E. and Glenn, A. J. (1976). "Vibrofloatation and Terra-Probe Comparision," *Journal of the Geotechnical Engineering Division, ASCE*, Vol. 102, No. GT-10, pp. 1059–1072.
110. Welsh, J. P. (1986). "In Situ Testing for Ground Modification Techniques," *Proc.* of In Situ '86, Geotechnical Division, ASCE, 23–25 June, Blacksburg, Va.
111. Lukas, R. G. (1980). "Densification of Loose Deposits by Pounding," *Journal of the Geotechnical Engineering Division, ASCE*, Vol. 106, No. GT-4, pp. 435–446.
112. Leonards, G. A., Cutter, W. A., and Holtz, R. D. (1980). "Dynamic Compaction of Granular Soils," *Journal of the Geotechnical Engineering Division, ASCE*, Vol. 106, No. GT-1, pp. 35–44.
113. Lukas, R. (1986). "Dynamic Compaction for Highway Construction Guidelines," Report No. RD-86-133, FHWA.
114. Stamatopoulos, A. C. and Kotzias, P. C. (1983). "Settlement-Time Predictions in Preloading," *Journal of Geotechnical Engineering, ASCE*, Vol. 109, No. 6, pp. 807–820.
115. Fellenius, D. H. (Oct. 1981). "Discussion on: Consolidation of Clay by Band Shaped Prefabricated Drains," *Ground Engineering*.
116. Hannon, J. B., and Welsh, T. J. (Jan. 1982). "Wick Drains, Membrane Reinforcement, and Lightweight Fill for Embankment Construction at Dumbarton," Presented to 1982 Transportation Research Board Annual Conf., Washington, D.C.
117. Hansbo, S. (July 1979). "Consolidation of Clay by Band-Shaped Prefabricated Drains," *Ground Engineering*.
118. Morrison, A. (Mar. 1981). "The Booming Business in Wick Drains," *Civil Engineering*, pp. 47–51.
119. Seim, C., Walsh, T. J., and Hannon, J. B. (July 1989). "Wicks, Fabrics and Sawdust Overcome Thick Mud," *Civil Engineering*, pp. 53–56.
120. Childs, K., Porter, D. L. and Holm, T. A. (Apr. 1983). "Lighweight Fill Helps Albany Port Expand," *Civil Engineering*, pp. 54–57.
121. Palermo, R. J. (1985). "Taking a Load Off an Aging Pier," *Engineering News-Record*, McGraw-Hill, New York.
122. Munfakh, G. A. (Jan. 1985). "Wharf Stands on Stone Columns," *Civil Engineering*, pp. 44–47.
123. Product literature, Reinforced Earth Company, Arlington, Va.

9

Design of Floating Structures

Floating structures are prominent features in many ports, serving as floating docks for small craft berthing and sometimes as piers or wharves for larger oceangoing vessels. There are also floating dry docks, breakwaters, mooring and navigation buoys, camel/separators, containment booms, and plant and equipment of a wide variety. This chapter is primarily concerned with basic design principles that apply in general to all floating structure types, with particular regard to floating pier applications. Basic principles of buoyancy, stability, motion response, and certain aspects of structural design are reviewed. Mooring and anchoring systems are of major importance in floating pier design; so their basic design principles are presented. Means of access to floating piers and ancillary systems such as ballast control, pumping, and flooding, and miscellaneous design features are reviewed. The final section of this chapter is devoted to floating docks for marinas and small craft facilities.

9.1 STRUCTURE TYPES AND APPLICATIONS

This section provides an overview of basic configurations and applications of floating structures commonly employed in the berthing and mooring of vessels. Floating dry docks and floating caisson gates that form the closure of basin-type dry docks, which also are common port floating structures, are discussed in Chapter 10.

Floating Piers and Platforms

Floating piers are usually of one of three basic hull configurations, as illustrated in Figure 9.1: the *single pontoon* (rectangular prism) or barge-type hull; the *catamaran* or multipontoon configuration with the deck spanning transversely between pontoon units; and the *semisubmersible* type, frequently employed in mobile offshore drilling units, with a deck superstructure supported on any number of vertical, usually cylindrical, buoyancy columns, which are themselves often connected below the waterline to continuous submerged pontoon-like buoyancy units.

The submersible hull-type configuration has as its principal advantage that it is largely transparent to wave motion because of its minimal waterplane area. The natural period of roll, for example, of a submersible unit usually is or can

DESIGN OF FLOATING STRUCTURES 359

Figure 9.1. Floating pier basic hull types.

be made to be longer than that of typical ocean waves. In addition to its relatively small change in displacement during the passage of a wave, the platform remains relatively motionless compared to a barge-type hull of the same displacement. Mooring loads also are lower for this configuration compared to hulls of equal displacement. Its disadvantages include a relatively low deck load capacity, sensitivity to moving weights on deck, and generally higher construction and operating costs. This structural configuration, however, would be ad-

vantageous at locations exposed to ocean waves where deck loadings are light, such as for an offshore tanker berth. The catamaran or multihull configuration may offer some cost savings or construction advantages over the single pontoon where an equal deck area is important but load capacity is not.

As it may not be feasible to construct a floating pier in one single length, individual pontoon units may be interconnected end-to-end by rigid hinges or semiflexible connections, or they may be joined together on site in the water by post-tensioning or welding to form a single unit that cannot be constructed or towed from its construction site in one piece. Alternative pontoon arrangements include individual pontoons spanned by a continuous deck system like a floating bridge, or a hinged deck system spanning between pontoon units, the latter applicable only at sites with little or no wave action.

Floating piers may offer several advantages over conventional fixed pier construction, depending upon local site conditions. They are most advantageous for these site conditions: very deep water, large tidal or river stage variations, difficult foundation and/or seismic conditions, remote locations lacking construction materials availability or equipment access, river or delta locations subject to migrating shorelines whereby the pier may need to be relocated. Other conditions favoring floating piers include temporary installations and those where speed of construction or rapid deployment is of prime importance, such as for military applications. Their principal disadvantages include higher maintenance and operating costs and susceptibility to wave action, resulting in down time due to unacceptable motions or to damage from storm waves.

In many instances, floating piers may comprise an entire marine terminal or port. Recent examples include the Falkland Islands "Flexiport," shown under construction in Figure 9.2, which consists of seven barges forming a 1000-foot-long floating quay with transit sheds that can accommodate Ro/Ro and general cargo. Shore connection is via a movable causeway on grounded pontoons (1). The Port of Valdez, Alaska, has a container terminal consisting of a 700- by 100-foot prestressed concrete pontoon with a 40-ton container crane and two shore access bridges. Extreme waves for design approach 8 feet in height, and the average tide range is 22 feet (2, 3). The Port of Valdez oil dock berth No. 1 loading dock consists of a semisubmersible-type tubular steel space frame supporting a 390- by 70-foot deck buoyed by 22-foot-diameter vertical cylinders (4). The Port of Iquitos, Peru, was upgraded in 1977 with a 614- by 50.5-foot wharf consisting of five rectangular steel pontoons added to an existing 285- by 30.5-foot pontoon that had operated with some intermittent upgrading for the previous 45 years. The wharf services general cargo vessels of up to 6000 GRT, and the river level varies by 35 feet (5).

Floating structures have been utilized as offshore terminals for the storage and transfer of hazardous cargoes such as the LPG storage facility in the Java Sea (6) (see Figure 1.6) and a proposed LNG terminal (7). The U.S. Navy has

DESIGN OF FLOATING STRUCTURES 361

Figure 9.2. Falkland Islands "Flexiport" under construction. Access bridge is supported on removable submerged pontoons. (Photo courtesy of MacGregor-Navire.)

carried out detailed studies for a two-level-deck floating pier to service naval vessels (8). In some instances, great savings can be realized by recycling existing barges for use as piers. An automobile unloading facility was built in the Port of Wilmington, Delaware, using existing 250- by 35-foot car floats to form a 515-foot-long wharf with two floating access ramp barges (9). Additional floating port case studies were given by Tsinker (4), who also provides an overall in-depth treatment of floating pier design and construction.

Link Spans/Transfer Bridges

Transfer bridges or link spans are required for the transfer of vehicular ferry and roll on/roll off (Ro/Ro) traffic from ship to shore. Although some larger oceangoing vessels are equipped with their own ramps that span directly from the ship to the quay or wharf deck, most ports that handle Ro/Ro cargo, and in particular those with large water-level fluctuations, have link span arrangements, which often are supported on floating structures. Floating link spans may be arranged for end-on, side-on, or quarter-point loading operations. Floating link spans can be readily moved and adjusted to suit tidal variations and vessels' threshold heights. Pontoons usually are fitted with pumping and flooding systems for ballast control, and some installations may have adjusting towers and

mechanical compensating devices on their mooring systems. Many link spans are moored directly alongside a pier or quay, as shown in Figures 9.3 and 9.4. Link span pontoons must be of ample size to minimize changes in draft and trim during cargo transfer. Normally pontoons should not change trim more than approximately 1° during routine cargo transfer operations.

Camels

Camels are floating separators placed between a vessel and quay or between two vessels in order to maintain a safe standoff distance. A camel may consist

Figure 9.3. Dockside movable link span. Pontoon-mounted transfer bridge for Ro/Ro Cargo transfer at Melbourne, Australia. (Photo courtesy of MacGregor-Navire.)

DESIGN OF FLOATING STRUCTURES 363

Figure 9.4. Submersible-type movable link span alongside quay at Douglas, Isle of Mann. (Photo courtesy of MacGregor-Navire.)

simply of a timber log or group of logs lashed together or of heavy timbers with struts and braces secured together in crib-work fashion. Floating fender units (see Section 5.5) also may serve as camels. Larger camels may be constructed of steel or timber frameworks with buoyancy tanks for flotation, or of steel pontoons. Sometimes small barges are adapted for use as camels. The width of a camel is determined by the ship's roll characteristics, freeboard, and presence of overhangs, such as the flight decks of aircraft carriers. The minimum length of a camel usually is on the order of 30 feet or as required to maintain acceptable contact pressures with the vessel's hull or to distribute the load against a sufficient number of pier fender piles. Consideration should be given to the stability of the camel against lifting and overturning. Larger camels have open deck surfaces or walkways, for which a minimum deck load of 20 psf or more should be considered. Where steel pontoons or buoyancy tanks are used, they should have internal watertight compartments or be foam-filled. The U.S. Navy, NAVFAC (10), has established some design criteria for camels and for fixed hanging-type separators.

Containment Booms

Containment booms are a common feature in most ports, especially at oil tanker berths. Booms must be readily deployable, and leave-in-place booms must be easy to move and resecure during vessel movements. Most booms today have synthetic fabric skirts coated with a hydrocarbon-resistant polymer, usually a polyurethane, with encapsulated buoyancy elements near their tops and continuous wire or other tension-member material incorporated in them. A typical in-

harbor boom has a total depth of about three feet with approximately one foot or less of freeboard. Booms often are put into hoop tension between mooring points, which may be increased by currents, wind, and tidal variations. Therefore, a tension capacity on the order of 10,000 to 15,000 pounds often is required (11). End and mooring connections must be strong but easy to secure in the water. Booms must be sufficiently ballasted and rigid enough to maintain their shape and stability under wind and current loads. An overview of containment boom design and use is given in reference (12).

Single Point Moorings

At offshore deepwater locations, tankers may be moored to single point mooring buoys (SPMs), and their cargo transferred between ship and shore through hoses and underwater pipelines or by lightering. There are several variations of SPMs, such as a single anchor leg mooring (SALM), consisting of a single bottom-anchored vertical chain, or a rigid strut riser attached to a large floating buoy or yoke arrangement, or a catenary anchor leg mooring (CALM), consisting of a buoy anchored by several mooring chains to independent bottom anchors. Figure 9.5 shows a VLCC transferring its cargo at an SPM, SALM terminal. This installation is described in some detail by Gruy et al. (13). Discussion of the design of these specialized facilities is beyond the scope of this book. A more detailed overview and design criteria can be found in references (14) through (17). Vessel moorings in general are covered in Section 9.7.

Floating Breakwaters

Floating breakwaters (FBWs) have been used with varying degrees of success, primarily in reducing the heights of short-period waves, of generally less than four seconds, at small craft facilities in exposed locations. The current state of the art of FBW design has been discussed by this author (18, 19), and many important papers on the subject are contained in the FBW conference proceedings (20, 21). Although FBW design is beyond the scope of this text, a few important principles that apply to floating structures in general will be reviewed here.

A rigid body moored normal to a wave train will reflect some amount of wave energy back seaward and dissipate a much smaller amount through friction and turbulence. The result is a smaller transmitted wave height on the leeward side of the structure although the transmitted wave has the same period as the incident wave. The incident wave height (H_i), the reflected wave height (H_r), and the transmitted wave height (H_t) are related by the following equation:

$$H_i^2 = H_r^2 + H_t^2 \qquad (9\text{-}1)$$

DESIGN OF FLOATING STRUCTURES 365

Figure 9.5. VLCC discharging crude oil at one of three SALM buoys at the Louisiana Offshore Oil Port (LOOP). (Photo courtesy of SOFEC, Inc.)

The ratio of the transmitted height to the incident height is called the transmission coefficient:

$$C_T = H_t/H_i \qquad (9\text{-}2)$$

Figure 9.6 shows the approximate variation of C_T with the ratio of the structure's width (B) to incident wavelength (L) for a rectangular object moored in regular waves. Note that the actual value of C_T also will vary with the relative water depth (d/D), the wave steepness (H/L), the mooring system stiffness, and other factors. It is evident that when B/L is relatively large, say on the order of 0.5 or greater, the FBW is an effective wave attenuator; and as B/L is reduced, more wave energy passes by the FBW. At a B/L of around 0.2 or less, the FBW essentially follows the wave contour with little or no wave attenuation. Reflected waves result in high mooring forces and cause standing wave patterns of increased wave height in front of the structure.

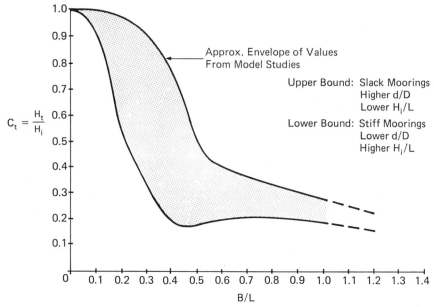

Figure 9.6. Representative values of wave transmission coefficient (C_T) versus beam to wavelength ratio (B/L) for a floating rectangular prism.

The mooring stiffness has an important effect on both wave attenuation and mooring loads (22). In general the stiffer the mooring and the more wall-sided and deeper-draft the structure, the greater the mooring loads and wave attenuation will be.

9.2 HYDROSTATICS AND STABILITY

A floating object displaces a volume of water equal in weight to its own weight (the Archimedes principle). For example, the displacement for a rectangular object is given by the product of its length (L_s), beam (B), and draft (D) and the unit weight of the water, γ ($= 64$ pcf for seawater). For an irregularly shaped object such as a ship, the displacement can be found by multiplying $L_s \times B \times D \times C_B$, where C_B is a block coefficient, used to account for the percentage of the volume of the ship's prism. Representative values of C_B and other form coefficients defining a vessel's shape were introduced in Section 2.2.

A floating object is in equilibrium when no external forces act on it, and its centers of gravity (c.g.) and buoyancy (c.b.) lie in the same vertical plane. If the body's equilibrium is disturbed by the application of a heeling moment (M_h), such as that induced by a moving weight on deck, a wind force, and so on, then the body will heel, or list, until a new equilibrium position is found,

whereby a righting moment (M_r) formed by the object's new c.b. acting with its c.g. balances the heeling moment. The righting moment is given by:

$$M_r = \Delta GM \sin \psi \tag{9-3}$$

where Δ is the object's displacement, GM is the metacentric height, and ψ is the angle of heel. Referring to Figure 9.7a, GM is the vertical distance between

a. No Internal Liquids

b. with Internal Water

Figure 9.7. Hydrostatic stability definition sketch: (a) pontoon without internal water; (b) pontoon with internal water.

the c.g. and the so-called metacenter (M), which is the point of intersection of the line of action of the object's buoyancy and heeled centerline for small ψ. A righting lever arm, $GZ = GM \sin \psi$, forms a couple between the object's c.g. and c.b. GM can be found from:

$$GM = BM + KB - KG \qquad (9\text{-}4)$$

where BM is the vertical distance between c.b. and M, and is equal to the moment of inertia of the waterplane area (Iwp) divided by the displaced volume (∇); KB is the distance from the object's bottom (or keel); and KG is the distance from the keel to the c.g. Note that in general for waterplane areas of arbitrary shape, the value of Iwp must be found by integration.

The reader is referred to the basic naval architecture texts cited in Chapter 2 for a more comprehensive theoretical development of floating body stability. For the stability of linked and hinge connected pontoons and catamaran hulls, reference (4) is very useful. It is important to note that if the pontoon has internal liquids (ballast water, leakage, and so on), then when it is heeled, the internal water will slosh (the "free surface effect") and come to a new equilibrium position with the c.g. of the internal water now shifted so as to increase the effective heeling moment (see Figure 9.7b). The effect of this internal free surface, which can be considered as a negative waterplane area, is to reduce the positive waterplane area. In this case, the negative waterplane area (I neg.) is taken about the neutral axis of the internal waterplane area.

Although there are no definitive guidelines for minimum acceptable values of GM for floating piers, a minimum value on the order of two to three feet or perhaps 10% of the structure's beam under the worst anticipated load conditions seems a reasonable assumption. High values of GM are desirable for initial stability, but there may be some instances where low values are sought in order to increase the period of roll, which is inversely proportional to the square root of GM.

The moment required to list the pontoon a given amount, say one inch, can be calculated from the GM as:

$$\text{MLI} = \Delta GM_T / 12B \qquad (9\text{-}5)$$

Where GM_T is the transverse metacentric height, and MLI is known as the moment to list one inch. Similar calculations can be carried out about the pontoon's centerline axis or midship section to obtain the longitudinal GM_L and the moment to trim one inch:

$$\text{MTI} = \Delta GM_L / 12L_s \qquad (9\text{-}6)$$

DESIGN OF FLOATING STRUCTURES 369

The side-to-side or end-to-end trim given by equations (9-5) and (9-6) gives the total amount of trim or list, which is equal at each end for a rectangular shape and must be proportioned by the distance to the center of flotation for an irregular waterplane area.

It often is convenient to know how much the float will sink under a given load. The load required to sink the pontoon one inch is termed the pounds per inch immersion, PPI, as given by:

$$\text{PPI} = WPA \times \gamma/12 \qquad (9\text{-}7)$$

where WPA is the water plane area and γ is the unit weight. The change in draft of the pontoon due to an added weight anywhere on deck can be found by superimposing the change in draft due to direct sinkage (PPI), heeling moment (MLI), and trimming moment (MTI). Maximum allowable angles of trim and list will vary with applications but in general will not likely exceed 1° or 2°, with perhaps up to 5° to 6° (1:10 deck slope) allowed under extreme conditions. Minimum freeboard requirements may vary widely with vessels to be serviced, but a reasonable amount should be maintained for reserve buoyancy and to prevent wave overtopping.

The overall hydrostatic and stability properties versus draft of any floating structure can be plotted in graphical form, as illustrated in Figure 10.43 for a floating dry dock. Floating piers and similar structures must be subdivided internally by watertight bulkheads to provide adequate stability with internal water and under damage conditions. Damage stability criteria should specify minimum required freeboards and trim/list with one or more compartments flooded. Floating pier intact freeboard requirements will vary but generally will range from four to six feet or more for the berthing of larger coastal and oceangoing vessels.

9.3 MOTION RESPONSE

The general problem of analysis of a moored floating body, which is most complex, was introduced in Chapter 6. This section considers basic concepts that are of practical application in the design of floating piers. Considering the six degrees of freedom of movement of an unrestrained or partially restrained floating body, described in Section 6.1 and illustrated in Figure 6.2, three motions—heave, roll, and pitch—have gravity as a restoring force and thus exhibit natural periods of free oscillation. Of these three modes, heave and roll (i.e., vertical rise and fall and side-to-side rotation) are of particular significance in floating pier design. Pitching motion (end-for-end rotation) usually is not important because of the relatively long length of a floating pier structure compared to the length of the incident waves. Exceptions would be locations exposed to long-

period swell propagating along the longitudinal axis of the pier, or piers such as landing floats with low aspect ratios of length to beam (L_s/B). In the latter case, a clear distinction between pitch and roll cannot be made. The remaining three modes of motion—surge, sway, and yaw—may be significant in piers moored with slack chains or cables, but they are seldom of critical importance in rigidly moored floating pier design.

The heave response normally is measured in terms of the ratio of heave amplitude to incident wave amplitude, the roll response as the structure slope to wave slope ratio and sway motion as the ratio of sway amplitude to wave amplitude. The trajectory of the center of mass of the structure under wave action varies with the wave length and height in relation to the water depth. In deep water and with steep waves, the motion ellipse is nearly circular, with heave and sway amplitudes being nearly equal. In shallow water and for low-amplitude waves, the wave particle orbits become elliptical, and heave motions are reduced, but sway amplitudes remain relatively large. The relative motion response of a floating structure also depends importantly upon the relative water depth to draft ratio (d/D), the structure beam to draft ratio (B/D) as it affects the virtual mass of the structure, and the beam to wave length ratio (B/L), as well as the wave direction and the degree of mooring restraint.

Acceptable limits to motion usually are dictated by the human response although certain cargo handling equipment and operations may have more severe limits. Rosen and Kit (23) arrived at the following limits in their study of acceptable motions within small craft harbors:

- Maximum linear accelerations: 1.3 fps^2
- Maximum angular accelerations: 2°/s^2
- Maximum peak roll to roll motion: 6°

Chapter 6 presented some general information on the acceptable motions of berthed ships versus type of operations.

Because lateral sway motions usually are sufficiently restrained by the pier's moorings, the remaining discussion will concern heave and roll motions, both of which can be greatly amplified when the natural period is near that of incident waves of sufficient height and length to excite resonant motions. Fortunately both heave and roll can be controlled to some degree by varying certain design parameters. Both heave and roll are affected by the amount of damping present, but only heaving is significantly affected by the added mass effect (introduced in Section 5.3).

Rolling may be affected by mooring restraint, which increases damping and thus reduces maximum amplitudes but does not significantly alter the natural period. For the general case of unrestrained rolling, roll motion characteristics can be solved from the following equation:

DESIGN OF FLOATING STRUCTURES 371

$$I_M \frac{\partial^2 \psi}{\partial t^2} + C_d \frac{\partial \psi}{\partial t} + M_r = 0 \quad (9\text{-}8)$$

where I_M is the structure's mass moment of inertia including the virtual mass of water about its longitudinal axis, ψ is the roll angle, C_d is the damping coefficient, and M_r is the righting moment = $\Delta GM \sin \psi$ (equation 9-3). Note that in general the added mass and damping can be neglected for unresisted rolling in calm water, especially at small values of ψ. This is not true in general for pitching motions, where the added mass in particular may have a significant effect in increasing the natural pitch period. The general solution to equation (9-8), neglecting the damping term, is:

$$\psi = \psi_0 \cos \omega_n t + \frac{V_0}{\omega} \sin \omega_n t \quad (9\text{-}9)$$

which gives the value of ψ as a function of time (t), with ψ_0 and V_0 the initial angle and angular velocity. The quantity ω_n is the natural circular frequency, and ω is the wave circular frequency ($2\pi/T$), where it can be shown that:

$$\omega_n^2 = \frac{\Delta GM_T}{I_M} \quad (9\text{-}10)$$

Here ΔGM_T is the spring constant and I_M is equal to $\Delta K_r^2/g$, where K_r is the radius of gyration and ω_n equals $2\pi/T_n$ by definition. By substituting into equation (9-10) and rearranging to solve for T_r, the natural period of roll is found to be:

$$T_r = 2\pi \sqrt{\frac{K_r^2}{gGM_T}} \quad (9\text{-}11)$$

which in feet-second units can be further reduced to:

$$T_r = 1.108 \frac{K_r}{\sqrt{GM_T}} \quad (9\text{-}12)$$

For ships K_r is usually in the range of .20B to .28B, and for a rectangular pontoon K_r is .29B. Therefore, the natural period in roll for a rectangular pontoon can be found from:

$$T_r = \frac{.32B}{\sqrt{GM_T}} \quad (9\text{-}13)$$

The angular acceleration in roll (α_r) affects personnel and equipment operations and should be kept as low as possible. For structures designed to be towed to remote sites, which may be subjected to much more severe wave conditions than would be expected during operations, the acceleration in roll may govern the design of appurtenances. It can be shown that:

$$\alpha_r = \frac{4\pi^2}{T_n^2} \psi \qquad (9\text{-}14)$$

where ψ is the peak roll angle in radians. The tangential force (F_t) acting on an object of weight W at a distance r from the roll axis is:

$$F_t = \frac{W}{g} \frac{4\pi^2}{T_n^2} r\psi \qquad (9\text{-}15)$$

Heave motions can be expressed in the form of a second-order differential equation similar to equation (9-8). Damping can be safely neglected for spud guide-type and most catenary chain-type moorings, but added mass effects cannot. The natural heave frequency (ω_h) can be solved in similar fashion to equation (9-10), where the spring constant is equal to the product of the waterplane area and the unit weight of water, and the mass is taken as the structure's displacement multiplied by an added mass coefficient (C_m) (see Section 5.3). Hence, the natural period in heave (T_n) for a rectangular pontoon is found from:

$$\begin{aligned} T_h &= 2\pi \sqrt{\frac{\gamma C_m(L_s BD)}{\gamma g(L_s B)}} \\ &= 1.108 \sqrt{C_m D} \end{aligned} \qquad (9\text{-}16)$$

The value of C_m may vary greatly with pontoon properties, (B/D) and relative water depth (d/D). Figure 9.8 shows T_h versus draft (D) for a range of B/D ratios for conditions with $d/D \geq 5$. Theoretical values of C_m were calculated on the basis of information provided in Blevins (24). The effect of increasingly shallow water is to increase C_m and hence also T_h.

As measured in terms of the heave amplitude to wave amplitude ratio, $a_h/a_w = 2a_h/H_i$, heave motion is highly dependent upon the structure beam to wavelength ratio, B/L. There is, then, a frequency-dependent response that can be subjected to a spectral analysis, as introduced in Section 6.6, to determine the structure's response in irregular seas. Figure 9.9 illustrates the heave response spectrum for a 30-foot-wide pontoon ballasted to a 7.5-foot draft and subjected to an incident wave spectrum with a spectral peak period, T_p, of 3.9 seconds

DESIGN OF FLOATING STRUCTURES 373

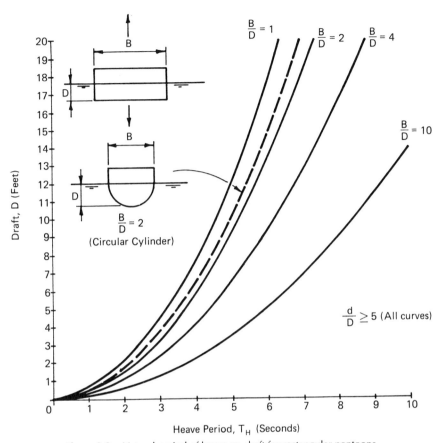

Figure 9.8. Natural period of heave vs. draft for rectangular pontoons.

and a significant wave height, H_s, of 3.0 feet. The response amplitude operator (RAO) or transfer function curve is based upon theory for a slack moored rectangular object in a beam sea condition with a value of 4.0 for B/D (22). The peak in the RAO near T_h is not so pronounced in actual model tests for in full-scale observations. In general, the heave response is negligible for $L < .75B$ and near unity for $L > 4B$. Note that this is a one-dimensional spectrum and does not account for the short-crestedness of real waves. Therefore, if the structure length, L_s, is much longer than the incident wavelength, the overall heave response will be greatly diminished. Applications of directional spectra accounting for wave direction and short-crestedness are beyond the scope of this text. The reader is referred to the references cited in Chapter 6 for further information.

Although waves are the usual source of excitation, it is important to consider

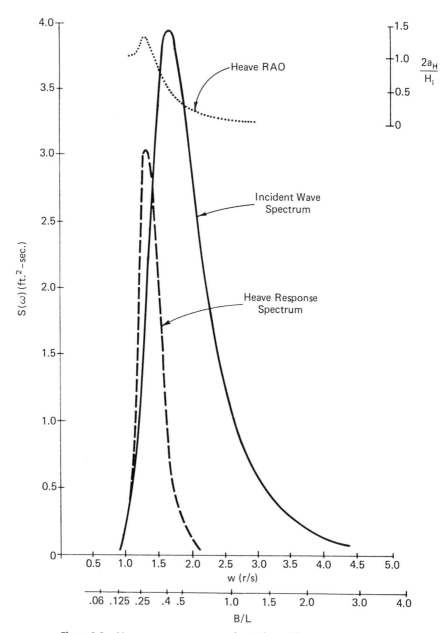

Figure 9.9. Heave response spectrum for 30-foot-wide pontoon—example.

the speed of traffic or traveling equipment that could initiate resonant motions. This is especially true with regard to the heave response of relatively small pontoons supporting link spans transfer bridges or floating bridge spans.

The maximum vertical acceleration (α_h) in heave is given by:

$$\alpha_h = \frac{4\pi^2}{T_h^2} \alpha_h \qquad (9\text{-}17)$$

Vertical accelerations are seldom a problem structurally but may be critical in determining separation forces and relative motions between access bridges and mooring compensation devices. Prolonged exposure to accelerations as low as approximately $0.05\ g$ ($1.6\ \text{fps}^2$) may affect personnel performance, especially if coupled with roll or pitch motions and depending upon the period of oscillation.

9.4 STRUCTURAL DESIGN

Floating structures such as piers and landing floats are designed to accommodate the same kinds of deck live loads, vehicular, and equipment loads, and vessel berthing and mooring loads as their fixed pier counterparts, with these main differences: reaction forces due to deck loads always will be resisted by a uniformly distributed buoyancy force over the structure's bottom, the structure is subject to floating body motions and hence dynamic effects not experienced by fixed structures, and horizontal forces are resisted at discrete mooring points and must be distributed through the hull structure. It is interesting to note that although floating structures themselves are isolated from seismic forces, very large forces may be developed at their mooring points. This is especially true for massive structures moored to rigid dolphins or piers in areas where strong ground motions are possible.

In general, floating structures must consider the following load categories:

- Deck loads that induce stillwater shear forces and bending moments and torsion effects.
- Wave-induced shear, bending, and torsion.
- Local hydrostatic pressures including dynamic effects of wave action and sloshing in tanks.
- Mooring loads due to vessel berthing and mooring and direct environmental loads against the structure.
- Thermal loads.
- Launching, dry docking, and towing loads.

Consideration also must be given to fatigue, impact, corrosion, and mechanical deterioration.

Contemporary floating piers are usually of steel or reinforced or prestressed concrete construction. Smaller floats (see Section 9.8) also may be constructed of timber, aluminum alloys, and various plastics such as fiberglass (GRP). There

are as yet no definitive codes or standards that govern the design of floating pier structures. However, some guidance is given in the rules of the vessel classification societies such as the American Bureau of Shipping (ABS) (25, 26) or their foreign country counterparts for steel structures, and in references (27) through (29) for concrete structures. For semisubmersible construction, reference (30) may be helpful. General guidance on design principles and structural design of steel floating structures may be obtained from basic naval architectural texts such as those cited in Chapter 2, of which reference (31) is especially helpful. For concrete floating structures, much useful information can be found in the proceedings of the few specialty conferences on the subject, such as references (32) and (33). Basic concrete and steel building codes also may be used for structural design, provided that appropriate load factors and allowable stresses are applied, and that due consideration is given to minimum thickness of members, continuity of welds, and other factors relating to possible fatigue and corrosion problems.

The overall hull strength of a floating pier is governed by its gross section modulus as determined from maximum stillwater bending moments due to deck and equipment loads and/or wave loading. Wave loadings govern the design of ships and are also likely to govern the design of floating structures subject to ocean towing or at sites where wave heights may exceed four to six feet, depending upon the ratio of structure length to wavelength (L_s/L). After being initially proportioned for the maximum longitudinal bending moment, the hull girder is checked for torsion and local combined load effects and strengthened as required. Because the framing of a rectangular pontoon typically is uniform throughout, the unloaded structure is uniformly supported over its length so that there is no stillwater bending moment associated with its dead load. Deck superstructures or moving equipment loads usually can be resolved into point loads in calculating the overall shear, bending, and torsion moments, the reaction to which is a uniform buoyancy force distributed along the pontoon bottom. Wave-induced bending moments are found by integrating the shear force (V_x) along the length of the pontoon:

$$M_x = \int_0^x V_x \, dx \tag{9-18}$$

where the shear force is obtained by integrating the net buoyancy force ($b - w$) along the structure length, with b the buoyancy force and w the weight per unit length of structure. Note that local concentrated loads can be deducted from the elemental buoyancy force to consider both stillwater and wave effects simultaneously. The shear force, therefore, is found from:

$$V_x = \int_0^{L_s} (b_x - w_x) \, dx \tag{9-19}$$

DESIGN OF FLOATING STRUCTURES

Assuming a sinusoidal wave form, b_x is given by the surface elevation (z) times γB, with z varying along the wavelength as:

$$z = \frac{H}{2} \cos \frac{2\pi x}{L}$$

The integration is carried out for the case of $L_s/L = 1.0$ with the wave crest at the structure midships (see Figure 9.10). This is called the hogging condition, as opposed to the sagging condition where the wave crests are at the structure ends. Sagging moments are found to be of equal magnitude but opposite sign

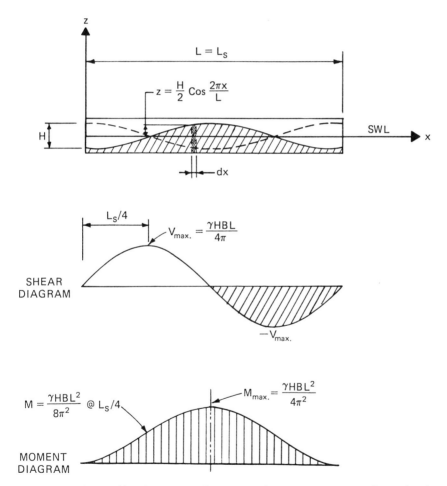

Figure 9.10. Shear and bending moment for a rectangular structure in waves with wavelength equal to structure length.

378 DESIGN OF MARINE FACILITIES

of hogging moments. The shear force is then:

$$V = \gamma B \frac{H}{2} \int \cos \frac{2\pi x}{L} dx$$

$$= \gamma B \frac{H}{2} \times \frac{L}{2\pi} \sin \frac{2\pi x}{L} + C_1$$

where the constant of integration $C_1 = 0$, given that $V = 0$, at $x = L_s/2$. The maximum shear force at the quarter points, $L_s/4$, is, then:

$$V_0 = \frac{\gamma HBL}{4\pi} \tag{9-20}$$

The bending moment is found from:

$$M = V_0 \int \sin \frac{2\pi x}{L} dx$$

$$= -\frac{V_0 L}{2\pi} \cos \frac{2\pi x}{L} + C_2$$

For the boundary conditions, $M = 0$ @ $x = 0$, $C_2 = V_0 L/2\pi$; and:

$$M_0 = \frac{V_0 L}{2\pi} \left(1 - \cos \frac{2\pi x}{L}\right) = \frac{V_0 L}{\pi} \text{ @ } x = L/2$$

$$M_0 = \frac{\gamma HBL^2}{4\pi^2} \tag{9-21}$$

The quantities M_0 and V_0 are known as the characteristic shear and moment, respectively, for $L_s/L = 1.0$. Worked example problems are provided by Tsinker (4). If similar calculations are carried out over a range of L_s/L, the maximum moment is found to occur at midships for $L_s/L = 1.12$ and is equal to $1.028 M_0$. The midship bending moment is zero for $L_s/L = 2.0$ and is generally small for $L_s/L > 2.0$ and for $L_s \ll L$. The maximum shear and moment in foot-pound units then are:

$$V\text{max.} = 5.24 \, HBL_s \tag{9-22}$$

$$M\text{max.} = 1.67 \, HBL_s^2 \tag{9-23}$$

Torsional stresses may be manifested as: wracking of sections in a primarily vertical or horizontal mode, warping out of planeness, and overall twisting ac-

tion. Maximum torsional moments usually are calculated on the basis of the structure's being supported along its opposing quarter points by wave crests. The angle of incidence of the waves and their length required to produce the maximum torque will vary with structure length to beam ratio (L_s/B). An excellent discussion has been provided by Muller in reference (32). According to Muller, the maximum torque moment in foot-pound units for a rectangular shape will be:

$$M_T = 2.53\ HB^2 L_s \qquad (9\text{-}24)$$

The wave height should be reduced by the factor $(L/L_s)^{1/2}$ to account for the reduced wave height of the shorter waves at an angle. The maximum torque moment usually will be found to be on the order of 20% of the longitudinal bending moment. Torsional stresses that are additive to direct and shear stresses associated with longitudinal bending are seldom a problem for the closed box sections of steel hulls. For concrete, especially nonprestressed concrete, combined shear stresses may be a problem. The following combinations of shear stresses should be checked (32):

1. Vmax. $+ 1/4\ M_T$max.
2. M_Tmax. $+ 1/4\ V$max.
3. $3/4\ (V$max. $+ M_T$max.$)$

Hull deflections (δ) due to bending are found by double integration of the moment curvature equation, and may be expressed semi-empirically by an equation of the form:

$$\delta = C_\delta \frac{ML_s^2}{EI} \qquad (9\text{-}25)$$

where E and I are the modulus of elasticity and moment of inertia of the hull girder, respectively. The coefficient C_δ is often taken as .09 for ship hull shapes (31). Pontoons usually have a depth (D_s) greater than approximately $L_s/20$ in order to practically meet strength and deflection criteria. For vessels, L_s/D_s must not exceed 15 (reference 25). For inland barges and floating piers, L_s/D_s generally should be less than 18 to 20.

Floating piers may be framed similarly to vessels either transversely or longitudinally (see Section 2.3) although transverse framing of open truss or rigid frames is most common (see Figure 9.11). Concrete hulls usually are modular rigid frames or vierendeel trusses with vertical sides and bulkheads haunched where they meet the deck and bottom. In either case, the deck and bottom form the girder flanges, and the sides and longitudinal bulkheads the webs in resisting overall shear and bending. The frames must resist overall transverse bending

380 DESIGN OF MARINE FACILITIES

Figure 9.11. Typical pontoon framing schemes.

moments, which are usually relatively small except in cases where very heavy concentrated loads act along the float centerline, as in the case of floating dry docks (see Section 10.4). The frames also act to distribute vertical buoyancy forces and horizontal hydrostatic pressures.

Internal watertight bulkheads also must resist hydrostatic pressures due to the differential head conditions between compartments. In addition to head pressures associated with the pier's maximum operating draft, a nominal equivalent head pressure acting above the deck level may be specified to account for the effects of overwashing waves or the impact contingency on the bottom and side shells. Japanese harbor standards (34) require that floating pier bottom and sides be designed for a 0.5 meter (1.6 ft) head above the freeboard deck level. Allowable stresses may be increased on the order of 75% to 90% of yield stress

DESIGN OF FLOATING STRUCTURES 381

in steel structures under extreme load pressures. Bottom and side shells are always in a biaxial or triaxial state of stress and must be adequately proportioned to meet primary, secondary, and tertiary stresses associated with the overall hull bending, the local secondary bending between bulkheads or frame elements, and the local tertiary bending between shell plating stiffening elements in steel structures (see Figure 9.12). Floating piers installed at freezing locations may need to be strengthened for ice forces. A general discussion of ice forces is provided in Section 4.5. For floating structures "beset" in solid ice, Noble (35) and Liljestrom (36) provide useful information.

Deck design usually will be governed by local wheel load or crane float re-

Figure 9.12. Schematic of global and local hull stresses.

quirements. Although floating piers usually are not used for the storage of cargoes, locally high uniformly distributed live loads may have to be considered for the temporary placement of heavy palletized or containerized loads. Orthotropic bridge deck type construction with either open or closed rib stiffening elements may be used in steel hull construction. Highway bridge design codes (37) are useful in determining wheel load distribution and deck framing requirements. Steel decks often are covered with cement concrete, bituminous concrete, or timber decking, the purpose of which is to provide a durable, skid-resistant, and easily maintainable wearing surface. Deck coverings should be kept as lightweight as possible. Tsinker (4) provides a more comprehensive discussion of deck coverings and details.

Allowable stresses and load combination factors for steel structures can be determined with guidance from references (25) and (26). References (27) and (28) give guidance for limit state design of concrete floating structures. Lightweight concrete with fresh unit weights of 120 psf to 125 psf can be reliably produced with compressive strengths in excess of 6500 psi (27). As with fixed concrete structures, concrete mixes should be sulfate-resistant and proportioned for low permeability. The amount of reinforcing steel cover is critical to the typically thin sections of concrete floating structures, and can be as low as $\frac{3}{8}$ to $\frac{1}{2}$ inch (27). The necessary cover depends upon crack control requirements (maximum crack widths of .01 inch should be considered for watertight elements), permeability, and the likelihood of degradation during normal service. Structures exposed to significant wave action should be designed to fatigue stress criteria. For a 20-year design life, it usually is considered that the structure will be subjected to 10^8 wave cycles, corresponding to an average wave period of 6.3 seconds. Other temporary design conditions include launching and towing stresses, for which higher allowable stresses normally are justified. Prior to the commissioning of a new structure, internal tanks or compartments should be tested for watertight integrity by hose test or by air pressurization.

9.5 BALLAST CONTROL AND ANCILLARY FEATURES

Many types of floating structures of interest herein may be equipped with pumping and flooding systems for ballast control, including tank interior ventilation and sounding and personnel access features. Floating structures also may be outfitted with mooring hardware, fender systems, crane trackage, handrails and curbs, cargo handling and ship services, and lighting and fire protection systems similar to those of fixed piers (see Sections 7.6 through 7.8). The design of pumping and flooding systems is beyond the scope of this text, but a few basic principles are presented below.

Pumping Systems

Floating dry docks, link spans, and some piers that are fitted with ballast systems generally require pumps capable of removing large volumes of water as rapidly as possible under relatively low but variable head conditions. Dewatering pumps most ideally suited to this purpose are usually of the axial mixed-flow or propeller type, operating at high specific speeds where the specific speed is defined as the speed in revolutions per minute (rpm) at which a given impeller would operate if reduced proportionately in size to deliver a capacity of one gallon per minute (gpm) against a total dynamic head of one foot. The total dynamic head (H_D) used for pump selection is the sum of the hydrostatic head (h) and the hydraulic system losses. The net positive suction head (NPSH) is defined as the effective pressure head that causes liquid to flow through the suction piping (if any) and enter the eye of the impeller. The NPSH required by a given pump is a function of the pump design and is determined by the pump manufacturer. The available NPSH must equal or exceed the required NPSH; and when the liquid source is above the pump impeller, it is equal to the sum of the equivalent head in feet of the barometric pressure plus the static head of liquid above the suction minus any friction losses in the piping minus the liquid vapor pressure.

There is a minimum height of water above the suction bell, referred to as the minimum submergence, below which the pump will begin to develop vortices and cease to operate efficiently. The depth of water remaining in the pontoon after the pump's "break suction" may create a problem for efficient operation, and it precludes inspection of the bottom. Therefore, pumps may be installed in depressed sump pits, or smaller stripper pumps may be installed as a remedy. In an efficiently designed sump, the height of the section bell above the bottom plate usually will be about one-half the suction bell diameter, and the distance from the side plate about one-fourth the section bell diameter, depending upon the proximity of the tank or sump sides (38). Flow velocities in the pump channel and at the suction bell must be properly controlled. Baffle plates and "umbrellas" fitted at the suction bell may help reduce vortexing and reduce the depth of residual water. Pump discharges normally are fitted with one-way flap valves to prevent water from entering due to wave wash or during deep submergence.

Pumps usually are driven by electric motors. The horsepower required is given by:

$$BHP = \frac{GPM \times H_D \times s.g.}{3960 \times \mathit{Eff.}} \qquad (9\text{-}26)$$

where BHP is the brake horsepower of the motor, H_D is the head in feet, GPM is gallons per minute, $s.g.$ is the specific gravity of the liquid, and $\mathit{Eff.}$ denotes the efficiency. Pump manufacturers publish performance curves of head requirements and efficiency versus capacity to aid in the selection of pumps operating over a range of head conditions. The final selection of the optimum pumping system often represents a compromise among initial cost, the cost and availability of electrical power, and the desired pumping time. Pumping systems for basin dry docks and for floating dry docks have their own particular requirements, as discussed in Chapter 10.

Flooding Systems

Flooding systems, like dewatering systems, are designed to meet some maximum allowable flooding time criteria. Consideration also must be given to varying head conditions, permissible flow velocities, and the size and the number of flood valves as they affect cost and operation.

The flow rate through a flood valve opening is given by:

$$Q = C_0 A_0 \sqrt{2gh} \tag{9-27}$$

where Q is the flooding rate, usually in cubic feet per second (cfs); C_0 is the discharge coefficient, which depends upon the flow velocity and contraction at the orifice and is usually taken to be on the order of 0.6 for circular orifices; A_0 is the area of the orifice; and h is the head, in feet. Maximum flow velocities generally should be kept below 25 feet per second (fps). The time rate of the flow under a varying head condition is found from:

$$Q/A_0 = \frac{dh}{dt}$$

Rearranging and substituting:

$$dt = \frac{A_{wp}}{C_0 A_0 \sqrt{2gh}} \, dh$$

where A_{wp} is the waterplane area of the tank, which is assumed to be constant; then:

$$\Delta t = -\int_{h_1}^{h_2} \frac{A_{wp}}{C_0 A_0 \sqrt{2gh}}$$

DESIGN OF FLOATING STRUCTURES 385

For a definite time interval:

$$t_2 - t_1 = \frac{2A_{wp}}{C_0 A_0 \sqrt{2g}} \left(\sqrt{h_1} - \sqrt{h_2} \right)$$

Therefore the total time, T, for an original difference in water level, H, for a rectangular tank is given by:

$$T = \frac{2A_{wp} \sqrt{H}}{C_0 A_0 \sqrt{2g}} \qquad (9\text{-}28)$$

On large-scale expensive projects where pumping and flooding times are critical, model tests may be carried out to determine actual coefficients and system hydraulic characteristics. Tank levels may be monitored remotely by use of liquid level gages, usually pneumatic manometer or electrical resistance-type devices. Where it is not critical to know the precise level of interior water at all times, sounding holes or wells may be installed for periodic checking. Flood valve operators may be compressed air, hydraulic, or manually operated. All systems should have a manual override capability in order to close valves in an emergency. The use of rising stem valve shafts allows operators to see at a glance and from a distance if valves are in an open or a closed position. Flood inlets should be fitted with screens to keep out trash and debris.

Tank Ventilation

All interior spaces must be adequately ventilated to remove noxious vapors and reduce condensation. Natural ventilation relying upon air density and temperature differences and wind usually is adequate for most uninhabited spaces. Ventilator openings are sized according to the tank volume and the number of air exchanges per hour required using semi-empirical relations. In some cases, the vent pipes may be used as a safety valve to limit the depth of submergence of the structure. When the bottom of the vent pipe is effectively closed off by the rising interior water level, a volume of air equal to the length of pipe below the deck times the tank plan area is trapped and compressed by the differential head pressure of water inside and out. The final depth of the air cushion inside can be calculated for the desired minimum freeboard by using the ideal gas laws, assuming an adiabatic process, which states that the product of the pressure times the volume remains constant, or $P_1 V_1 = P_2 V_2$.

Tank Access

Tank access for inspection and maintenance usually is provided through watertight gasketed manholes with welded steel vertical ladders. For very deep tanks, intermediate landings may be required. Manholes should be at least 18 to 21 inches in diameter. All sealed tanks and enclosed spaces must be tested for harmful vapors and adequate oxygen content prior to personnel entry, as the corrosion process can consume all available oxygen; so manhole covers should have appropriate safety markings.

9.6 ACCESS BRIDGES AND GANGWAYS

Access to shore for vehicles and equipment most commonly is provided by single span articulated bridges or ramps of similar construction to highway bridges. If the floating structure is far from shore, a series of bridge spans supported on intermediate pontoons forming a floating bridge such as shown in Figure 9.2 may be used. Other forms of floating pier access include sliding wedge structures such as those used along steep river banks with large water-level variations and vertical lift-type bridges. Smaller lighter spans limited to personnel access only usually are called gangways. This section reviews some basic requirements of typical single span access bridges and gangways. The reader is referred to Tsinker (4) for further description of floating bridges and alternative access arrangements.

Access Bridges

Design of access bridge structures follows the same principles as basic highway bridge design (37), where the governing loads include the same vehicles and equipment in traveling condition as used in the pier deck design. Other important design considerations include: lightweight, free-draining decks with good traction; simplicity and ease of maintenance; and special attention to end connections and bearings. For long spans where the bridge dead load reactions are high, a separate pontoon may be employed for the sole purpose of supporting the bridge so as not to affect the trim and capacity of the main pier. Alternatively, fixed adjusting towers with counterweight arrangements may be employed to keep bridge reactions to the minimum required for positive grounding on the pier deck. Fixed adjusting towers further allow the bridge to be lifted clear of the floating structure when not in use, which may be beneficial for pontoon maintenance, under severe weather conditions, or for security reasons.

Access bridges may have hinge pin connections at both shore and pier ends, or at either end with the other end having a sliding bearing. Generally, a short transition plate or apron plate will be required at both ends to span the gap created as the bridge pivots. Hinged bearings should be slotted to allow for

some lateral motion even though the bridge may be stayed to fixed points with relatively taut cables. Slotted, energy-absorbing, sliding bearings may be used for bridges subjected to longitudinal movements due to vessel impact. At exposed locations where the floating structure is subject to notable heaving and pitching motions, the vertical separation force at the pontoon end must be calculated based upon the dynamic characteristics of the bridge, pontoon, and counterweight system if applicable. If it is found that the pontoon could accelerate downward faster than the bridge could follow because of its large angular inertia, a vertical restraint must be provided or system characteristics changed. For the general design of vertical lift access bridges, reference (39) is most useful.

General design criteria for Ro/Ro ship-to-shore connection of interest in access bridge design can be found in references (40) and (41). Figure 9.13 illus-

NOTE: Dimensions given in meters (M) are per ISO Ro/Ro standards. Actual angles & dimensions may vary with specific facility type requirements.

Figure 9.13. Pontoon access bridge design criteria.

trates some general design criteria. Minimum roadway widths of 5 meters (16.4 ft) and 9 meters (29.5 ft) are recommended for one- and two-lane traffic, respectively (40). Minimum vertical clearance under worst normal tide conditions should be 7 meters (23 ft). The maximum bridge slope should not be steeper than 1:10 (5.7°) up or down with the pontoon at its maximum normal freeboard at the highest normal water level (e.g., MHW or MHHW) or with the pontoon at its minimum normal draft at the lowest normal water level (e.g., MLW or MLLW). The change in gradient between deck surfaces should be limited to allow small wheeled cargo handling equipment to cross the transition point without difficulty. In general, this requires a limiting angle of approximately 6.5°. Further recommendations can be found in reference (40).

Gangways

Gangways, sometimes referred to as brows, especially when used for direct ship to shore personnel access, may be constructed of timber, steel, or aluminum alloy. Marine grade aluminum alloys are by far the most common construction materials, and prefabricated aluminum truss-type gangways are readily available in lengths up to 60 feet or more. Beam type construction is usually suitable up to approximately 25-foot lengths. Walkway widths are usually from two to six feet, although for general purposes a minimum width of $3\frac{1}{2}$ feet is recommended. Maximum gangway slopes of 1:3 with the floating structure at its normal (MLW or MLLW) position and 1:2.5 at extreme low water usually are recommended in small craft harbors (see general references for Section 9.8). Gangways typically are fitted with handrails a minimum of 3.5 feet high designed for a 20 plf lateral force applied along the top rail. Shore end connections should allow some freedom of movement in both directions. Bottom ends usually are fitted with wheels or rollers that ride on the pontoon deck. Gangways wider than six feet should be designed for a uniform load capacity at 100 psf, and those less than six feet wide for a minimum of 40 psf live load. NAVFAC (10) stipulates that all shipboarding brows, three-foot minimum to five-foot brows as required, be designed for a uniform live load of 75 psf and be tested for a 150 psf live load. In addition, the brow is to be designed for an impact load, with one end being dropped five feet onto an unyielding surface. The maximum vertical deflection under drop of the brow is not to exceed $\frac{1}{240}$ of the span. Gangway surfaces should be skid-resistant and preferably fitted with transverse cleats at one-foot centers, approximately 1 inch wide by $\frac{1}{4}$ to $\frac{1}{2}$ inch high.

9.7 MOORING SYSTEM DESIGN

A mooring system generally consists of the connection at the floating structure, a tethering cable, a strut or linking mechanism, the anchoring device or struc-

Figure 9.14. Schematic of mooring connections to fixed structures.

ture, and the soil mass that holds it. Moorings can be broadly categorized as fixed, whereby the structure is secured directly to a fixed pier, dolphin, or cantilever pile, or free, with the structure tethered via cables or chains to anchors in the bottom soil. The design of fixed structures was covered in Chapter 7, and some typical means of connection to fixed structures are illustrated in Figure 9.14.

Fixed mooring connections must allow free vertical movement of the floating structure while keeping it securely in position. The connection also must permit some degree of list and trim of the structure. The simplest type of connection is the hoop and guide pile arrangement common in marinas. Spud and gripper arrangements commonly are used to moor floating dry docks and pontoon structures, and the gripper connection usually is articulated to allow for trimming

and listing of the dock. Floating dry docks alternatively may be moored by a combination of spud guides and shear spuds, providing only lateral and longitudinal restraint, respectively.

Free moorings may be further classified as free swinging, where a vessel is moored at a single point, usually to a buoy which supports the mooring chain or chains, or spread moored, where a vessel or floating structure is secured from movement in both principal directions. Figure 9.15 illustrates some free mooring arrangements. Almost all harbors have anchorage areas where vessels can anchor awaiting their turn at berth. At some locations substantial single buoy moorings are provided for this purpose. The use of SPMs at offshore deep draft tanker terminals was introduced in Section 9.1. Conventional tankers or barges may be moored offshore or "in-stream" in rivers in a relatively fixed position to a series of mooring buoys, and their cargoes pumped or lightered ashore. This is one of the simplest forms of marine terminal. Floating piers, which invariably are shore-connected, can be primarily spread moored but often have cables, chains, or struts connecting them to anchors along shore at their shoreward end. Mooring chains usually are cross-connected, leading to opposite sides of the pontoon in order to allow deeper draft alongside. Tsinker (4) and Tsinker and Vernigora (42) provide further discussion of anchor cable arrangements.

The remainder of this section is devoted to some basic principles common to the design of most free mooring systems. Particular attention is given to the catenary chain mooring, with which all harbor engineers should be familiar. Wire rope and fiber lines also may be used as components of floating structure moorings. Their general properties were discussed in Section 6.2.

The Catenary Equations

A suspended chain or cable will assume a characteristic curve shape under its own weight, known as a catenary. The catenary action of a mooring chain is beneficial in reducing the moored object's motions and in acting as a shock absorber in reducing peak mooring loads. Because a chain or cable only can act in tension, it produces a horizontal force, equal at any point along its length, that is a function of the chain's unit weight and its extension length. Figure 9.16 illustrates basic catenary geometry where the following equations apply:

$$F_V = w_s l_s \qquad (9\text{-}29)$$

$$F_H = w_s c \qquad (9\text{-}30)$$

$$T = w_s y \qquad (9\text{-}31)$$

and F_V, F_H, and T are the vertical and horizontal force components and tension respectively, w_s is the submerged unit weight of the cable, and l_s is the total

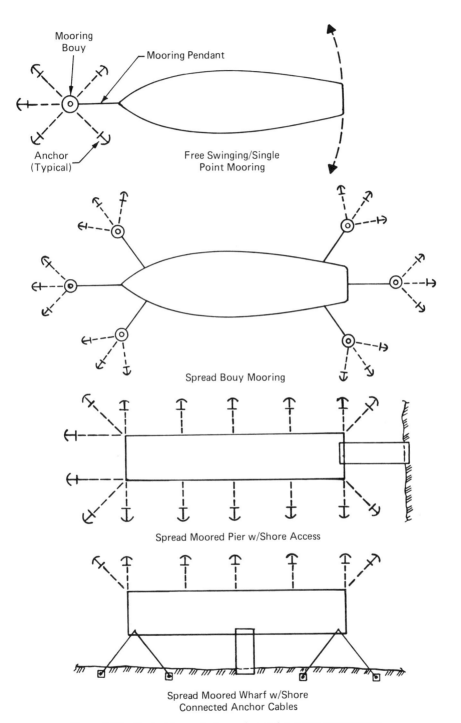

Figure 9.15. Typical free swinging and spread mooring arrangements.

392 DESIGN OF MARINE FACILITIES

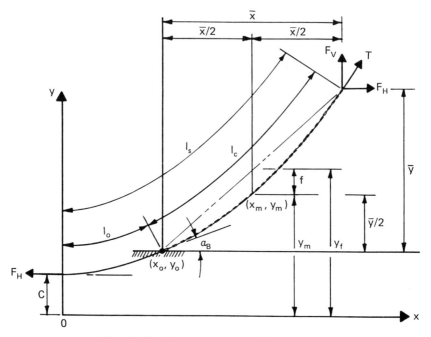

Figure 9.16. Catenary geometry definition sketch.

curve length from the mooring connection to its hypothetical point of horizontal tangency and intersection with the y-axis. Note that the curve shown actually is only one-half of a true catenary, which would continue symmetrically to a suspension point at $(-x, y)$. The distances c and y are constants of the curve. Note that if x were reduced such that T became tangent at y, then c would equal 0, and $F_H = T$ would be tangent at the origin. We are usually interested, however, in only a portion of the catenary acted upon by a horizontal force (F_H), where it assumes some angle (α) with the bottom at the anchor point. For the case of 0 slope at the anchor, the cable length, l_c, equals l_s. In general, it is desirable to keep α minimal; in particular, when drag type anchors are used, α should be 3° or less under design load conditions.

The following equations are used in catenary computations:

$$y^2 = l_s^2 + c^2 \tag{9-32}$$

$$y/c = \cosh\left(\frac{x}{c}\right) \tag{9-33}$$

$$l_s/c = \sinh\left(\frac{x}{c}\right) \tag{9-34}$$

$$\sqrt{l_c^2 - y^2} = 2c \sinh\left(\frac{\bar{x}}{2c}\right) \qquad (9\text{-}35)$$

$$\frac{\bar{y}}{l_c} = \tanh\left(\frac{x_m}{c}\right) \qquad (9\text{-}36)$$

$$x = x_m + \frac{\bar{x}}{2} \qquad (9\text{-}37)$$

Worked examples of the use of the catenary equations and a computer program for their solution are provided by NAVFAC (43).

It often is desirable to know the cable sag, as for checking vessel keel clearance. The sag (f) from a taut line measured vertically at $\bar{x}/2$ is given by:

$$f = \frac{w_s(\bar{x})^2}{8F_H} \qquad (9\text{-}38)$$

Weights called sinkers often are attached to mooring chains near their midlengths in order to reduce the angle at the anchor and sometimes also to increase under-keel clearances for berthed vessels. In this instance, the final chain geometry is calculated by assuming the intersection of two catenaries at the sinker location where the submerged weight of the sinker is taken as an equivalent length of chain. A series of clump weights may be used in lieu of a single larger sinker. Sometimes wire rope and/or fiber lines may be used in combination with chain. Such systems are referred to as multicomponent mooring systems. A direct solution of the behavior of such systems is not possible; iterative techniques with assumed equilibrium positions must be followed. Reference (44) gives an overiview of multicomponent mooring systems design. In very deep water and where strong currents exist, the hydrodynamic resistance of the mooring cable becomes important. The book by Berteaux (45) and tables by Pode (46) are very useful in such applications.

General Design Criteria

Mooring system design criteria and loads on floating port structures are determined in a manner similar to that described for berthing and mooring loads in Chapters 5 and 6. Free moored structures, however, have greater freedom of motion than rigidly moored structures; so a wider range of dynamic response to environmental forces is possible with free moorings. Headland et al. (47) performed a dynamic analysis of a spread moored floating dry dock under various combinations of steady and time-varying wind and regular and irregular waves and concluded that the presence of an unsteady wind field gives consid-

erably greater peak loads than those given by a static wind analysis. Also, the relatively low natural frequencies of spread moored systems make them prone to high loading under low frequency excitation. Wind spectra have been developed for offshore structure design applications (48), and means of calculating the effects of low frequency vessel motions on offshore drilling and semisubmersible units have been given by the American Petroleum Institute (API) (49). Although not directly applicable to floating piers per se, references (43) and (49) offer some guidance for the design of catenary chain systems.

In general, for chain moorings, the load in the mostly highly loaded chain should not exceed 33% of the mean break load (MBL) of the chain under maximum design operating conditions or up to 50% to 60% of MBL under extreme environmental conditions. This is exclusive of any corrosion allowance, and further reduction should be taken in cases where the chain is subject to sharp changes in directions such as at fairleads and hawsepipe connections. In many instances, however, the chain may be sized for its weight to help limit excursions under load with tidal variations, and much greater factors of safety will be realized. Anchor efficiencies drop off greatly with increased chain angle, and limiting angles usually are within the range of 3° to 6° above the horizontal. Floating pier movements are a function of water depth, tide range, and mooring system stiffness, and allowable surge/sway movements usually are determined by access bridge requirements. The mooring chain scope also affects relative movements with tide changes and bottom chain angles. Although a greater scope is favorable, it must be traded off against cost and spatial restraints. Minimum scopes are usually on the order of 2.5:1 to 3:1 unless substantial uplift resistance can be provided at the anchor.

Although some floating piers may be considered to have shorter design lives than the nominal 25 years often assumed for the design of fixed structures, it is still recommended that extreme conditions associated with the 100-year event or a .999 probability level be designed for, with the exception of temporary or expendable structures intended to be removed from the site in less than one to two years, for which a 25- to 50-year recurrence interval may be appropriate.

Mooring System Components

Mooring system components include anchors, chains and/or cables, connecting hardware, sinkers, and structure end connections. The structure end connections should allow a convenient means of adjusting and resecuring the mooring chains, but at some installations where anchors can be accurately located, chains may be connected directly to fixed mooring eyes on the pontoon hull or deck. Chains may be led around submerged fairleads at the bottom of the pontoon and up to the pontoon deck, or directly to the deck over a fairlead and chafing plate, and may be connected to pad eyes directly or to a chain "wildcat" or

windlass or tackle arrangement for ready adjustment. Alternatively chains may be led through the pontoon hull and secured with chain claws that straddle the chain link and hawsepipe opening. Mooring hardware includes: shackles, connecting links, equalizer plates, end connectors, and so on, whose description is beyond the scope of this text. The reader should consult references (50) and (51) for a detailed description of mooring chain and wire rope hardware and its application.

Chain is either of open link or stud link construction, of mild carbon or high alloy steel, depending upon strength requirements. Open link chain normally is made by automatic electric welding processes, whereas stud link chain may be made by welding, forging, or casting. Open link chain is classified typically as proof coil, buoy, BBB, conveyor chain, and so forth, according to its link proportions. Open link chain commonly is used in smaller sizes, below one inch, but stud link chain is preferred for permanent moorings and is almost universally used in sizes of one inch and greater. Chain strength is based upon actual break tests of a sample, and individual lengths (usually 90 foot shots) are certified by proof tests of around two-thirds of breaking strength. Working load limits may be applied to chains and are usually on the order of one-half of proof load. Complete tables of chain strengths and weights are published by the various chain manufacturers.

The ultimate strength of a chain link is approximately 1.55 times the theoretical tensile strength of one of its side wires, and the elastic stretch of a ferrous stud link chain is approximately 2.65 times the elastic stretch of one of its side wires (51), resulting in an elongation of about 1% at proof load. According to API (49), the elasticity (T/δ_c), where T is the chain tension in pounds and δ_c is the stretch in feet, is given by:

$$T/\delta_c = 1.2 \times 10^7 \, D_c^2/l_c \qquad (9\text{-}39)$$

where D_c is the nominal diameter (wire size) in inches, and l_c is the chain length in feet. The submerged weight of chain is equal to 87% of its weight in air. The general engineering properties of wire rope and fiber lines were introduced in Section 6.2

Anchors

Anchors can be loosely classified according to their primary mode of developing lateral resistance within the bottom soil. Figure 9.17 illustrates representative anchor types.

Gravity type anchors such as concrete blocks depend upon their submerged weight and the effective coefficient of friction with the bottom soil, which for a concrete block usually is assumed to be between 0.5 and 1.0. Thus its effi-

Figure 9.17. Classification of basic anchor types—schematic.

ciency or ratio of horizontal holding power to its dry weight in air is on the order of 0.25 to 0.5. Such massive objects may gain some added resistance due to bottom suction in cohesive soils and passive soil resistance if well embedded, as by jetting into the bottom. Pekin et al. (52) have conducted experiments on gravity anchors in sand for large-scale applications. Anchor chains themselves will contribute to the holding power of a mooring; according to API (49), the holding power of a chain or wire rope resting on the bottom is equal to the length of cable on the bottom times its submerged weight per unit length times a friction factor of 1.0 for chain and 0.6 for wire rope. Gravity anchors also may be pinned to the bottom with short needle piles driven through the anchor block or via projections cast into their bottoms. Gravity anchors may be cast in shapes that help them to penetrate the bottom and gain additional passive resistance within the soil, such as the self-burying type described by Tsinker and Vernigora (42).

Drag type anchors must be dragged into position in order to set their flukes and gain holding power by means of activating shear stresses within the bottom soil. Drag type anchors can also be divided into traditional *ship type anchors*—such as the mushroom, kedge, and stockless, which hook into the bottom rather than penetrate it and thus still depend upon their weight for holding power—and contemporary *burial type drag anchors*. Holding power ratios range from around two to three for mushroom types up to five to seven for stockless types under the most favorable soil conditions. Contemporary ship anchors are mostly of the burial type, designed to penetrate the bottom and engage as large a portion of a soil mass as possible.

The capacity of burial type anchors depends upon the soil type, depth of penetration, fluke area, fluke to shank angle, and the anchor's overall shape and configuration as they affect its stability and ability to penetrate the soil. Holding power ratios of from 15 to 60 or more times the anchor weight may be attained. Reference (50) contains much physical and descriptive data, and reference (53) contains comparative data on anchor holding power and means for calculating soil resistance for contemporary burial type anchors. Carchedi et al. (54) have carried out experiments on a generic burial type drag anchor in sand with variable fluke angles, shank lengths and shapes, and surface roughness for different soil densities and bottom slopes. They found anchor holding power to be directly dependent upon the depth of burial, which is in turn dependent upon soil properties, bottom slope, and anchor configuration. Sandy soils generally are considered to be good holding ground because of their high friction and easy penetration, but anchors must be stabilized against rolling over as they approach their tripping moment, where they rise on the tip of the flukes during burial. The optimum fluke to shank angle in sand is considered to be 32° and for silts and mud 50°. Mud also may provide a good holding bottom for anchors with large fluke areas. Stiff penetrable clays provide good holding, whereas softer clays may be disturbed during penetration.

Embedment anchors are forcibly driven into the bottom either by underwater pile hammers or surface equipment with followers or by propellants such as explosives. Stake piles usually are driven with chains attached and provide excellent holding power where site conditions permit. In addition, they can be accurately located unlike drag anchors, which must be dragged an uncertain distance before they "fetch-up." Reese (55) presents a design methodology for anchor piles. Other embedment types include umbrella piles, which open their flukes upon reaching the required penetration depth, and the vibratory type (56), which vibrates itself into the bottom like a vibratory pile hammer. The suction pile type anchor (4) is driven into the bottom by pumping water from within its cylindrical body, which is open at the bottom and sealed at the top.

The final choice of the best anchoring scheme for a given project depends upon the nature and depth of the bottom material, holding power required and scope as it affects chain slope, need to accurately locate anchors or limitations on dragging anchors into position, need to recover anchors, and first cost and long-term maintenance. General discussion of anchors and anchoring techniques is provided by Hinz (57) and Puech (58).

Free Swinging Moorings

Free swinging or single point moorings allow a vessel to "weather vane," bringing itself into alignment with the wind and/or current, thus generally minimizing its total mooring load. Preliminary designs of free swinging moorings

traditionally have been prepared by using static analysis methods with final designs based on physical model tests. Recent progress in numerical analysis of single point moorings, however, has led to an increase in the use of such models for both preliminary and final designs. A brief description of static and dynamic analysis methods is provided in the following paragraphs.

Static analysis methods presume that a vessel secured by a free swinging mooring will assume an equilibrium position under wind and current loading. Though wind and current can be coincident, the largest mooring loads generally occur under oblique wind and current attack. In calculating pendant loads, account must be taken of the increased projected area with yaw angle and of the horizontal and vertical components of the mooring pendant. When wind and current act together, the vessel will lie at an odd equilibrium position that can be found only by trial-and-error calculations. It is noteworthy that the maximum mooring force for combined wind and current may not occur with the maximum wind or current speed (43).

Experience both in the field and in physical model tests has shown that vessels secured to single point moorings often do not assume an equilibrium position. Instead, vessels tend to oscillate around the equilibrium position even under steady wind and current attack. This phenomenon, commonly called fishtailing or kiting, can result in mooring loads considerably higher than those computed by static analysis methods. Numerical methods for the dynamic analysis of single point moorings are based on the time domain approach discussed in Chapter 6. Wichers (59) presents a good summary of time domain analysis for single point moorings, classifies single point mooring response as either dynamically unstable or dynamically stable, and presents a criterion for establishing stability conditions under wind and current attack.

Figure 9.18 summarizes typical results from a computer simulation of a naval vessel secured to a single point mooring under wind and current attack, presenting time histories of vessel yaw and mooring hawser loads as well as a plot of the probability of exceedence of hawser loads for the simulation. It should be noted that the yaw time history reflects a fishtailing period of roughly $4\frac{1}{2}$ minutes.

Particular caution must be exercised in relatively shallow water at sites exposed to ocean swell or long period waves where surge motions may be significant. Note too that design mooring conditions will vary with the intended use of the system, for example, occasional transient or harbor of refuge under storm conditions.

The U.S. Navy classifies standard moorings for given holding power and swing radius conditions (43) and (60). The watch circle described by the vessel's swing radius takes up a great deal of space, as described in reference (60) and in the following discussion on moorings for yachts and small craft.

A typical mooring arrangement usually consists of a single stake pile or grav-

Figure 9.18. Example computer output for 7000 DT vessel at single point mooring subject to combined wind and current. (Provided by John R. Headland, P.E., NAVFAC.)

400 DESIGN OF MARINE FACILITIES

Probability of Exceedance for Hawser Force

Figure 9.18. (Continued)

ity anchor or an arrangement of three to six drag-type anchors in a radial pattern connected to a heavy bottom chain and thence to a riser chain via a ground ring or spider plates and a mooring buoy, which supports the weight of the chain and transmits the load to the vessel's mooring lines. A representative free swinging riser-type mooring is illustrated in Figure 9.19.

Some buoys are fitted with their own, usually braided nylon, pendants supported by smaller pickup buoys to facilitate pickup by the vessel. Contemporary mooring buoys usually are constructed of a rigid closed cell foam core encapsulated in an elastomeric or fiber-reinforced shell with a central tension member or hawse tube arrangement for connecting the mooring chain. Commercially made mooring buoys are available with up to approximately 100 tons of pull-through load capacity, with net buoyancies of 25 tons. Mooring buoys alternatively may be constructed of steel plate with internal watertight compartments for damage control. The minimum freeboard under load should be on the order of 18 inches. Other buoy applications include: marker buoys (usually spherical), utility floats, chain or pendant buoys, and navigation buoys (50).

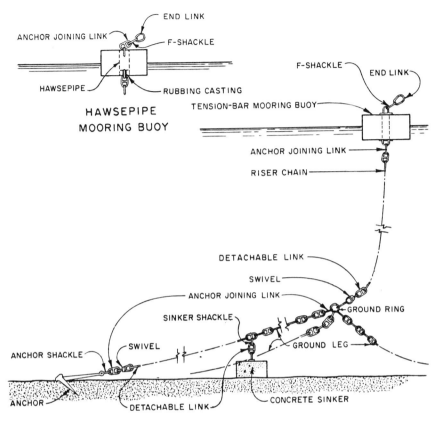

Figure 9.19. Representative free swinging riser-type mooring. (After NAVFAC (43).)

9.8 MARINAS AND SMALL CRAFT FACILITIES

Floating Dock Systems

Floating docks are a prominent feature of most marinas with tide ranges or seasonal water level changes of three to four feet or more, and are almost universally employed where the tide range exceeds five to six feet. Floating docks also are commonly utilized as landing floats and increasingly as floating breakwaters. Floating dock systems may be moored by cantilever piles, piers, or pile groups, struts to fixed points along the shore, or cables or chains to underwater anchors. A common design requirement for marina floats is that they maintain some minimum freeboard or change in draft under a uniformly distributed live

load, often given in the range of 20 to 30 psf. At rough water locations, this requirement should be given special consideration to assure that adequate reserve buoyancy is obtained.

Design live loads in the range of 25 to 40 psf should be considered at exposed sites, and live loads of 40 to as much as 100 psf for large public landing floats and commuter boat service and commercial small craft facilities. Corresponding minimum freeboards under full live loadings are usually within the range of 8 to 16 inches, and greater for live loads exceeding around 40 psf. Concentrated live loads on the order of 500 pounds acting anywhere also must be accommodated by the float's structure. For solid prismatic-type floats, the wave bending moment for design waves of one foot or greater height usually will be much larger than the stillwater bending moments associated with concentrated live load requirements. Torsional stresses in waves also may be critical, especially for landing floats with large beam-to-length ratios (B/L_s). Equivalent lateral loads for design due principally to wind on berthed vessels and a contingency for normal berthing and mooring loads usually range from 60 to 120 plf for most marina applications. At exposed locations and for commercial small craft, these loads may be much higher. Gaythwaite (61) has reviewed general design criteria for small craft facilities at exposed locations.

There is a large variety of commercially available floating docks on the market. Figure 9.20 illustrates some typical generic types and their usual types of construction and behavior under wave action. The layout and design of marinas and small craft facilities in general is covered in references (62) through (66).

Small Craft Moorings

Free swinging moorings are a common and expedient means of mooring yachts, especially at exposed locations and sites where the construction of marinas is not feasible. The general geometry of a typical yacht mooring is illustrated in Figure 9.21. Moorings usurp a relatively large water area, so a central focus of mooring design is to minimize the swinging radius or watch circle at low water while providing adequate scope (S_c) under high water conditions. The minimum value of S_c should be based upon an appropriate design water level (DWL), considering maximum seasonal water levels or storm conditions. It is desirable to maintain a condition of zero slope of the mooring chain at the anchor under such conditions. The watch circle radius is equal to the sum of: the water depth times ($S_c - 1$) plus the pendant length plus the vessel length plus any desired safety margin or contingency. The maximum value of S_c should be taken at MLW or MLLW. At many locations, where the water area is at a premium, less than the full swinging clearance may be provided. In such cases, it is essential to place vessels with similar hull shape characteristics together and to isolate dissimilar types (for example, deep draft sailboats and high superstruc-

DESIGN OF FLOATING STRUCTURES

FLOAT TYPE	TYPICAL CONSTRUCTION	BEHAVIOR IN WAVES
UNITIZED FLOATATION	Metal or wood deck frame w/ wood decking supported on individual bouyancy units, usually of closed cell foam encapsulated in rigid plastic shell.	Often springy underfoot & lively in waves. Rel. low mooring & conn. load Good survivability in waves if soundly built. Low reserve bouyancy.
CATAMARAN	Metal deck frame w/ metal or wood decking supported on cont. bouyancy tanks often of metal pipe sections w/ floatation fill.	Rel. solid underfoot. Mod. to low motion resp. in waves. Moderate mooring & conn. loads. Cross structure req's. careful design. Mod. to low reserve bouyancy, may lose righting moment rapidly w/ heel angle.
RAFT	Wood deck over bolted wood frame supported on heavy timbers laid parallel & w/ additional bouyancy provided by rigid foam billets battened between timbers.	Rel. massive, solid underfoot. Timber flexes allowing good survivability in waves if well constructed. Billets may wash loose under wave action. Moderate mooring & conn. loads & reserve bouyancy.
SOLID	Monolithic, box hull of steel or reinforced conc. construction w/ foam fill.	Solid underfoot, slow motion response, massive w/ rel. large mooring & conn. loads which may become problem under sustained wave action. Monolithic hull is inherently strong. Excellent reserve bouyancy.

Figure 9.20. Comparison of generic floating dock types. (After Gaythwaite (61).)

ture power boats respond very differently to variable wind and current combinations). Where this is not possible, fore and aft moorings or other restrictive mooring arrangements may be used (65).

404 DESIGN OF MARINE FACILITIES

Figure 9.21. Free swinging yacht mooring geometry. (After Gaythwaite (61).)

In laying out new mooring areas, the engineer can design for classes of moorings with a given holding power and watch circle radius based upon site-specific design wind speeds, water depths, and so on. Vessels then can be classed according to their holding power requirements and swinging characteristics, in terms of their displacement or other characteristics that correlate with wave surge loads and windage as well as the *LOA*. In areas affected by currents, an equilibrium position must be calculated by trial and error, based on relative current and wind directions. It likely will be found that vessels lie at odd angles, and that the resultant mooring force is not a simple addition of wind and current vectors.

Wind loads are calculated by using the drag force equation on the forward projected area yawed to 30° to get the longitudinal component of wind force, and then by applying a surge factor (C_{SF}) to account for wind gusts and swinging and for vessel surge, pitch, and heave motions in waves. The surge factor is normally within the range of 1.25 to 2.0 (57), depending upon the degree of exposure. Drag force coefficients typically range from 0.6 to 1.2. Caution should be exercised at certain shallow water locations where long period waves may be present, as the wave particle orbits become more elliptical and may result in large surge, fore, and aft motions. Van Dorn (67) provides useful information

for calculating wind loads on sailing yachts and also an excellent discussion of the dynamics of anchored yachts. The American Boat and Yacht Council (68) gives recommended permanent mooring capacities for pleasure boats ranging from approximately 1500 pounds for a boat of 25 feet *LOA* to 6000 pounds for a vessel of 60 feet *LOA*.

The use of chains is preferred for permanent moorings because of their weight and catenary action, strength, and abrasion resistance. In areas of large tide range, it is recommended that a short length of heavier ground chain be used, permanently secured to the anchor block and of sufficient length to be brought to the surface at low tide for periodic replacement of the smaller riser chain, which terminates at the mooring buoy. Mooring buoys can be designed to support the full weight of chain at high water, or at relatively shallow water sites the pendant simply can be secured to a pickup buoy, which need only support the submerged weight of the pendant. The pendant should be made of nylon line for its shock-absorbing characteristics and preferably be of a length equal to two to three times the height of the vessel's bow chocks above water. The pendant should be wrapped with chafe protection where it passes through the vessel's chocks. Twisted nylon generally possesses better chafe resistance than the braided types.

Galvanizing a chain, which typically reduces its strength about 10%, may add from six months to perhaps a year or more to its life, after which time the zinc coating goes into solution in seawater. The relative cost of galvanized chain should be compared with the cost of an extra corrosion allowance on ungalvanized chain.

REFERENCES

1. ——— (Dec. 1983). "Link Span Design for Floating Falklands Port," *The Dock and Harbour Authority*.
2. ——— (July/Aug. 1982). "Floating Container Terminal, Valdez, Alaska," *PCI Journal*.
3. Zinserling, M. H. and Cichanski, W. J. (May 1982). "Design and Functional Requirements for the Floating Container Terminal at Valdez, Alaska," *Proc. OTC*, Houston, Tex.
4. Tsinker, G. P. (1986). *Floating Ports: Design and Construction Practices*. Gulf Publishing, Houston, Tex.
5. Tanner, R. G. et al. (1983). "Modernization and Upgrading of the Dock and Associated Facilities at Iquitos, Peru," *Proc. ASCE*, Ports '83, Spec. Conf., New Orleans, La.
6. Anderson, A. R. (Jan. 1973). "A 65,000 Ton Prestressed Concrete Floating Facility for Offshore Storage of LPG," *Marine Technology*, Vol. 15, No. 1, SNAME, New York.
7. Anspach, G. et al. (May 1980). "Floating Terminal for Storage and Sendout of LNG," *Proc. ASCE*, Ports '80, Spec. Conf., Norfolk, Va.
8. Jahren, C. T. and Davis, D. A. (May 1986): "Construction Approaches to the Navy's Floating Pier Concept," *Proc. ASCE*, Ports '86, Spec. Conf., Oakland, Calif.
9. Felsburg, R. E. (Oct. 1977). "Port of Wilmington Automobile Unloading Facility," Port Engineers Notebook, *American Seaport*.
10. NAVFAC (1980). *Piers and Wharves*, DM-25.1. Dept. of the Navy, Alexandria, Va.

11. Hubbell, P. et al. (Sept. 1978). "Development of New U.S. Navy Standard Oil Containment Boom," Int. Oil Pollution Prevention Conf., Hamburg, W. Germany.
12. Smith, M. F. (1975). "Planning, Equipment and Training for Oil Pollution Control," Slickbar, Inc., Southport, Conn.
13. Gruy, R. H. et al. (1979). "The LOOP Deepwater Port: Design and Construction of the SALM Tanker Terminals," *Proc. OTC*, Paper No. 3562, Houston, Tex.
14. ABS (1975). *Rules for Building and Classing Single Point Moorings*. American Bureau of Shipping, Paramus, N.J.
15. Benham, F. A. et al. (1977). "Guidelines for Deepwater Port Single Point Mooring Design," U.S. Coast Guard, Report No. CG-D-49-77.
16. USACE (Dec. 1976). "Report on Design of Single Bouy Mooring Terminals," Dept. of the Army, Washington, D.C.
17. Pedersen, K. I. (Jan. 1975). "Mono-Mooring Terminals," from "Harbors, Ports, and Offshore Terminals" Conf. Ed. Program UCLA, Berkeley, Calif.
18. Gaythwaite, J. W. (Spring 1987). "Floating Breakwaters for Small Craft Facilities," *Journal of Civil Engineering Practice, BSCE*, Vol. 2, No. 1.
19. Ibid. (1988). "Practical Aspects of Floating Breakwater Design," PIANC, Bull. No. 63, Brussels, Belgium.
20. Kowalski, T., ed. (1974). "1974 Floating Breakwaters Conference Papers," MTRS, No. 24, Univ. of R.I. Sea Grant, Kingston, R.I.
21. Adee, B. H. and Richey, E. P., eds. (1981). "Proceedings of the Second Conference on Floating Breakwaters," University of Washington, Seattle.
22. Isaacson, M. d. St. Q. and Fraser, G. A. (Sept. 1979). "Effects of Moorings on Floating Breakwater Response," *Proc. ASCE*, CEO IV, Spec. Conf., Vol. I, San Francisco, Calif.
23. Rosen, D. and Kit, E. (1985). "A Simulation Method for Model Study of Small Craft Harbors," *Proc. ASCE*, 19th Coastal Engineering Conf.
24. Blevins, R. D. (1979). *Formulas for Natural Frequency and Mode Shape*. Van Nostrand Reinhold, New York.
25. ABS (1988). *Rules for Building and Classing Steel Vessels*. American Bureau of Shipping, Paramus, N.J. (issued annually).
26. ABS (1973). *Rules for Building and Classing Steel Barges for Offshore Service*. American Bureau of Shipping, Paramus, N.J.
27. ACI (1988). "State of the Art Report on Barge-like Concrete Structures," ACI 357.2R-88, Detroit, Mich.
28. FIP (1986). *Design and Construction of Concrete Ships*, Federation Internationale de la Precontrainte, FIP Recommendations. Thomas Telford, London.
29. NKK (1975). "Provisional Rules for Prestressed Concrete Barges." Nippon Kaiji Kyokai, Tokyo.
30. ABS (1985). *Rules for Building and Classing Offshore Mobile Drilling Units*. American Bureau of Shipping, Paramus, N.J.
31. Taggart, R., ed. (1980). *Ship Design and Construction*. SNAME, New York.
32. Gerwick, B. C., Jr., ed., (Sept. 1975). "Proceedings of the Conference on Concrete Ships and Floating Structures," UCLA, Berkeley, Calif.
33. ———— (1980). *Proceedings of Concrete Ships and Floating Structures Convention*. Thomas Reed Publications, Rotterdam.
34. ———— (1980). *Technical Standards for Port and Harbour Facilities in Japan*. The Overseas Coastal Area Development Institute of Japan, Bureau of Ports & Harbours, Ministry of Transport, Tokyo.
35. Noble, P. G. (1983). "Design of Submerged & Floating Structures for Ice Forces," in "Design for Ice Forces," Report by the Technical Council on Cold Regions Engineering, ASCE, New York.

36. Liljestrom, G. (1983). "Design of Ships for Ice Environments," in "Design for Ice Forces," Report by the Technical Council on Cold Regions Engineering, ASCE, New York.
37. AASHTO (1983). *Standard Specifications for Highway Bridges*. AASHTO, Washington, D.C.
38. Dicmas, J. L. (May 1967). "Development of an Optimum Sump Design for Propeller and Mixed Flow Pumps," *Proc.* ASME, Fluid Engineering Conf., Chicago.
39. AASHTO (1988). *Standard Specifications for Movable Highway Bridges*. AASHTO, Washington, D.C.
40. ISO (1983). "Roll-on/Roll-off Ship to Shore Connection—Interface between Terminals and Ships with Straight Stern/Bow Ramps," ISO-6812-1983, Geneva.
41. ICHCA (1978). "Ship and Shore Ramp Characteristics," ICHCA, London.
42. Tsinker, G. P. and Vernigora, E. (1980). "Floating Piers," *Proc. ASCE*, Ports '80, Spec. Conf., Norfolk, Va.
43. NAVFAC (1985). *Fleet Moorings*, DM-26.5. Dept. of the Navy, Alexandria, Va.
44. Ansari, K. A. (Mar. 1979). "How to Design a Multi-Component Mooring System," *Ocean Industry Magazine*, Houston, Tex.
45. Berteaux, H. O. (1976). *Buoy Engineering*. Wiley, New York.
46. Pode, L. (Mar. 1951). "Tables of Computing the Equilibrium Configuration of a Flexible Cable in a Uniform Stream," David Taylor Model Basin Report #687, U.S. Navy.
47. Headland, J. R., Seelig, W. N., and Chern, C. (1989). "Dynamic Analysis of Moored Floating Drydocks," *Proc. ASCE*, Ports '89, Spec. Conf., Boston.
48. Ochi, M. K. and Shin, Y. S. (1988). "Wind Turbulent Spectra for Design Consideration of Offshore Structures," *Proc.* OTC, No. 5736, Houston, Tex.
49. API (May 1987). "Analysis of Spread Mooring Systems for Floating Drilling Units," API, RP-2P, Washington, D.C.
50. NAVFAC (Apr. 1986). *Mooring Design Physical and Empirical Data*, DM-26.6, Alexandria, Va.
51. Myers, J. J. et al., eds. (1969). *Handbook of Ocean and Underwater Engineering*. McGraw-Hill, New York.
52. Pekin, O. et al. (1987). "Gravity Anchors Dragged on Sand and Implications for OTEC," *Proc.* 6th Int. Offshore Mechanics and Arctic Engineering Symposium.
53. NCEL (Mar. 1985). *Handbook for Marine Geotechnical Engineering*. Naval Civil Engineering Lab., Port Hueneme, Calif.
54. Carchedi, D. R. et al. (Mar. 1984). "An Experimental Study of Drag Anchors and Implications for OTEC," *Proc.* Offshore Technology Conf., OTC #4768, Houston, Tex.
55. Reese, L. D. (1973). "A Design Method for an Anchor Pile in a Mooring System," *Proc.* OTC, #1745, Houston, Tex.
56. Shaw, S. H. (Jan. 1972). "New Anchoring Concept Moors Floating Dry Dock," *Ocean Industry Magazine*.
57. Hinz, E. (1986). *The Complete Book of Anchoring and Mooring*. Cornell Maritime Press, Centreville, Md.
58. Puech, A. (1984). *The Use of Anchors in Offshore Petroleum Operations*. Gulf Publishing, Houston, Tex.
59. Wichers, J. E. W. (1988). "A Simulation Model for A Single Point Moored Tanker," Pub. No. 797, Maritime Research Institute Netherlands, Wageningen, The Netherlands.
60. NAVFAC (1981). *Harbors*, DM-26.1. Dept. of the Navy, Alexandria, Va.
61. Gaythwaite, J. W. (1989). "Design of Small Craft Facilities at Exposed Locations," *Proc. ASCE*, Ports '89, Spec. Conf., Boston.
62. ASCE (1969). "Report on Small Craft Harbors," American Society of Civil Engineers, Manual No. 50.
63. Dunham, J. and Finn, A. (1974). "Small Craft Harbors: Design, Construction and Operations," USACE, SR#2.

64. NAVFAC (1981). *Ferry Terminals & Small Craft Berthing Facilities*, DM-25.5. Dept. of the Navy, Washington, D.C.
65. PIANC (1976). "Final Report of the Int. Comm. on Sport and Pleasure Navigation," Annex to Bull. No. 25.
66. State of California (1984). "Layout and Design Guidelines for Small Craft Berthing Facilities," Dept. of Boating and Waterways.
67. Van Dorn, W. (1974). *Oceanography and Seamanship*. Dodd Mead, New York.
68. ABYC. *Standards and Recommended Practices for Small Craft*. American Boat and Yacht Council, Amityville, New York.

10

Introduction to Dry Docks

by Paul M. Becht, P.E., Vice President and James R. Hetherman, P.E., Senior Engineer, Crandall Dry Dock Engineers, Inc.

Dry docks are structures that allow complete dry access to a vessel for maintenance, overhaul, and repairs or for new construction and launching. Dry docks also may provide a means of transferring vessels to or from dry land for temporary storage, lengthy overhauls, or the launching of newly constructed vessels. They are the workhorses of ship repair facilities and may be used in lieu of traditional building ways at shipyards devoted to new construction.

There are various types of dry docks, including those that physically lift the ship from the water, such as floating dry docks, marine railways, and vertical lift systems, and traditional basin dry docks that dewater an enclosed space around the vessel. This chapter is intended as an introduction to the various types of dry docks and their basic principles of design and operation. The first section presents a brief review of the capabilities and major features of each type of dry-docking facility and factors affecting the selection of the appropriate facility. Following sections are devoted to the basic design of the principal dry dock types. The final section deals with mobile straddle lifts and other miscellaneous means of dry-docking small craft.

10.1 GENERAL CHARACTERISTICS, FEATURES, AND FACTORS AFFECTING DRY DOCK SELECTION

There are four basic types of dry-docking facilities: graving or basin docks, marine railways, floating dry docks, and vertical lifts. Vertical lifts may be further divided into two groups, the shiplift and the mobile marine lift.

Basin dry docks are excavations with a gate at one end opening to the waterway. Marine railways consist of a cradle that travels on inclined tracks to launch and retrieve vessels. Floating dry docks are structures capable of sinking below the water's surface to receive a vessel and lift it out of the water. Vertical lifts

410 DESIGN OF MARINE FACILITIES

are ship elevators designed to be lowered into the water to receive a ship and then lift it clear of the water.

Several characteristics and features are desirable for the efficient operation of a dry dock, regardless of the type of facility chosen (1, 2):

1. Adequate space is necessary in and around the dry dock for ease of personnel and material movement with a vessel in the dock.

2. Fast and efficient access is needed to and from the dry dock and the vessel in the dry dock. Access for vehicular traffic is very desirable. The introduction of traveling staging or dock arms for use in basin and floating dry docks has helped to make ship repair more efficient. The staging, as shown in Figure 10.1, is a mobile platform supported on a hydraulically adjustable hinged arm that cantilevers from an electric or diesel-powered carriage running on rails attached to the dry dock side walls and/or coping. The rails run the entire length of the dock. High pressure water blasting, shot blasting, and painting can be performed from the platform.

3. Adequate light and ventilation are necessary to ensure good working conditions.

4. Support facilities are required, such as electrical and mechanical services. Outlets for water, compressed air, steam, and welding and cutting gases often

Figure 10.1. Traveling staging on wing wall of floating dry dock.

are located at service stations for ready access to the vessel. The service stations usually are placed along the wing and pontoon decks of floating docks, in recessed galleries or at the coping of basin docks, and in surrounding ground areas for marine railways and lifts. Electrical outlets are provided to supply power to the vessel in the dry dock and for repairing the vessel. Ship services and utilities are discussed in Section 7.8.

5. Material handling systems are needed for heavy items and equipment. Cranes are an essential dry dock feature, making ship repair and outfitting more efficient. Cranes may operate on the wing walls of a floating dry dock or on the adjacent pier and/or land area to service all types of dry docks. More discussion of cranes may be found in Section 4.3 and reference (3).

6. An efficient method should be provided for moving a vessel in and out of dry dock. Docks often are equipped with tensioning winches, capstans, and other line handling hardware, which allow the dry dock crew to control the vessel as it enters the dock and position it correctly over the blocks.

Electric or electro-hydraulic capstans are most often used to handle the lines. Some of the larger dry docks utilize centering/haul-in trolleys that travel on monorails fitted along the inboard side of each side wall on a floating dock or basin dock. These trolleys have lines led from winches that can center the vessel in the dock and haul it into its docking position.

7. A proper blocking system must be provided. Blocks are used to support the weight of the ship while positioning it at a convenient height to give work access underneath and leave much of the bottom area free for cleaning, repair, and painting. They also provide stability to prevent the ship from tipping over. The blocking can be considered as a mattress that provides support yet yields elastically to account for irregularities in the fit of the ship.

The most common blocking scheme uses a row of fixed, pre-positioned keel blocks with a row of stabilizing bilge blocks on each side. The bilge blocks may be pre-positioned or brought into position after the ship's keel has touched the keel blocks. The latter case is preferred where the ship has considerable shape or where clearance for sonar domes or propellers is needed to permit the ship to enter the dock. Many systems have been developed for this purpose, including electrical, hydraulic, and compressed air systems and chains with winches. Because of its simplicity and durability, the chain/winch system is most often used.

Most blocks are made of composite construction with concrete or steel bases and a timber or rubber cap piece to provide the necessary cushioning against the ship's hull. Timber may be added to the base of the block to provide cushioning against irregularities in the dock floor.

Removal of blocks to provide access to hull areas in need of repair is accomplished through the use of sand frames set on top of the blocks' concrete base, as shown in Figure 10.2. A confined layer of sand in the frame supports the top

412 DESIGN OF MARINE FACILITIES

Figure 10.2. Block with sand frame for easy removal.

pieces of wood. Provision is made to remove the sand by water jet, allowing the wood to drop so the block can be removed. Other, more sophisticated systems are available for block removal, but this simple system is perhaps the most common and successful one. More information on blocks and blocking concepts can be found in references (4) and (5).

Selection of the appropriate type of dry-docking facility will be influenced by many factors, including:

1. The dimensions, weight characteristics, and general features of the vessels to be serviced by the dry dock.
2. Conditions at the site of the dry dock and the associated land facilities, including available land area, available area in the water, and proximity to navigable channels or open water, tides, currents, topography, and soil conditions.
3. The purpose of the dry dock. Will it be used solely as part of a new building program, for long-term vessel repairs, for short-term vessel repairs, or for some combination of all these purposes?

4. The near- and far-term goals of the shipyard and the potential for future extension of the dry-docking facilities.
5. Financing. Investors may be more willing to provide working capital for one type of facility than for another.

These considerations will be discussed as they relate to each type of facility in the sections that follow. Additional discussion of these factors can be found in references (1) and (6).

Factors that affect the operational aspects of dry-docking include characteristics of the ship, features of the dock, and, in some cases, the two together (4).

A ship is necessarily strong to resist stresses encountered at sea. The stiffness that allows the vessel to resist excessive flexing in waves also permits the distribution of weight concentrations due to machinery, fuel, cargo, and so on, fairly uniformly to blocks along its length, which support the ship when it is dry-docked. The ship hull girder stiffness usually is sufficient to permit overhang of the ship beyond the block length both fore and aft. As a general rule, the safe overhang at either end (or both ends) may be up to one and one-half times the molded depth of the ship (7). Figure 10.3 shows a floating dry dock with a large vessel overhanging at one end.

Figure 10.3. Floating dry dock with large vessel overhang. (Crandall Dry Dock Engineers file photo.)

Ships are designed to be stable, floating upright in the water, and they generally have a good degree of stability when dry-docked. Certain conditions, such as the effect of free liquids in a ship, can threaten the docked stability, however. Grounding instability, occurring when the ship first touches the blocks, can be crucial, especially for ships with a large amount of trim. Ship stability and attitude considerations as they pertain to the various dry dock types are discussed in the sections that follow.

The weight and weight distribution of the ship must be known to ensure safe docking in questionable or limiting cases. A method for determination of the ship's weight distribution is presented in Section 10.4.

As with a ship, the strength properties of the dry dock must be considered prior to dry-docking. The load distribution through the dock structure is significant for railway dry docks, vertical lifts, and, to a lesser extent, floating dry docks. The inherent stability, practically speaking, is of concern only with floating dry docks.

Basin dry docks rarely are damaged by overloading. The foundation and floor, if properly designed, should safely support any ship that can fit into it, even with high load concentrations. The size and the docking capacity of basin dry docks are practically unlimited, with several basins capable of docking ships over 1,000,000 DWT having been built in recent years.

A heavy load concentration, or a ship that is too heavy, can cause damage to a marine railway's cradle, its track, and/or its foundation. A railway's safe capacity is certified in terms of total load and load per unit length, neither of which should be exceeded. The largest railway in service today is rated at 5200 long tons lift capacity, but 8000 long tons lift capacity is within the capability of these systems.

A major difference between a floating dry dock and other kinds of dry docks is that the forces between the blocking and the ship are a function of the floating dock's buoyancy as well as the ship's load. These forces can be somewhat controlled, however, by adjusting the buoyancy through variations in ballast tank water levels. As with railways, floating dry dock capacities are given in terms of total load and load per unit length. The stability of the ship–dock system is a critical factor, and must be considered in the design and the operation of a floating dock. The largest floating dry docks currently in operation in the United States have lifting capacities of 81,000 long tons.

Vertical lifts have no inherent questions of stability, but adequate strength is critical because a failure can lead to a free-gravity fall of the ship. The ship's weight is distributed by the platform to the peripheral lifting means—wire ropes, chains, jacks, and so on. Depending on the ship's position on the platform and its weight distribution, some hoists carry more load than others. A lift's ability relates to the total capacity and the maximum load per unit length, which is a function of the number, size, and spacing of the lifting units. Shiplifts have

been built as large as 25,000 tons lift capacity, and their capabilities have increased steadily over the years.

10.2 BASIN DRY DOCKS

A basin, or graving, dry dock is simply an excavation or depression in the earth with one end in free communication with the sea. For dry dockings, the seaward end is sealed off with a gate.

One of the earliest graving docks was constructed in the late 1400s at Portsmouth, England, but it was not until the middle to late 1600s that the basin dry dock came into prominence.

Some of the earliest basin dry docks were simple excavations lined with timber, having a brick or concrete floor and fitted with a gate bearing against a sill to exclude the tide. The vessels entered at high tide, and the entrance gate was closed. As the tide receded, the dock emptied itself through a tidal sluice, and the vessel settled on blocks that had been prepared for it. At locations with little or no tide, the water had to be pumped out of the basin. As ships became larger, dry docks became important civil works. Solid masonry was adopted for lining the walls, and large pumps were added for rapid dewatering of the dock. A history of basin dry docks and further description of some of the early designs can be found in references (2) and (8).

A present-day dock, when compared to its predecessors, differs significantly in terms of dimensions, materials, details of construction, and sophistication of the equipment, but the basic principles of operation and design are still the same. The general design of basin dry docks is covered in references (3) and (9) through (11).

General Features and Characteristics

The basic features of a basin dry dock, shown in Figure 10.4, include the floor, side walls, head wall, and dock gate. There are three types of dry dock structures: mass gravity, ground anchored, and underdrain (or pressure-relieved). Various methods of construction utilize underdrains and massive stone or concrete slabs in combination with concrete or masonry retaining walls.

In general, basin dry docks are best suited for large vessels. Advances in materials and methods of construction have continually expanded the dimensional limits of basin dry docks. Shipbuilding and ship repair basins over 1200 feet long, 200 feet wide, and capable of handling vessels of all types and under all conditions, are common among recently completed docks. The construction of some of these very large dry docks is discussed in references (12) through (15).

416 DESIGN OF MARINE FACILITIES

Figure 10.4. Basic features of a basin dry dock.

The operation of a basin dry dock is quite simple. Once the ship and docking blocks have been properly prepared, the basin is filled with water until the water level inside it is equal to the level in the harbor. Then the gate is opened, and the ship is moved into the dock. Warping into dock usually is accomplished by using head lines and spring lines on capstans or winches with assistance from harbor tugs. The dock gate is closed, and the ship is properly positioned before the pumps are energized. Next the water level is lowered, and the ship settles gently on the blocks. During dewatering the ship's position is carefully monitored, and the tension in the springing and breasting lines is adjusted accordingly. An undocking evolution is similar, with the order of events reversed.

In a basin, the vessel weight usually is not of concern. If the ship can fit in the basin, it can be safely dry-docked. Nevertheless, each docking evolution should begin with a careful investigation of all aspects of the ship.

Unlike a floating dock, the floor of a basin is not adjustable. A vessel docked with a substantial trim will result in an extremely high load on making initial contact with the blocking, and the ship could become unstable. Vessel trim can be accommodated to some degree by shoring or building the blocks on an incline, but, as a general rule, an acceptable trim usually is one foot for every 100 feet of ship length. Beyond this limit special preparations for blocking should be considered.

Dewatering a basin dock involves moving huge quantities of water through large-capacity pumps. Basins are different from other types of docks in that energy consumption in basins during an operation is inversely proportional to

ship size; that is, the smaller the vessel docked, the greater the volume of water that must be pumped out, and the greater the amount of energy consumed.

Because the dock floor is far below yard level, natural light, air for ventilation, and access are somewhat restricted. This situation tends to adversely affect working conditions; however, cranes traveling along the coping on each side of the dock or large gantry cranes spanning from side wall to side wall can easily move heavy equipment and large structural modules in and out of the basin.

Site Considerations

Site selection for dock construction involves careful evaluation of many factors, including topography, hydrography, geology, and meteorology. As most of these factors have a direct impact on construction and operation of the dock, each must be carefully considered before a final site can be chosen. Locating the perfect site is extremely rare; so a series of trade-offs must be made between type and cost of construction, as well as operating and maintenance costs. Some preferences for dock location and chief concerns regarding construction are summarized below, but for a more detailed discussion on this subject see reference (16).

A basin should be located so that large ships can enter it without interfering with navigation. Normally a dock is oriented perpendicular to the general shoreline. Providing a turning basin in front of the dock entrance is advisable to allow ships to align themselves with the axis of the dock before entering. At sites where there is a strong current parallel to shore, the dock may be oriented at a skew angle to the shoreline to avoid cross currents as a ship enters the dock. Similarly, orienting the dock parallel to the prevailing winds is desirable for maneuvering ships into dock. In general, a location in a protected harbor or within the confines of a breakwater is essential, as ships should be sheltered from strong winds, waves, and currents during docking. Also, during the planning stages a sheltered anchorage or standby berth should be considered in conjunction with the dock because there usually is some time between a ship's arrival in the yard and its going into dry dock.

Silting and scouring at the dock entrance and its effect on the gate and gate recess may warrant special consideration. The potential for excessive silting could indicate high future costs for maintenance dredging.

Tides and waves at the site have a direct impact on construction, as they influence the elevation at the top of the gate and along the coping. Soil types and their strength as well as groundwater levels greatly influence the kind of structure suitable for the site, the depth to which it can economically be built, and the type of construction methods that can be employed. The groundwater level is the most critical factor in the structural design of the floor and walls.

The design of a basin dry dock involves a wide variety of soil engineering problems, as discussed in Section 8.8. For this reason a thorough subsurface exploration program must be implemented prior to the design of a new dry dock.

Finally, supporting shop facilities should be in close proximity. The availability of a large amount of lay-down space around the dock is a definite advantage, and some type of crane service is essential for moving material into and out of the basin.

Dimensions of a Basin Dry Dock

The key dimensions of length, width, and depth depend on the type of ships to be docked and whether the basin is to be used strictly for new construction, vessel repair, or a combination of both. The effective length of a basin is the minimum horizontal distance measured along the centerline between the head wall and the dock gate. This is illustrated in Figure 10.4. The effective length should be at least 10 to 15 feet longer than the overall length of the maximum design ship. To allow for propeller and shaft work, an additional 25 to 100 feet should be provided. Docks that are extremely long can be partitioned off by an inner gate (Figure 10.5) or gates, which provide operational flexibility and savings on pumping costs in docking smaller vessels. It is quite conceivable for one, two, or possibly more small vessels to occupy the dock simultaneously.

As shown in Figure 10.6, the entrance width is the clear distance between the permanent dock fenders or wall structure at the dock entrance. The width at the entrance should be at least 6 to 10 feet wider than the widest ship to be docked. The clear width within the basin normally is greater than the entrance width. Most basins built today have an effective length to width ratio of between 5:1 and 7:1, in approximate proportion to large contemporary vessels (3). The inside width is established from the beam of the design vessel plus an allowance for clearance on each side of the vessel for working space. Figure 10.7 shows a cross section with minimum clearances noted.

The required depth at the entrance should be decided only after careful study

Figure 10.5. Shipbuilding dock with intermediate gate.

INTRODUCTION TO DRY DOCKS 419

Figure 10.6. Cross sections showing entrance widths.

of all available water level data in conjunction with the vessel's drafts. Ships needing repair or routine maintenance, or those coming into dock in a damaged condition, require a deeper dock than vessels under construction, which generally are moved out of the basin once the main hull is watertight and before final outfitting is complete. Where appropriate, the designer should include an allowance for ship hull appendages such as sonar domes or propellers, which sometimes extend below the vessel's keel. In areas where tide is minimal, the mean low water level (MLW) should be used in determining the depth of the entrance. Where docking or undocking delays cannot be tolerated while waiting for certain tide levels, as may be the case with warships, then the mean lower low water level (MLLW) should be used in determining the entrance depth. For commercial vessels, dry docks in tidal harbors may rely on high tide for docking

Figure 10.7. Minimum clearances for construction or repair work.

and undocking operations. In all cases, the depth over the sill must include at least a one-foot allowance for clearance.

The depth to the floor of the basin usually exceeds the depth to the sill by a minimum distance equal to the height of the keel blocks (see Figure 10.4). At the head wall the depth to the basin floor may be slightly less than at the sill if the dock floor is built on an inclination.

The depth of a basin has the single greatest impact on overall project cost. Extending the depth of the dry dock considerably increases construction costs, in proportion to the third power of the increase in depth (e.g., doubling the depth will increase construction cost by a factor of eight) (3).

Types of Basin Dry Docks

Of the three basic types of basin dry docks—mass gravity, anchored, and underdrain (or pressure-relieved)—each relies on a different method to counteract the effects of buoyancy due to the surrounding groundwater.

In a mass gravity design, the structure has sufficient mass to overcome the upward pressure of the groundwater acting on the underside of the floor when the dock is empty. Gravity docks can be constructed as a monolithic frame or with a floor slab separate from the side walls. They also can be built with a

Figure 10.8. Mass gravity dry dock structures (I).

Figure 10.9. Mass gravity dry dock structures (II).

very heavy floor slab, which in itself is sufficient to resist the uplift. In docks where the floor slab is separate from the walls, the floor usually is made as a homogeneous slab or inverted arch supported by the walls. In this case the walls can be steel sheet pile walls resisting pullout by virtue of soil friction, or they can be the reinforced concrete retaining type of wall. In the latter case, the thickness of the slab is reduced under the walls, thereby taking advantage of the walls' weight. Four arrangements of mass gravity docks are illustrated in Figures 10.8 and 10.9.

The mass gravity type of dock generally is preferred for docks of significant width; but for docks of great depth, difficulties arise during attempts to lower the groundwater if the slab is to be cast in the dry. Moreover, with an increase in depth, the thickness of the floor slab also increases considerably. For docks of great depth, use of the gravity-type dock becomes very labor-intensive and uneconomical.

In an anchored dock design, the hydrostatic uplift is opposed by the permanent weight of the structure plus some type of anchorage. The soil conditions must be such that they can provide bearing capacity for the floor slab loads and allow for the development of tension in pilings and/or anchors embedded in the soil. Where pilings are used, they may play a dual role, acting in tension when the dock is empty, and in compression when there is a ship in dock. Two types of anchored docks are shown on Figure 10.10. To avoid some of the disadvantages of a mass gravity design, the anchored type of dock relies on a relatively light floor slab. This type is less expensive to build and can be constructed in a shorter period of time, compared to a gravity dock.

In the design of gravity and anchored-type docks, special consideration must be given to ensuring an adequate safety margin against uplift. According to PIANC (9), a factor of safety of 1.1 is recommended for comparing the downward weight of the dock to the uplift forces. For this calculation, it is recommended that the designer assume the highest possible level for the groundwater and the lowest expected density for concrete with no ship or other loading on the dock floor. In anchored docks, the hydrostatic load usually is increased by a factor of 1.3 for the design of the anchoring system.

In an underdrain or pressure-relieved design, the groundwater around the dock is lowered sufficiently that no uplift pressure develops. A precise knowledge of the soil parameters is very important. Data on the continuity of the impervious layer and its exact depth as well as horizontal and vertical permeability coefficients must be determined. A pressure-relieved dock may be fully drained, partially drained, or situated on impervious soil.

In docks where the pressure is relieved at the floor slab and the walls, an external drainage system is installed, which allows the groundwater level to be lowered. The drainage system is a series of pipes and culverts connected to the dock's pumping station. In docks where the pressure is relieved under the floor slab only, water flow to the drainage system is controlled by a sheet pile cutoff

Figure 10.10. Anchored-type dry dock structures.

wall installed below the dock walls; the cutoff wall extends to the impervious layer. This type of dry dock is illustrated in Figure 10.11. Piezometers are recommended for fully or partially relieved dry docks to ensure that pressure relief is being maintained.

Also, depending on soil conditions, lowering of groundwater levels may be accomplished with a shallow dewatering system such as sand drains, a deep pressure relief system such as deep wells, or a combination of both. A recent application of these methods for a basin dry dock at the Long Beach Naval Shipyard is described in reference (17). Basically, the system was designed to induce drainage from the backfill soils behind the dry dock walls through sand drains, into the underlying aquifer. The water subsequently was pumped from the deep wells to relieve hydrostatic pressure on the base of the dry dock. Some ground conditions do not permit lowering of the water table, as this can lead to consolidation of clays and silts and result in unacceptable settlements of surrounding structures.

Variations on methods of construction for the three basic types of docks can be found in references (3) and (9).

Dock Floors

Today, practically all dry dock floors are built of reinforced concrete, but before the advent of reinforced concrete, brick arch and stone block floors were pro-

Figure 10.11. Dry dock with underdrain system.

vided, both for structural strength and for mass. The floor slab is designed for several types of loading conditions including the concentrated loading from a ship and, depending on the type of dock, the full hydrostatic uplift on the underside of the floor. The two main means of supporting the floor loads are: direct bearing on the ground and end bearing or friction-type pilings. Direct bearing on the ground requires the slab to have sufficient strength to spread the load to the soil, and the soil to have adequate bearing capacity to support the load without excessive deformation or settlement. The analysis is similar to that for a beam on a continuous elastic foundation.

Dock floors may be built horizontal or with an inclination. An inclined floor can slope longitudinally toward the dock entrance or, transversely, from the centerline to the side walls. The floor generally slopes at an inclination of about $1:300$, which enhances drainage of the dock during dewatering.

In docks where the floor is horizontal, better drainage is achieved by inclined dewatering channels or troughs recessed into the slab. Transverse dewatering channels are oriented perpendicular to the axis of the dock, directing water flow toward the side walls where channels run longitudinally, allowing water to flow to sumps and into the main intake culverts.

Control joints may be required in the floor for shrinkage. Expansion joints rarely are required, as the temperature of the slab remains fairly constant because it is in contact with the underlying groundwater.

Dock Side Walls

Originally docks were designed with sloping walls having many offsets called altars, as shown in Figure 10.6. Today most docks have vertical walls. The side walls are designed for a variety of loadings, including earth pressure and surcharge, groundwater pressure, water pressure from inside the dock, and loadings from equipment and fittings such as cranes, traveling dock arms, ship hauling gear, bollards, and services. In addition to these loadings, the walls must be designed for the hydrostatic loads transmitted up through the dock floor. In most locations earthquake loads must be considered.

The main types of dock wall construction include mass concrete, reinforced concrete, sheet piling, and caissons. The use of mass concrete was prevalent during the early twentieth century and may still be economical in some instances, but it generally is not common in modern construction. Many of the docks built within the last 40 to 50 years were built with reinforced concrete walls. Some of the more common configurations are shown in Figure 10.12.

When sheet pile walls are used, slightly heavier sheeting than that required by design is recommended to allow for corrosion. Special considerations are necessary if the walls are to carry vertical loads from cranes and building columns; and the watertight integrity of the installed sheeting at the joints is also of concern. If a groundwater cutoff wall is required, then sheet piling can be used to great advantage by terminating the sheeting in an impervious layer some distance below the dock floor. This effectively reduces the amount of water to be pumped from an underdrain system.

In some cases where real estate is at a premium, it may be possible to extend the basin out into the channel by using interlocking floating caissons for the walls. Once in place, the caissons are filled with earth to secure them.

The three basic wall profiles are illustrated in Figure 10.13. The uniform vertical profile requires that all equipment, piping runs, electrical raceways, and so on, be set inside the closed gallery with all service outlets located on the dock coping. The vertical profile with recessed or cantilever galleries just below the coping provides a place for service lines and service outlets. In all cases the walls may be inclined as well as vertical, and the wall profile should be such that it makes total use of the full floor width. The layout of the dock wall should be closely coordinated with the many services and other equipment required along the coping (see Figure 10.14).

Structural Analysis of Basin Docks

The finite-element method of analysis is the most preferred method for analyzing the complex structure of a basin dry dock. Whereas previous analyses were accomplished by using strength of material concepts and simplified theories of

VERTICAL, WITH HIDDEN GALLERY

VERTICAL, WITH RECESSED GALLERY

TWO TYPES WITH CANTILEVERED GALLERIES

INCLINED VERTICAL

Figure 10.12. Dock wall construction of reinforced concrete.

elasticity, this method considers the interaction between the soil and the structure as well as the stress/strain characteristics of the structure itself. A detailed description is beyond the scope of this text, but the reader can consult reference (18), which describes the method and includes examples of the results of a

INTRODUCTION TO DRY DOCKS

Figure 10.13. Typical wall profiles.

typical analysis. Such an analysis provides a thorough evaluation of the safety of a dry dock subjected to static soil and hydrostatic pressures, ship loadings, and earthquake-induced dynamic loading.

Pumping and Flooding Systems

Pumping and flooding systems control the emptying and filling of the dry dock, which must be accomplished in a gradual, controlled manner. The dewatering process must allow for the ship to ease down onto the blocks, and the flooding

Figure 10.14. Arrangement of services at dry dock coping for recessed gallery.

must be controlled to the point where there is no large inrush of water, which could disturb the blocks.

Pumping and flooding times vary from dock to dock, but the times usually are independent of dock size. Normal pumping times for a basin dry dock are in the range of 2 to 3 hours, but some docks require up to 12 hours or more for dewatering. Normal flooding times, on the other hand, are about half as long. In general, for dry docks engaged primarily in repair work, the less time required to pump and/or flood the dock, the better. This translates to shorter docking times and thereby decreases vessel turn-around time and idle time for shipyard workers. For basins where the emphasis is on new construction, longer pumping and flooding times are acceptable. The basics of the dewatering and flooding systems are illustrated in Figures 10.15 and 10.16, respectively.

The three main types of dewatering pumps used in basin dry docks are centrifugal, mixed-flow, and axial flow pumps. Centrifugal pumps may be installed with either a horizontal or a vertical shaft, but the mixed and axial flow pumps are vertical installations. Pump types and guidelines for proper pump selection are described in detail in references (19) and (20). Final pump selection must take into account dewatering time, power consumption, initial cost, and operating cost. The number of main pumps usually ranges from two to five. A single pump never should be used because a breakdown would render the dock inoperable.

Dry dock flooding systems utilize several types of valves, the most common being: sluice valves, which are used in the flooding culverts; equilibrium valves, for filling the dock; and gate valves and butterfly valves, which can be used for isolating valves and flood valves in the gate. Sluice valves, found mostly in small, older docks, can vary in shape and in materials of construction. Their shape is determined by the shape of the culvert; the materials of construction can include timber, cast iron, or steel. Equilibrium, or cylindrical, valves consist of a large vertical cylinder supported in vertical guides. The valves are located in compartments with direct access to the sea, and the valve bottom consists of a seal set over the flooding culvert (see Figure 10.16). When the valve is lifted, water flows radially into the culvert. Because the cylinder is extremely large, its weight is counterbalanced so that a relatively small force is required to open the valve. Isolation valves in the suction and discharge culverts provide a means for repairing or performing maintenance on a pump without affecting the rest of the dewatering system. Valve operators for the larger-size valves are electric, hydraulic, or compressed air. All should have a provision for manual operation in the event of system failure. More information on culvert sluices and valves can be found in reference (21).

A separate building or chamber called a pumping station, built into the wall of the dock, contains the dewatering pumps and valve operators, as well as other smaller pumps and equipment for discharging rainwater and providing

INTRODUCTION TO DRY DOCKS 429

Figure 10.15. Dry dock dewatering system.

Figure 10.16. Dry dock flooding system.

fire-fighting or ballast water, or in the case of an underdrain-type dock, pumps for the discharge of groundwater. A typical pumping station is illustrated in Figure 10.17. Local control of the pumps and valves is provided in the pumping station. In addition, remote operation from a central control house at yard level is found at most facilities.

The preferred location for the pumping station is within one of the side walls near the entrance to the dock, to minimize the lengths of suction and flooding culverts. The dock's side wall requires special detailing for the pump house. In the case of parallel twin docks, the ideal location for the pumping station is in the wall between the docks, as one large pumping station is less expensive than two smaller ones. For docks fitted with an inner gate, it is preferable to have the pumping station near the inner gate sill or recess, to serve both sections of the dock efficiently.

INTRODUCTION TO DRY DOCKS 431

Figure 10.17. Typical dry dock pumping station.

The arrangement and size of the pumping station depends mainly on the type of pumps selected. If horizontal shaft pumps are chosen, they must be mounted as close to the floor level as possible because of their limited suction height. The pump motor is coupled to a shaft extension from the pump and is mounted on the same level as the pump. This arrangement requires a large floor area for each pump/motor unit.

The most recent docks have utilized vertical shaft pumps. The drive motor is on a common vertical shaft with the pump and is mounted directly above it. This arrangement has the advantage that it significantly reduces the amount of floor space required, resulting in considerable construction savings (see Figure 10.17).

In a modern pumping station it is necessary to have a reliable source of electrical power, sufficient lighting, a communication system consisting of a telephone or radio and loudspeaker system, and adequate ventilation and heating. Also if there are auxiliary pumps in continuous operation for handling groundwater seepage, it is imperative that an independent power source be available in the event of a power failure.

All pumping and flooding operations are controlled from the control center. The center is usually in a special tower or building conveniently located with respect to the basin. In addition to the pump and valve controls, there are gages for monitoring water levels in the dry dock as well as the major suction chambers. The control center has electrical status boards and indicators, an alarm indicating system for fire, equipment failure, main power source failure, and so on, a communication system, fire pump controls, and controls for all auxiliary pumps. The control console usually has a mimic diagram showing schemati-

432 DESIGN OF MARINE FACILITIES

cally the arrangements of culverts, sumps, pumps, and valves as well as their open/close or run/non-run status.

Dry Dock Gates

The dry dock gate provides the necessary separation between the basin and the harbor. These gates vary widely in terms of design, configuration, and means of operation. There are five basic types of gates: floating, hinge, sliding, mitre, and flap gates. Examples for each type are illustrated in Figures 10.18 through 10.22. The type, shape, and overall dimensions are influenced by a variety of factors:

- Width of the entrance.
- Depth of the entrance.

Figure 10.18. Typical free floating gates.

INTRODUCTION TO DRY DOCKS 433

Figure 10.19. Dry dock hinged gate.

- The head of water to be retained during various phases of operation.
- The reverse head of water in docks capable of being superflooded.
- Tides, winds, and waves as they affect the gate's height.
- The speed of operation. For repair docks a gate operating time of 10 to 15 minutes is preferred; for building basins, an acceptable time is about 30 minutes.
- Cost of construction, including associated civil works and operating mechanisms.
- Depth available below the sill just outside the dock entrance, and the possibility of siltation.
- Availability of parking space for the gate against a pier or quay wall.
- Cost and ease of maintenance.
- Labor and mechanisms required to operate the gate.
- Power source required to operate the gate.
- Access across the top of the gate for pedestrians and/or vehicles.

Figure 10.20. Sliding caisson gate.

- On-site versus off-site location for construction of the gate.
- Special considerations such as ice loads.

Whatever the type of gate chosen, it is imperative that it make a tight seal against the dock entrance. Wood and hard rubber are the most common materials used for gate seals. The seal must be accurately shaped to provide a good fit; its cushioning ability makes up for irregularities and unevenness along the face of the concrete.

Buoyancy is an important consideration for all gates, as they should have sufficient buoyancy to aid their removal from and placement in the sill.

Floating Gates

Free floating gates are the most widely used. They can be constructed of steel, reinforced concrete, or prestressed concrete. A floating gate is completely independent of the dry dock. Once it is free of the sill, it is towed to a pier or quay for mooring, out of the way of ship traffic. A floating gate is particularly well suited to long spans and is used primarily in cases where heavy vehicular traffic will pass over it. Floating gates can be made reversible for ease of main-

INTRODUCTION TO DRY DOCKS 435

Figure 10.21. Mitre gate.

tenance, and they can be constructed at a remote location and towed into position for immediate use. Their main disadvantages are that the operating time is long, and they require a large crew to operate winches and capstans and handle lines during gate removal and placement.

Floating gates can have a variety of cross-sectional shapes; and in elevation, the sides may be vertical or inclined, as shown in Figure 10.18. Internally, the gate is subdivided into watertight compartments for buoyancy, stability, and ballast control. The ballast control system consists of pumps and valves mounted in the gate.

The stability of a floating gate must be carefully calculated. As the amount of water ballast changes, the center of gravity of the gate shifts, as does its GM (see Section 9.2). To preclude any instability, most floating gates contain concrete ballast near the bottom. The stability of floating gates is discussed in more detail in reference (22).

436 DESIGN OF MARINE FACILITIES

Figure 10.22. Flap gates for basin dry docks.

Hinge Gates

Hinge gates are modified versions of floating gates, made with a vertical hinge on one side as shown in Figure 10.19. The hydrostatic principles of the hinge gate are similar to those of the floating gate. The gate is opened by means of wire ropes and winches. An alternative method is to connect a boom operated by a hydraulic ram capable of pushing or pulling the gate. The hinge gate has

many of the same advantages and disadvantages of the floating gate. For maintenance, the gate can be disconnected from the hinges and moved to another location for topside repairs or drydocking.

Sliding Gates

Sliding gates typically are floating pontoons or caissons made of steel or concrete. When the dock is opened to the sea, the sliding gate is shifted transversely into a special pocket or recess built into the side wall of the dock, as shown in Figure 10.20. Sliding gates are similar in construction to floating gates, and can be used for outer as well as intermediate gates. They are easy to maneuver, requiring only a small crew, and are designed for fast opening and closing. Their main disadvantage is the need for a special storage chamber built into the side wall of the dock; such a chamber can add significantly to the overall cost of construction. The usual configuration is rectangular, but some are built in an arched shape. The typical width-to-height range is 1:4.5 to 1:6.5, in order to allow for traffic on top of the gate (3).

The gate slides on a system of wheels or rollers on rails, and also can slide on a smooth surface. In either case, the gate is deballasted to minimize the reaction on the sliding tracks. Of these designs, those without rails provide better watertightness, as the entire bottom of the gate presses against the sliding surface to form a tight seal. The gate can be built with meeting faces on both sides if there is the possibility of reverse head pressure. Despite the fact that the gate is deballasted prior to opening, considerable breakaway force is required, especially if the gate has been sitting for a long time. Wire cables connected to electric or hydraulic winches normally are used for moving the gate, although hydraulic rams sometimes are installed to provide the initial breakaway force.

Mitre Gates

The mitre gate is split at the middle and consists of two leaves, each rotating about a vertical hinge at the side wall of the dock. The vertical axis is supported by an upper and a lower bearing, and the entire hinge assembly is recessed into the wall of the basin. The leaves are shaped so that they form an arch across the dock entrance to resist water pressure. Watertightness is achieved by mating pieces of greenheart timber or hard rubber on the mitre posts at the centerline, as well as timber or rubber at the wall hinges and along the bottom at the sill. A mitre gate is fast-operating, requiring only a small crew, and the gates are designed with some buoyancy to minimize the load on the hinges and eliminate the need for bottom rollers.

Upon opening, the gate leaves enter recesses formed in the side walls at the

dock head. Thus the entrance width can be as wide as the main dock chamber, as shown in Figure 10.21.

Some disadvantages include the gate's inability to accommodate reverse loading or roadway traffic across the top. Also maintenance of the structure is difficult, and it is generally considered unsuitable for large entrance widths.

Flap Gates

A flap gate consists of a single leaf spanning the entire width of the dock, hinged horizontally on two heavy trunnion bearings at the level of the dock floor. Operation of the gate is simple and speedy, as it can be raised or lowered in about five minutes by electrically driven winches. When open, the entire gate is below the sill level of the dock, thus posing no limitation on the water depth at the sill. It can be accommodated in the waterway or channel adjoining the dry dock and does not require a special recess in the floor or wall structure. The gate is designed to be semibuoyant to reduce the load on the wire rope used for opening and closing the flap. The design of a flap gate is greatly influenced by the entrance height-to-width ratio. The gate may be designed with a deep box girder forming the top portion of the gate spanning the entire entrance and with the lower portion a series of stiffeners spanning vertically between the sill and the girder. Alternatively, the gate may be designed as a series of heavy beams supported on three sides—the two side walls and the sill.

For very wide entrances the simple flap gate is too heavy and uneconomical. In this situation a strutted flap gate may be considered. When closed, the gate is supported at intervals by a series of inclined struts designed to take the horizontal water pressure. Another alternative is the cantilever flap gate. This gate is arranged to cantilever from the sill just above where it is hinged (see Figure 10.22). The tendency in dock construction has been to eliminate complex devices and simplify the process of opening and closing the dock chamber. The flap gate has become the most economical and the most reliable type of dry dock closure. A more detailed treatment of the subject of dry dock gates can be found in references (3) and (22) through (25).

10.3 MARINE RAILWAYS

The marine railway dry dock is a mechanical means for lifting a ship out of the water to an elevation above the highest tides. Operating on the basic principle of the inclined plane, the marine railway is comprised of a cradle or carriage that is lowered into the water along a sloping track on a system of rollers or wheels, until the vessel to be dry-docked can be floated over it. Once the vessel has been centered, the cradle then is hauled up the track until the vessel grounds out on blocking. The operation is complete when the cradle deck is clear of the water, leaving the ship high and dry. The railway is designed so that the dry-

docked vessel is positioned at or above the level of the yard, allowing materials and staging to be easily moved into position from the adjacent yard area. Also, because the cradle is virtually open on all sides, there is free circulation of air and good illumination, allowing vessels to be repaired quickly and efficiently. Figure 10.23 shows the major components of a marine railway.

The marine railway was developed during the mid-1800s, originating from the English-type slipway. The features introduced on the first marine railway in East Boston, which differentiated it from the slipway, included higher keel blocks and a deck over the cradle, docking platforms for men to walk along while handling lines during a docking, and steam power instead of the power of men and horses.

A marine railway of modern design is fast-operating, reliable, and extremely durable in the harsh environment of the sea. In general, it is economical in both installation and operation when compared to other facilities in the same capacity range.

The marine railway is best suited for a capacity range of 100 to 8000 tons. Theoretically, marine railways beyond 8000 tons are possible, provided that the foundation conditions are satisfactory. The type of ships to be docked and any special restrictions of the site usually dictate the final dimensions of the railway. Figure 10.24 gives the usual dimensions for marine railways.

Marine railways can be either end-haul or side-haul designs. Most modern railways are end-haul because that type of operation is generally safer and less complicated than the other. Also, the end-haul type requires only about one-third as much water frontage as the side-haul type.

Site Considerations

The marine railway is best suited to sites with gradually sloping bottoms. In areas where siltation is a problem, the natural bottom at the offshore end of the track should be below the track grade. A site in a sheltered harbor with natural protection from strong winds, waves, and currents is essential. For some locations, the formation of ice is a serious concern, as ice can severely hamper the operation of a marine railway. Because the trackage is fairly long, a sufficient area must be available to avoid infringement on the channel. Favorable subsurface conditions are essential for supporting the anticipated loadings. The capacity of the foundation will determine the per foot or lineal capacity of the railway.

Track and Foundation Design

The load that the ship exerts is transmitted directly to the track and foundation through the cradle and roller (or wheel) system. The design loadings for the

Figure 10.23. The major components of a marine railway.

INTRODUCTION TO DRY DOCKS 441

LIFTING CAPACITY in tons of 2000 lbs.	LENGTH OVER KEEL BLOCKS in feet	LENGTH OVER DECK in feet	WIDTH OF CRADLE in feet	CLEAR WIDTH in feet	DEPTH FORWARD in feet	DEPTH AFT in feet
100	78	78	30	26	6	11
200	90	90	32	26	6	11
300	102	102	34	28	7	12
400	115	115	36	30	7	12
500	128	128	38	32	8	13
600	140	150	40	33	8	13
800	160	172	42	35	9	14
1000	180	195	44	37	9	14
1200	200	215	46	39	10	15
1500	220	235	50	42	11	16
2000	240	255	54	46	12	17
2500	270	285	56	48	12	17
3000	300	320	60	52	13	18
3500	320	340	65	57	13	18
4000	340	360	70	62	14	18
4500	360	380	72	63	14	18
5000	380	400	74	65	15	19
6000	400	420	76	67	16	20
7000	420	450	77	69	17	21
8000	440	470	78	72	18	22

NOTE:—
These dimensions may be modified to suit requirements.
The length of the track varies with the slope which conforms to the natural conditions of the site.

Figure 10.24. Typical dimensions for marine railways.

track vary along the length of the track. For example, the inshore portion must be designed for the dead load of the cradle plus the live load of the ship. The offshore portion of the track is designed for the dead load of the submerged cradle and no live load. Moving inshore, the design load steadily increases from

442 DESIGN OF MARINE FACILITIES

Figure 10.25. Profile of a curved track.

the point at which the vessel grounds on the first block until the vessel is completely out of the water, and the full live load and dead load are realized.

Depending on the natural shoreline, availability of space, river or tidal currents, variations in water level, and desired lifting capacity, the declivity of the track may vary from one in ten to one in thirty. In most cases, the slope of the track is carefully selected to minimize or completely eliminate dredging.

The track must be constructed with a smooth gradient. A trend in railway design has been departure from a straight gradient and development of a track constructed on a vertical circular curve (see Figures 10.25 and 10.26). This geometry causes the cradle to rotate as it travels along the track, making it possible for the keel blocks and cradle deck to be on a declivity when submerged, to be closely aligned with the trim or drag of a vessel. It also allows the cradle to be horizontal in the full-up position, which is necessary in the case

Figure 10.26. Photograph of track built on a vertical curve.

of transfer to shore. A secondary benefit of the curved track design is that it can provide necessary water depths while minimizing the overall track length. With respect to water depths, the optimum design provides for docking at mean low water, especially at locations where the tide range is small.

The earliest, small-capacity marine railways used mud sills or sleeper ties laid on a bed of gravel. As load concentrations increased, the need for greater foundation capacity became apparent. Today, pilings of timber, steel, or concrete are used. When there is rock available in close proximity to the track, concrete footings can be used to support the track. The footings must be designed with the proper aspect ratio for stability, and spaced to be compatible with the lineal load capacity of the railway.

Early track designs were made entirely of timber. Today a variety of material is used, including steel, concrete, and timber. The most common track arrangements are composite structures of concrete and steel or concrete and timber. The composite design consists of a reinforced concrete section above the low water line and steel or timber for the submerged portions. New steel track can be fabricated in the dry in sections 40 to 60 feet long. The sections then are floated into position and submerged where they are secured to steel piles. Compared to the all-timber track and foundation, this type of structure lends itself much better to heavy waterfront construction methods. The finished product is free from marine borer attack, and with good protective coating and cathodic protection systems is very durable in seawater.

The track structure may be composed of two, three, or four ways, depending on the vessel type and size. Most marine railways today are built with a two-way track arrangement.

Cradle Design and Construction

For durability and strength, the cradle superstructure on a modern railway is built of steel, including the transverse beams, runners, columns, and uprights that support the docking platforms. The cradle deck and blocking are made of wood. The cradle structure must have strength and stability to support the ship, and yet at the same time be flexible in longitudinal bending and in torsion to accommodate any irregularities that may occur in the track and/or the ship. Usually the cradle runner structure is a simple running log for the inshore portion and a vierendeel truss for the aft portion. The aft portion of a typical steel cradle under construction is shown in Figure 10.27.

The gage of the cradle runners must be wide enough to provide stability against overturning from wind, current, or seismic loads. As a general rule, the gage of a railway is approximately one-half the beam of the widest vessel and/or one-third the width of the cradle.

The cradle platform is comprised of a series of long and short transverse

444 DESIGN OF MARINE FACILITIES

Figure 10.27. Photograph of built-up cradle under construction.

beams. The short beams are equal in length to the runner gage, and are designed to take loads from the keel blocks only. The long beams extend symmetrically beyond the cradle runners and are designed to support the bilge blocks as well as the keel blocks. Beams may be spaced at any increment, but the generally accepted pattern calls for long beams at 12 feet on center with short beams in between at 4-foot or 6-foot spacing.

A divided cradle is one that is built in multiple sections, usually two, that can be operated as a single unit or detached for independent operation. Its design provides operational flexibility and facilitates cradle maintenance. The divided cradle allows for docking one large ship or two smaller ships. In the latter case, it is possible to have a ship docked on the forward section, for long-term repairs, while the aft section serves ships requiring minor repairs and maintenance. Figure 10.28 shows the aft portion of a cradle operating independently from the bow portion. Maintenance of a divided cradle is somewhat easier than that of a one-piece cradle because it is a relatively simple matter to float away the bow cradle and then haul the after cradle completely clear of the water.

Vertical steel beams, called uprights, are fitted at the end of the long cradle beams. The uprights serve several functions. They support a continuous plat-

Figure 10.28. Photograph of a divided cradle.

form that is used by docking personnel for handling lines during the positioning of the ship. Winches that operate sliding bilge blocks usually are mounted on stands spaced along the docking platform. Also, the uprights are fitted with timber fendering and designed to resist a modest amount of load from ship impact during a docking operation.

Blocking Considerations

Railway dry docks typically are fitted with a standard all-timber blocking system that is three to four feet high. In laying out the block line on a cradle, it is advisable to have the top of the keel blocks parallel to the ship's keel when contact is made; otherwise, tipping forces develop that, in severe cases, could cause grounding instability. The line of the keel blocks can be different from the line of the track. Thus the slope of the track may be gentle or steep to suit local conditions, yet the vessels are lifted on practically an even keel.

Cradles equipped with releasing bilge and keel blocks, sliding retractable bilge blocks, and special high blocking towers permit many types of vessels to be accommodated with minimum time lost and minimum manpower. For example, Figure 10.29 shows sliding bilge blocks towers used on a large marine railway for docking naval vessels.

Figure 10.29. Photograph of high bilge blocks on a marine railway.

Hauling Machinery and Chains

Hauling up the inclined track is accomplished by means of powerful machinery and chains, which pull the cradle and its superimposed load on a system of rollers or wheels. Some of the smaller railways utilize wire rope winches with multipart wire rope in place of chain. The following discussion focuses on the chain haul system and associated machinery.

Developed especially for railway dry docks, hauling machines consist of an electric motor that drives a speed reducer and a train of gears. The gears turn one or more toothed chain wheels, or sprocket wheels, driving the hauling chain. A schematic of a typical haling machine is shown in Figure 10.30. Unlike the cradle, track, and foundation, where local concentrated loads can greatly affect the design, the hauling machine and chains are sensitive only to the total live and dead loads. The load in the chain is entirely a function of the cradle load, the gradient, and the friction, and does not come from the driving motor or engine. Basically, the load in the chain may be expressed as $W \sin \theta + WC_f$, where W is the total weight of the maximum design vessel plus the weight of the cradle and chains, θ is the angle of inclination of the track in degrees, and C_f is the coefficient of friction.

INTRODUCTION TO DRY DOCKS 447

Figure 10.30. Schematic of a railway hauling machine.

For a roller system, the design value for the coefficient of friction ranges between .015 to .020. (Lutts (26) reported that in actual tests completed at the Boston Naval Shipyard on marine railway No. 11, the average coefficient of friction for an existing roller system was calculated to be about .007.) For a wheel system, a coefficient of about .05 is used for design.

The horsepower required is a function of the pull times the speed. Whenever possible, the machine is designed to have a hauling speed of about one vertical foot per minute. For example, if a railway is built on a gradient of 1:22, then the ideal hauling speed would be 22 feet per minute.

With regard to safety features, automatic brakes protect against runaway cradles, which may be the result of human error or electrical failure. Once it is in the full-up position, cradle latches secure the cradle. Also, to protect the machine as well as personnel around it, electrical interlocks are incorporated into the control system to prevent its inadvertent operation.

Chains are far more satisfactory than the best wire rope because of their enormous tensile strength, durability, economy, and ease of connection with special hauling shackles. Railways built today utilize Class 3 or oil-rig-quality chain. This welded, high-strength, alloy chain is about 50% stronger than cast steel chain, and has a working load that is about 40% of the chain's breaking load. Hauling chain can be easily gaged for stretch or checked with calipers to determine its remaining strength. Chains will last up to 20 years in a salt water environment with little or no maintenance.

From the cradle the chain goes to the chain wheel where it turns 180° and then is stretched out along the track for storage. The bitter end is connected to a smaller-sized down-haul or backing chain, which passes through an underwater sheave secured to the track and back to the cradle to form an endless loop. This is illustrated in Figure 10.31. The endless loop system provides a

Figure 10.31. Chain arrangement for a two-chain machine.

448

INTRODUCTION TO DRY DOCKS

Figure 10.32. Marine railway roller system.

means of pulling the cradle down the track if the friction force caused by silt and light debris on the track cannot be overcome by gravity.

Rollers and Wheels

For medium- and large-capacity railways the cradle runs along the track on a system of rollers. A typical roller system is similar to a longitudinal roller bearing and demands the same precision, clearance, and perpendicular alignment. Present-day roller systems consist of cast iron, ductile iron, or cast steel rollers, held in position by steel frames consisting of angles with welded-in malleable iron bushings to receive the roller pintles, as shown in Figure 10.32.

The roller width and spacing are dictated by the maximum lineal load. The minimum roller width for a modern railway is 6 inches, whereas the largest railways to date have 14-inch-wide rollers. Each roller frame holds about 12 individual rollers spaced at approximately 12 to 18 inches on center. Frames are connected together, forming a continuous length of rollers on each way.

Small railways of 250 tons capacity or less usually are built with wheels. Wheels of the fixed-axle type have at least $2\frac{1}{2}$ times the friction of rollers, and so require heavier chain and slightly more powerful machinery than rollers.

Side-Haul Railway Dry Docks

A marine railway side-haul arrangement is most commonly used at river sites, where there are wide ranges of water levels. Periods of low water may last several months, but may be interspersed with shorter periods of high water. The typical vessels navigating such rivers are shallow draft, flat-bottomed craft of light construction. The typical side-haul railway, as shown on Figure 10.33, is constructed on a relatively steep gradient, providing sufficient vertical movement to allow docking at all river stages. In general, the side-haul is expensive to build and is not very well suited to oceangoing vessels. The side-haul railway should be chosen only if careful study indicates that other types are unsuitable.

Figure 10.33. Side-haul railway.

A side-haul typically consists of several cradles, each handled by one or two parts of chain or wire rope. The principal technical disadvantage is that the load on any one cradle section is statically indeterminate and varies according to the location of the ship's longitudinal center of gravity. Consequently, it is necessary to use oversize chains or wire ropes to allow for overload. Furthermore, the heavily loaded wires may stretch, causing differential movement of the cradle. This may be overcome by using chains driven by sprocket wheels running at the same speed (27, 28).

Vessel Transfer on a Marine Railway

Adding to the versatility of marine railways is the ability to transfer vessels from the cradle to berths on shore or vice versa. The advantage gained in having the cradle deck horizontal in the full-up position is fully realized when transfer is involved. Ship transfer systems have become extremely popular because of needs created by the prefabricated, modular-type assembly line techniques used in modern ship construction. Also, the economic benefits of handling several ships for long-term repair jobs as well as the possibilities for winter storage are among the reasons why owners choose to incorporate a transfer berth or berths in the original construction, or to make allowance for such installation at a future date.

The relatively low cost of a transfer system that uses a single railway dry dock as the basic lifting and launching facility makes transfer very advantageous for large-scale developments. The cost varies greatly, depending on the arrangement, number of berths, and selectivity required; but, in general, one transfer berth, including a transfer cradle, will cost from 25% to 30% of the cost of the railway itself.

The principles of rolling and hauling used for the inclined railway can be applied to the transfer of a vessel on a horizontal plane. The difference is that no component of gravity needs to be overcome; only the friction forces must be. The effort required to move a ship horizontally is only a fraction of what it takes to pull it up an incline.

The typical transfer system consists of a pair or a series of ways bearing on concrete footings or pilings. The ways support the transfer cars, which in turn

INTRODUCTION TO DRY DOCKS 451

Figure 10.34. Marine railway with longitudinal transfer.

support the ship. Blocking is provided between the transfer car and the ship to give head room and clearance for working on the hull. It also serves to distribute the vessel weight as much as possible. The transfer cars are fitted with wheels or travel on a roller system.

The choice between side transfer and longitudinal transfer is largely a matter of convenience with respect to the various conditions of the site. Longitudinal transfer off the end of the cradle has proved to be very economical, especially if only a single track is used. However, it does require that the entire hauling machine be depressed in a pit below the level of the transfer rails. When full selectivity of storage space is desired, then two-directional transfer is required. In this case, the cost of end transfer is about equal to the cost of side transfer. With two-directional transfer, the second direction of movement is accomplished at a lower level than the initial transfer. This permits the car for one direction to be superimposed on the car for the other direction. Figures 10.34 and 10.35 show typical transfer schemes for a marine railway.

The ability to transfer vessels on a marine railway has enabled some yards

Figure 10.35. Marine railway with side transfer.

Figure 10.36. Marine railway at Punta Arenas, Chile with longitudinal transfer.

to abandon the use of greased ways for launching ships and to adopt the safer method of controlled launching, with the added advantage of being able to construct the vessel from a horizontal work plane. A modern longitudinal transfer system is depicted in Figure 10.36.

For additional information on the features, characteristics, and general guidelines for design of marine railways, see references (11), (27), (29), and (30). For information on the care and maintenance of marine railways, see reference (28).

10.4 FLOATING DRY DOCKS

Floating dry docks are floating structures with sufficient dimensions, strength, displacement, and stability to lift a vessel from the water. They typically consist of two main parts, pontoons and wing walls (see Figure 10.37). The pontoon (or pontoons) is the prime supporting body, which must displace the weight of the vessel and the dry dock. It also must be able to withstand the transverse bending caused by the ship's weight along the centerline opposed by the water pressure from beneath. The wings provide stability while the pontoons are submerged, as well as space for equipment on the wing deck and inside the wing structure. Part of the wing structure is used for ballast water, needed to sink

INTRODUCTION TO DRY DOCKS

Figure 10.37. Major features of a floating dry dock.

and control the depth of submergence of the dock. Continuous wing walls serve as longitudinal girders. Most dry docks are rectangular in shape, although some are shaped more like ships.

Floating dry docks may range in size up to 100,000 long tons lift capacity, with lengths in excess of 1000 feet and widths greater than 200 feet. They have been constructed of wood, iron, steel, reinforced concrete, and a combination of wood and steel. Nearly all modern floating dry docks are of all-welded steel construction.

The earliest reported use of a floating dry dock occurred around 1700 in the tideless Baltic Sea. In order to affect repairs to his vessel, a ship's captain acquired the hulk of a larger vessel, the "Camel," removed the decks and bulkheads, and cut off the stern. After floating his own vessel into the larger hull, he bulkheaded the stern and removed the water from inside the larger vessel. The "Camel" floated with the smaller vessel inside, high and dry and with the hull accessible for repairs.

For some time after this, floating dry docks were either made from, or made in the shape of, vessel hulls. Gradually the dry docks were made in U-shapes and rectangular or box shapes toward the middle and latter parts of the eighteenth century. At first only wood was available for construction, but the introduction of iron and eventually steel as construction materials provided for improved capabilities and increased sizes in floating dry docks.

As vessels grew larger and heavier, floating dry docks followed suit, with a variety of designs. Maintenance of the dry dock by virtue of a self-docking capability was an important feature. Naval powers, particularly England and Germany, required mobility of their dry docks to service their large fleets scat-

tered across the globe. These two factors contributed heavily to the various types of floating dry docks developed through the years.

Today, floating dry docks continue to be important parts of many ship repair facilities. Through the use of improved protective coating and cathodic protection systems to extend useful life before repairs or drydocking are required, modern floating dry docks are most often built as non-self-docking, one-piece steel units.

Floating dry docks are the most flexible type of repair facility available, in that they can accommodate vessels of all sizes, up through supertankers and VLCCs, and they represent an asset that can be readily relocated as conditions require. They are able to operate with a list and/or a trim to accept vessels floating at almost any attitude. Because these dry docks float adjacent to a pier or dolphins, they can be located beyond pierhead lines and do not occupy valuable waterfront real estate.

Types of Floating Dry Docks

Floating dry docks may be categorized in any of several ways. They may be divided into those that are intended to be self-docking and those that are made as one-piece units and require another dry-docking facility for underwater maintenance and repairs. The self-docking designs generally utilize removable sections that will fit on the remaining portions of the dry dock. Through a cycle of self-docking operations, the entire dry dock can be serviced.

The first such structure was developed by Rennie in the mid-1700s, utilizing sectional pontoons with continuous wing structures. An individual pontoon unit could be disconnected from the wing structure and lifted by the remaining pontoons for maintenance and repairs. This was an important feature as it helped to extend the serviceable life of a dock. This type of dry dock is still in use.

Variations of the Rennie concept include dry docks that are completely sectional with several one-piece wing and pontoon units joined together (see Figure 1.5). An individual unit can be disconnected from the adjacent unit(s) and lifted by the remaining units for service. Other types utilize two relatively short end pieces and a long center section as an operating unit. The short end units can be dry-docked on the longer center unit, and the two end units can be used to lift the center section. Several of these docks are still in existence from their World War II days.

Earlier sectional concepts utilized removable connections along the vertical face between the wing structure and the pontoon units, which permitted the wings to lift the pontoons and the pontoons to lift the wings for service. This design concept, as well as several others, is shown in Figure 10.38.

All sectional dry docks exhibit inherent weakness at the connections between

INTRODUCTION TO DRY DOCKS 455

Figure 10.38. Types of floating dry docks. (After Amirikian (46).)

sections. The strongest dry dock with the most efficient use of building material is the one-piece, continuous box dock.

Floating dry docks also may be differentiated by their shape (i.e., ship-shaped or trough-shaped). The ship-shaped floating dry dock is best exemplified by the U.S. Navy's ARDM (formerly ARD) dry docks of World War II vintage. This is essentially a floating graving dock with a closed bow and a stern gate. The vessel to be dry-docked must fit completely within the confines of the dock. Lift is provided by the displacement of the dry dock structure and the evacuated cavity within that structure. Although it is not self-propelled, its shape allows it to be readily towed to advance bases. The pontoon has less depth than a trough-shaped dry dock of similar lift capacity because the working or pontoon deck is below the outside water level, requiring less operating depth and possibly less dredging at the operating site.

The trough-shaped dry dock, which must have its working deck above water level, requires a deeper pontoon and operating site for an equivalent capacity, compared to the ship-shaped type. Because the ends are open, a docked vessel may exceed the length of the dry dock and overhang the ends of the pontoon. These docks can be relocated, but tow preparations for this type of facility usually are more involved than those required for the ship-shaped dry dock.

Another way to categorize dry docks is by the materials of construction. Originally all floating dry docks were constructed of wood. Iron, then steel, as well

as reinforced concrete, gradually replaced wood as the favored construction material. Composite construction, usually consisting of wooden pontoons and steel wings in a Rennie or sectional design, produced a sound dry dock capable of many years of useful service.

Wooden dry docks were limited in their dimensions and load carrying capacity; so as ships became larger, the wooden dry dock became less practical. Wood was used extensively in World War I and World War II because of a desperate need for dry docks and a shortage of steel, but no wooden dry docks have been built since then. Difficulties in obtaining good-quality wood in the dimensions required and the fact that working with large timbers is a lost art, as well as the increased size of dry docks and the capabilities of steel construction, have led to a discontinuance of wooden dock construction. Wood is a natural construction material for salt water because of its life expectancy in that preserving environment; as long as the wood was kept wet by occasional operation and marine borers could be repelled, it would be preserved. However, exposure of some areas of the wing walls to rainwater and the destructive capabilities of marine borers in the pontoons led to the demise of most wooden dry docks.

Composite construction, incorporating wooden pontoon units and steel wing walls, utilized the advantages of wood's self-preservation in seawater and the strength and survivability of the steel above it. Many of the all-wood dry docks were repaired by replacing the decayed wing walls with steel wing boxes. This helped to greatly extend the useful life of the facility, and a few of these structures are still in operation today. No new composite docks have been built, for reasons cited previously.

Reinforced concrete was used for construction of several floating dry docks for the U.S. Navy during World War II, largely because of the lack of steel during the war period. These docks were smaller in size, with lift capabilities between 1000 and 2800 long tons. Reinforced concrete has not been used much since, although a large floating dry dock project using it was started in Genoa, Italy in the early 1980s (31). As of this writing, that dry dock has yet to be completed and put into operation. Concrete dry docks have a high dead-weight-to-lift ratio, requiring deeper pontoons than steel docks for equivalent lift capabilities (see Figure 10.39).

Additional information on the history and types of floating dry docks can be found in references (2) and (32) through (35).

Today, the use of all-welded steel construction provides an efficient relationship between the weight, strength, and lift capacity needed for modern ship repair facilities. These docks generally are built by shipbuilders or fabrication yards dedicated to building floating dry docks using modern shipbuilding techniques. Figures 10.40 and 10.41 show more modern floating dry docks. Several

INTRODUCTION TO DRY DOCKS 457

Figure 10.39. Comparative water levels and head conditions for dry docks constructed of different materials. (After Crandall Dry Dock Engineers (33).)

of the equipment items mentioned in Section 10.1, such as blocks, cranes, and traveling stages, are evident in these photographs.

Principles of Operation

A floating dry dock works on the basis of the Archimedes principle, displacing a volume of water equal in weight to its own weight. The dimensions and arrangement of the structure, and the presence of the pumping and flooding system, permit the weight of the dry dock to be varied so that the dock's pontoon deck may be submerged to dock or undock a vessel. Water is added to the ballast tanks, sinking the dock until the desired water depth is attained. The vessel is then placed in, or removed from, the dry dock. Water is then evacuated from the ballast tanks, and the dry dock rises from the water until the pontoon deck is above the water level.

The capacity of a floating dry dock may be limited by its buoyancy, its stability, and/or the strength of the structure. Generally the buoyancy will be the

458 DESIGN OF MARINE FACILITIES

Figure 10.40. Photograph of modern floating dry dock, Halifax, Nova Scotia. (Photo courtesy of Marine Industries Limited, Sorel, Quebec.)

limiting condition because sufficient stability and strength will be designed around the buoyancy. There are, however, exceptions to this general rule, involving both stability and strength.

The dimensions of the wing walls and the compartmentation of the pontoon provide the stability for a ship–dock system, and the intact stability must be assessed for all phases during a docking operation. Figure 10.42 shows the important elements for determining stability for a ship–dock system. Critical stability occurs after the vessel's keel comes out of the water but before the pontoon deck breaks the water's surface. As the dry dock rises and lifts the vessel from the water, the vertical center of weight (VCG) of the ship–dock system rises. Positive waterplane inertia is provided by the wing walls and the waterplane of the vessel being lifted. When the vessel's keel breaks the water's surface, the center of gravity of the system is very high, and the positive inertia, now being provided only by the dock's wings, is at a minimum. Until the pontoon deck breaks the water surface, stability is at a minimum. This is the critical phase of intact stability, and the dimensions of the wing walls and the ballast tanks must be coordinated with the dock's design vessel(s) to assure positive

INTRODUCTION TO DRY DOCKS 459

Figure 10.41. Photograph of modern floating dry dock, Portland, Oregon. (Photo courtesy of Port of Portland, Portland, Oregon.)

stability characteristics. Information on the hydrostatic and stability characteristics of a floating dry dock are provided on the curves of form and stability curves, as shown in Figures 10.43 and 10.44. Refer to Section 9.2 for a discussion of the stability of floating bodies.

The change in the stability characteristics of the dry dock due to the large waterplane of the pontoon versus the small waterplane of the wing walls can have a dramatic effect on the operation of the dock. This ''multiplication effect'' is calculated as the ratio of the metacentric height (GM) of the ship–dock sys-

Figure 10.42. Important dimensions for ship–dock stability calculations.

Figure 10.43. Hydrostatic curves for floating dry dock.

tem before and after the pontoon deck breaks the water's surface. Any list present in the dock's attitude as it passes through this phase will be compounded by this ratio (e.g., a 6-inch list will become a 24-inch list for a dock with a multiplication effect ratio of 4:1 as it submerges). This is of more concern as a vessel submerges than it is during the lift of a vessel because the effect is the inverse of the ratio. The dock operator may slow the flooding rate and/or operate the dock on a longitudinal trim to reduce the transition rate and minimize the impact of the multiplication effect during the operation. The dimensions of the dry dock will determine the ratio and the impact of this phenomenon.

Relatively new requirements for ship repair yard operators desiring to drydock U.S. Navy vessels in floating dry docks demand stability analysis for the dry dock in a damaged condition. The impact of this requirement on the design of the dry dock is that a structure is used with more compartmentation and larger wing walls than would be required for intact stability considerations alone. Buoyancy, structural, or intact stability limitations may be secondary to the restrictions on safe docking capacity dictated by these damage stability requirements (36, 37).

The second situation in which buoyancy limitations are not the controlling factor in design involves those dry docks designed with the capability of transferring vessels between the dock and the shore. Often when the purpose is to

Typical Dock-Ship Stability for Various Ship KG's

Figure 10.44. Intact stability curves for floating dry dock. (From Crandall (4).)

use the dock to transfer vessels onto shore, it is necessary to lift vessels to a higher freeboard than would be required for a conventional dry-docking, in order to get the pontoon deck to yard grade. To accomplish this, additional pontoon depth is needed. For economy of design in the strength and weight of the structure, the structural capacity of this extra-deep pontoon may be less than the actual buoyant capability. Thus, the structure is the limiting consideration in the capacity of the dry dock.

Compartmentation of the pontoon provides for more precise control of the dry dock, in addition to enhancing the stability. Differential deballasting of the dry dock's ballast compartments permits a more equal and opposite reaction to the loading imposed by the vessel, reducing stresses in the dry dock and vessel structure during a dry-docking. Also the dock's attitude can be adjusted by differential deballasting to accommodate a vessel's list and/or trim.

The load distribution of a vessel can be computed as shown in Figure 10.45. The analysis treats the vessel as a rigid beam and requires the vessel's displace-

Figure 10.45. Determination of vessel load distribution.

ment, the location of the longitudinal center of gravity, and the location and length of the keel bearing on the blocks. More detailed information, such as that provided by a typical naval architect's weight distribution, usually is not required for standard dockings because of the elasticity of the blocking system between the vessel and the dry dock, as well as the strength and rigidity of vessels and dry docks. More involved docking situations such as those created by damaged ships or in cases where the ship is to be cut while in dry dock require a more detailed analysis. More complex ship hull configurations, resulting in large overhangs and unsupported lengths of the vessel in dry dock, have created concerns about the validity of this simple analysis. More complex analysis procedures are described in references (38) through (42).

Ballasting and dewatering of the dry dock is accomplished through an arrangement of pumps and valves. Pumping or flooding operations should be completed in one to three hours. A floating dry dock may be described as pump-controlled or valve-controlled.

A pump-controlled dock will use a large number of small pumps and valves, as each ballast compartment typically has its own dewatering pump and flood valve. The rate of dewatering for each ballast tank is controlled by starting and stopping the pump. A cross connection is provided to join adjacent ballast compartments together so that if one pump fails, the adjacent pump can be used to dewater two ballast tanks. The pump-controlled arrangement can be beneficial in areas where sediment accumulation in the ballast tanks may be significant. The ballast tanks can be flushed by flooding and dewatering at the same time to stir up the accumulated sediment and discharge it. This type of arrangement may not be practical for pontoons with three or more ballast tanks across their width because of difficulties in reaching the inner tanks and the redundancy needed in the piping system.

A valve-controlled dry dock utilizes a few large pumps to dewater the ballast

Figure 10.46. Pump and valve arrangement for pump-controlled dock.

compartments through a manifold arrangement of piping and valves. The rate of dewatering is controlled by adjusting the valves to each tank while the pump runs continuously. A few large flood inlets also are used to supply ballast to the many tanks via the piping arrangement. This type of arrangement is well suited to docks with three or more ballast tanks across their width. Figures 10.46 and 10.47 illustrate typical pump and valve arrangements for the two concepts.

Design Considerations

A dry dock generally is intended to service a specific vessel or a specific group of vessels that determines the design criteria. Any vessels whose requirements do not exceed those of the design vessel may be serviced in the dry dock.

The area of the pontoon deck must provide sufficient working space under and around the vessel for access by shipyard workers and their equipment. The

464 DESIGN OF MARINE FACILITIES

Figure 10.47. Pump and valve arrangement for valve-controlled dock.

wing deck and safety deck areas must accommodate the equipment to be mounted on or in these areas. The pontoon must be deep enough to provide the required lift capacity for the design vessel, and the wing walls must be sufficiently high to achieve the proper draft, so that vessels can enter and leave the dock over the blocking.

The wing walls must provide the required inertia and buoyancy for conditions of minimum stability and full submergence. The compartmentation of the pontoon must provide adequate stability and buoyancy control for the safe operation of the dock.

The transverse strength of the pontoon must adequately support the concentrated vessel loading along the keel line, and distribute the load across the full

INTRODUCTION TO DRY DOCKS 465

Figure 10.48. Basic design criteria for floating dry dock.

width of the dock to the buoyancy below. Other situations for the distribution of vessel load on the pontoon deck also must be considered, including the loading imposed by side blocks and the effect of buoyant loading with no vessel loading from above (see Figure 10.48).

Longitudinally, shear forces and bending moments must be investigated to ensure that resulting stresses and deflections are within acceptable limits. For a one-piece dock, this loading can be carried by the wing and pontoon structure. For Rennie-type docks, all longitudinal stresses must be carried by the continuous wing walls. Sectional docks cannot carry longitudinal stesses unless the sections are pinned or otherwise connected. In this case, the capacity of the connections is likely to be very limited, and careful ballasting will be required.

The shell plating and the internal watertight bulkheads must be designed to withstand the head pressures experienced during the operation of the dock. A determination of the maximum head pressure encountered during a docking or undocking operation will reveal the worst shell plate loading. The maximum pressure between ballast tanks will be a function of the intended submergence of the dock, the dimensions of the wing walls, and the draft and beam of the vessel to be dry-docked.

Local loading conditions also must be investigated, including keel and bilge block loads and other loadings on the pontoon deck such as vehicular traffic for forklifts or trailer trucks. Foundations for equipment to be mounted on the wing deck and safety deck, such as winches, capstans, ventilators, cleats, and so on, must be designed into the structure.

The mooring arrangement, and means of access by both vehicular and pedestrian traffic, must be considered. The dry dock may be anchored to chains in the middle of a harbor, with access via launches or an access ramp. Often, to make the dry dock more accessible, the dock will be attached to a pier or dolphins close to shore by a fixed mooring arrangement.

Design guidance can be found in the rules of several classification societies, such as the American Bureau of Shipping (43, 44) and Lloyd's Register of Shipping (45), which have prepared rules for the classification of dry docks by their organizations. Additional information on the design of floating dry docks is available in references (4), (33), and (46), and (47).

Features of Floating Dry Docks

Although they are not specifically required for operation, several features considered in dry dock design offer practical advantages. Its accessibility can be a major advantage for this type of facility, as compared to a basin dry dock. The pontoon deck is at or near yard grade and, with adequate access via pedestrian and vehicle access ramps and crane service, can serve as an extension of the yard itself. The rapid and efficient movement of personnel and materials onto and off the dock can make ship repair efforts more efficient.

Remote pump and valve controls and ballast indication systems greatly increase the operators' knowledge of the status of the operation and reduce the manpower needed to control the pumps and valves. A central control station, often a control house located atop one wing wall, contains the water level indicators, valve operators, and pump controls needed to monitor and operate the dry dock. Draft indicators can be displayed and their data interpreted to indicate the deflection, list, and trim condition of the dock.

Alarm systems allow monitoring for fires in the safety deck spaces, and communication systems permit communication between the control house and all areas of the wing and safety decks.

Transfer Systems

Floating dry docks can be equipped to accommodate the horizontal transfer of vessels to and from landside facilities. This may be done in either the longitudinal or the transverse direction. Typically the dry dock will operate in a deepwater or sinking berth, and be translated to the shallower transfer area. The transfer area may include one or more transfer positions, and often will include some means of supporting the dry dock at a fixed elevation.

Support may be provided for the entire dry dock, such as an underwater grid, or just for the end of the dock that mates to the land side. The end support can take the form of a shelf along the waterfront upon which the end or the side of the dry dock rests; it is intended to align the dock at the proper elevation with the landside transfer ways. The support need only be capable of supporting a seating load, and is not required to carry the full weight of the dry dock. The differential ballasting capabilities of the dry dock are used to counteract the weight of the dry dock and the moving load. Some floating dry docks rely solely on ballast control to maintain proper elevation, but utilize long transition beams from the shore to the dry dock to accommodate errors in the elevation of the dry dock and the shore.

The transfer capacity generally will be less than the lift capacity because the pontoon deck must be lifted to yard grade, thus reducing the amount of lift applicable to vessel loading. End transfer is readily accomplished because of the open ends of the floating dry dock; a side transfer is considerably more cumbersome than end transfer, as the operation involves removal of the wing boxes on the side where the transfer will be made. Figure 10.49 schematically shows both types of operations, and Figure 10.50 shows an end transfer operation in progress.

Several methods and mechanisms are available for the movement of vessels from shore facilities to or from a floating dry dock, including skidding, roller systems, wheeled systems, air or water bearings, and walking beams.

The horizontal skidding of vessels resting on wooden blocks and cradles along greased ways is an ancient concept. Methods in use today are similar to it, with the exception of the equipment used to propel the vessel. Hydraulic jacks, used in tandem with wedged grippers that clamp onto the flanges of the transfer way, are capable of continuous motion.

Roller systems may use hardware similar to that of marine railways. Recirculating rollers and spherical ball units also may be used as the transfer mechanism. Roller frame systems are capable of considerable load, but have the disadvantage of requiring extra length in the roller frames in consideration of the differential movement between the frame and the vessel. Recirculating rollers eliminate this problem, but tend to concentrate loads to reduce the number of units required for a transfer. Spherical ball units are capable of movement in any direction, but have not been widely used.

SIDE TRANSFER WITH DOCK
RESTING ON GRID

END TRANSFER WITH END
OF DOCK ON SILL

Figure 10.49. Schematic of end and side transfers from a floating dry dock. (After Crandall (48).)

Air and water bearings are fairly recent developments in transfer systems. Several pallet-type units are placed beneath a cradle in a pattern that counteracts the weight of the ship to be moved, and the flow of air or water creates a pressure between the pallet and the surface of the transfer area, floating the cradle with the vessel. Movement is relatively easy to achieve because of the low friction involved, although this could be detrimental in high winds or on surfaces with very small slopes. The need for a very smooth surface is another drawback of such systems.

Walking beams often are used as a transfer mechanism for subassemblies but only rarely used for complete ships. These units are self-contained and capable

INTRODUCTION TO DRY DOCKS 469

Figure 10.50. Photograph of end transfer operation on a floating dry dock. (Crandall Dry Dock Engineers file photo.)

of movement in any direction; they will traverse slight grades and smoothly graded surfaces that have adequate bearing capacity. Their disadvantage is that they are rather complex.

The most commonly used transfer method is the wheeled system, consisting of continuous cradles or of individual cars, usually of four wheels each. Continuous cradles usually are equipped with two or more rows of flanged wheels that travel on flat or crane rails. Movement is provided by an external source, such as winches with wire rope or a tractor. Individual cars usually have four wheels each, and require a double track. Often these cars are equipped with vertical jacks and are electrically self-propelled, offering additional capabilities for vessel construction. Individual strongbacks or cradle beams usually are used with these cars to provide flexibility. Although the continuous cradles will remain with a vessel during its stay on shore, individual cars can be removed from under the strongbacks and used elsewhere. They also must be removed prior to submergence of the dry dock so they will not receive water damage. Wood and/or rubber blocking may be used to support the vessel on the cradle or strongbacks. References (27), (49), and (50) provide further information on transfer systems.

10.5 VERTICAL LIFTS

A vertical lift dry dock or shiplift is simply a ship elevator, consisting of a horizontal platform and lifting equipment. The platform, which supports the

470 DESIGN OF MARINE FACILITIES

Figure 10.51. Plan of a typical shiplift.

ship, is fabricated from transverse and longitudinal steel beams. The platform support is transmitted from the ends of the beams, through cables, chain, or hydraulic components, to the lifting machinery, which is a series of jacks or hoists mounted on a pier structure on each side of the platform.

The number and capacity of the lifting units determine the overall lifting capacity of the system, whereas the spacing of the units determines the per foot capacity. The dry dock is designed to vertically lift a total load equal to the dead load of the platform plus the superimposed live load. A schematic plan of a typical shiplift is shown on Figure 10.51.

Although various types of do-it-yourself systems have been developed, including some using buoyancy principles with positive stabilizing and support structure, the two most common types of vertical lifts are mechanical and hydraulic. Figures 10.52 and 10.53 illustrate the major components of each of these types.

Vertical lift systems are the newest innovations in large shiplift systems; the practical versions of this concept are about 30 years old.

Features of the Vertical Lift System

The horizontal, land-level attitude of the platform ideally suits vertical lifts for transfer of vessels to shore, and most vertical lifts are designed with extensive transfer systems (see Figure 10.54). As many as 40 land berths have been served by a single vertical lift.

Material can flow naturally to a vessel on a shiplift because of the uncluttered area around the platform. The absence of vertical walls, which are present in floating dry docks and basin dry docks, permits easy access to the ship.

INTRODUCTION TO DRY DOCKS 471

Figure 10.52. Components of a mechanical-type lift.

The speed of operation for a vertical lift is greater than that of most other types of dry dock; the average time for dry-docking a ship in this manner is 20 to 40 minutes. Furthermore, the total number of operating personnel required for a typical docking is somewhat less than for other types of dry docks of equal capacity. The remote operations center that houses the hoist controls and load

Figure 10.53. Components of a hydraulic-type lift.

Figure 10.54. Typical transfer berth arrangement.

readouts is in an enclosure nearby, with good visibility of the entire docking operation.

Finally, a vertical lift is particularly well suited for lengthening, and can be expanded by the addition of platform sections along with an appropriate number of hoists supported on extensions of the existing piers.

Mechanical Lift Systems

Typically, mechanical lift systems use electrically driven hoists with multipart wire rope cables attached to structural beams through a system of running sheaves. Most of the lifts use galvanized wire rope, which should be inspected frequently, as it is subject to fatigue from axial and bending stresses. Also, immersion in salt water leads to corrosion between the strands that is difficult if not impossible to detect. Therefore, hoisting ropes should be regularly greased to minimize wear and to prevent corrosion. Also it is recommended that a schedule be established for cable testing and/or replacement. Some systems

utilize galvanized rope with a plastic sheathed core. The plastic, which has a high compressive strength, forms a water-resistant coating and improves the fatigue life of the wire rope (51).

The motion of individual hoists must be synchronized as closely as possible to ensure that all platform lift points cover the same distance in the same period of time. Seeholzer has stated that "synchronous running merely serves to prevent uncontrolled changes in rope loads during lifting or launching, taking into account that these loads are not uniform from rope to rope due to varying load distribution of the ship" (51). Synchronization is achieved through the use of AC induction motors that operate at a constant speed regardless of load.

Two of the most popular mechanical-type shiplift systems are the Schiess-Defries Lift-Dock and the Synchrolift system. The Syncrolift system was patented in 1958 and has been perfected for lifts of 30 tons up to nearly 30,000 tons. Further description of the Lift Dock and Syncrolift systems can be found in references (51) and (52).

The largest vertical lifts built to date are Syncrolift systems. The largest Syncrolift, in terms of overall platform size, is the shiplift system at Todd Pacific Shipyards, Los Angeles Division (see Figure 10.55). It has a gross lifting ca-

Figure 10.55. Photograph of Synchrolift installation. (Courtesy of Todd Pacific Shipyards Corporation.)

pacity of 26,400 long tons. The live load lifting capacity for docking directly on the platform (i.e., with no transfer cradles) is 22,200 long tons. The maximum allowable load per foot for docking on the cradles is 33 long tons per foot. The platform measures 655 feet by 106 feet and has a lifting speed of 9 inches per minute. The available vertical travel is 54 feet, and the maximum draft over the cradle is 32 feet.

Todd's Synchrolift is an ideal example of a vertical lift's suitability for transferring vessels to shore. The system is designed for end transfer from the platform onto a transfer table that allows side transfer to a number of work bays. This scheme allows optimum utilization of the available land and provides a means for servicing several ships at one time. Further information on Todd's Syncrolift system can be found in reference (53).

Hydraulic Lift Systems

Hydraulic shiplift systems are not nearly so popular as the mechanical systems. Several types of hydraulic jacking systems have been proposed and built, some utilizing long thrust jacks and others utilizing short thrust jacks, the latter being more common. A long thrust jacking system is designed to lift the platform in a single stroke, whereas the short thrust jacking system relies on a number of repetitive strokes. In general, lifts that use automated short thrust jacks are simpler, more reliable, and less costly than the long thrust type. Information on the general characteristics, advantages, and disadvantages of long thrust jacking systems may be found in reference (54). The remainder of this section will focus on short thrust jacking systems.

Two types of hydraulic lift systems using the short thrust jacks are the Kramo jacking system, produced by Kramo Limited of England, and the Hy-Chain lift system, produced by the joint venture of AG Weser Seebeckwerft, F.A.G. Kugelfischer, and Georg Schafer & Company (55, 56).

In the Kramo system the load is suspended from the jacking units by six- or eight-inch square steel bars. The jacking units are mounted on a pier at yard level, and the lower end of the bar is attached to the platform. Operation of the jacking units raises the steel bar, thereby lifting the platform. The upper end, or free end, of the steel bar must be guided or removed as it projects upward from the jacking unit during the lifting operation.

As shown in Figure 10.56, each jacking unit is fitted with a set of equalizing cylinders. In operation, the suspended load is transferred from the load rod to the jacking unit, through the equalizing cylinders, and into the jack's support beam attached to the pier. A spherical bearing incorporated into the main support beam that houses the jacking unit ensures that only axial loads will pass into the jacking unit, and also provides sufficient articulation to allow the platform to adopt a modest angle of trim.

The standard hydraulic pumps allow for a climbing rate of between 7 and 12

Figure 10.56. Kramo jacking system. (Reproduced with permission of Kramo Ltd. (55).)

inches per minute. All jacking units are operated from a central control panel by a single button. At any point during the lifting or lowering of the platform, the operation may be stopped and any single jacking unit adjusted by means of its own individual control.

A typical jacking unit, which is made in standard sizes from 100 up to 400 tons, is comprised of two steel head blocks with double-acting hydraulic rams between. Each head block contains a pair of steel wedges whose faces have grooved teeth. When a vertical load is applied, wedge action forces the grooved teeth against opposite-side faces of the square lifting bar, causing the teeth to bite into the bar and thus provide a virtually fail-safe holding. A 400-ton jacking unit is shown in Figure 10.57.

The Hy-Chain lift system is somewhat similar to the Kramo system in that it relies on a hydraulic locking and lifting unit, but instead of square steel load bars it uses a specially designed chain made from heavy rectangular link bars joined together with steel rod stock.

Design and Construction of Vertical Lifts

The waterfront space requirements for a vertical lift are similar to those of a floating dry dock. The actual space required is the area of the shiplift platform

476 DESIGN OF MARINE FACILITIES

Figure 10.57. Photograph of a 400-ton jacking unit. (Courtesy of Kramo Ltd.)

plus some additional space on each side for the lifting mechanism support platform. The lift may be installed perpendicular or parallel to the shoreline, either fully or partially outside the low water line or completely above the high water line. Although a vertical lift does not require space out in the water as the track structure of a marine railway does, it is not very well suited to long sloping sites where the shipyard is at a high elevation. The vertical lift has proved to be economical for existing deepwater berths where dredging is minimal.

The lifting force required is determined by the intensity of the load from the ship. Because of load concentrations along the length of a vessel, certain jacks or hoists carry more load than others. Thus for a vertical lift system the lifting capacity required is not determined by the total vessel weight, but by the maximum load concentration applied along the entire length of the platform. The net capacity could then be as much as twice the ship's total displacement. This is particularly true of dry docks designed for certain surface warships with very high load concentrations aft. Most lifts today are fitted with load cells at the lifting points that display the total weight and the load distribution on the platform. In the Syncrolift shiplift, load cells automatically preclude the possibility of hoist overloads.

For those systems designed for transfer, the ship's movement onto and off the platform may be transverse or longitudinal, with the direction determining the type and construction of the platform. In general, the weight of a platform designed for longitudinal transfer is about one-third greater than that of one designed for side transfer because heavier longitudinal beams are required for support of the transfer rails in the former case.

There are two basic types of platform: rigid and articulated. In a rigid platform, the longitudinal beams are continuous, spanning across the transverse beams. In an articulated platform, the longitudinal beams are framed intercostally with the transverse beams. The longitudinal beams are lighter in this case, and are framed to act as statically flexible elements between the main transverse beams. Relatively speaking, any type of platform is flexible when compared to the very stiff structure of a ship's hull; but, in general, the rigid-type platform ensures a linear distribution of load over the platform due to its ability to redistribute load to adjacent beams. Either type of platform can be designed to operate as two or more separate sections for independent dockings of small ships. Furthermore, one section can be vertically offset from another for clearance of a hull projection such as a propeller or a sonar dome.

The wire rope hoisting cables should be sized for a design load or working load equal to the nominal load (i.e., dead load plus live load) times an impact factor of at least 50%. The factor of safety, which is the ratio of the breaking load of the cable divided by the working load, should be on the order of 6.0 to 8.0.

The hoisting units for wire rope systems consist of small AC synchronous motors, reduction gearing, a wire rope drum, and a brake. The brake is designed to stop the unit automatically in the event of a power failure. The size of the motor and the reduction gearing are generally selected to allow for a lifting speed of 6 to 12 inches per minute.

The lifting units usually are mounted on pile-supported timber or concrete piers on each side of the shiplift platform. In addition to supporting the lifting units, the piers are designed for rubber-tired cranes or provided with rails for traveling cranes. The elevation at which the electric hoists are mounted must be carefully selected. They should be well above extreme high water levels (i.e., storm tides) to avoid electrical damage from flooding.

Further information and guidelines for the design of structural and mechanical elements of a vertical lift may be found in reference (57).

10.6 STRADDLE LIFTS AND BOATYARD EQUIPMENT*

The introduction of straddle cranes into marine service in the late 1950s revolutionized the small craft hauling and storage industry. Boatyards were enabled to readily haul, store, and load yachts and small craft onto trailers with complete flexibility. Straddle hoists are currently available in 15- to 500-ton lifting capacities. Generally, however, even the largest yachts (to 60-foot length or larger) can be lifted on hoists of 50-ton or smaller capacity (see Figure 10.58). Larger-

*Sections 10.1 through 10.5 were contributed by Mr. Becht and Mr. Hetherman; Section 10.6 was written by Mr. Gaythwaite.

Figure 10.58. A 50-ton-capacity boatyard straddle lift with a small fishing vessel. (Courtesy of Marine Travelift™, Inc.)

capacity lifts allow the docking of commercial and service craft, including tugs and barges, and may be an attractive option to small marine railways. For larger vessels, of 100 tons or more, special attention should be given to the vessel load distribution and the placement of slings (58).

Straddle lifts require two parallel runways, usually on pile-supported finger piers, of sufficient length and water depth between them (see Figure 10.59). Typical clear widths between piers range 15 to 20 feet for lifts up to 60 tons capacity to over 30 feet for the larger commercial vessel machines. The vessel is supported fore and aft by slings, usually of reinforced nylon fabric, and hoisted by wire ropes reeved in several parts and equalized for load balance. Most contemporary machines are equipped with one top cross-beam that can be retracted (or is omitted entirely) to allow hauling of vessels with masts or high superstructures. Runway girders may be of heavy timbers drifted together for the smallest sizes, combinations of timber and steel plate or shapes, or reinforced or prestressed concrete. They should be designed for stiffness to minimize deflections and assure favorable load distribution to the piles. Wheel loads for a given piece of equipment must be obtained from the manufacturer. As a rule of thumb, single wheel or single leg dual wheel loads are usually on the order of one-half of the machine's rated capacity, to which a 15% impact factor should be applied in the girder design. Maximum tire inflation pressures are on the order of 115 psi. A lateral load equal to a minimum of 10% of the wheel

INTRODUCTION TO DRY DOCKS 479

✱ DENOTES DIMENSION USUALLY SPECIFIED BY LIFT MANUF.

Figure 10.59. General features of typical straddle lift pier.

load should be applied simultaneously with the vertical loads to the overall structure stability. Each finger pier should be capable of resisting a minimum longitudinal thrust due to braking of 10% of the combined live load of the machine plus the vessel.

Runway widths should be obtained from the equipment manufacturer. Usually 3 to 4.5 inches inside clearance from the tire to the curb plate is required, and adequate allowance must also be made for minor steering corrections and possible misalignment of the hoist frame. A four- to five-foot walkway width should be provided outside of the runway clearance. The walkway and the runway girder may be one continuous member, or for larger machines some cost savings may be attained by constructing the walkway of lighter (timber) framing. In this case, a minimum uniform live load capacity of 60 to 100 psf is recommended for the walkway. Maximum travel speeds are on the order of 125 feet per minute, and maximum grades should be kept to 5% or less. Turning radii and operating clearances must be obtained from the manufacturer.

Other types of small boat lifting and handling equipment include forklift trucks, bulkhead-mounted forklifts, and crane-type davit hoists. Special forklift trucks have been developed for dry-stacking of small boats (typically power boats under 25 feet). These lift trucks have extended forks that may be elevated

Figure 10.60. Jib-boom hoist for small craft.

to 30 feet above ground, and some have negative lift capability for extending their forks 10 to 12 feet below grade to lift boats from the water. Such trucks usually have maximum capacities on the order of five to six tons at an 8-foot load center. Maximum wheel and axle loads should be obtained from the manufacturer and designed for in accordance with Section 4.2. Less common are bulkhead-or pier-mounted forklifts that rotate about their base at deck level. Capacities usually are limited to less than 10 tons.

Crane-type hoists with fixed bases may be of various types, the most common probably being the jib-boom type depicted in Figure 10.60. The cantilevered boom can rotate on its pedestal base or column support, and usually is equipped with a remote electric hoist that can be fixed at its outboard end or mounted on a trolley that can move along the boom. These hoists usually are limited to five tons or less capacity, and less than a 10- to 15-foot outreach. Hydraulic pedestal-mounted cranes with fixed or telescoping booms, as designed for shipboard use, also may be employed. Their capacity usually is limited to less than 10 tons and a 10- to 20-foot maximum outreach. Derrick-type cranes, consisting of a vertical post (mast) with a long boom hinged near its base and secured to the mast via a wire rope topping lift, occasionally are seen. Hoisting is accomplished by a wire rope winch, usually reeved in parts to reduce the line pull. The fixed-base pedestal and derrick-type hoists create large moments about their base, which usually require substantial foundations or below-deck framing.

Additional discussion of boatyard equipment can be found in references (59) and (60).

REFERENCES

1. Crandall, P. S. (1964). "Characteristics and Relative Merits of Railway and Floating Dry Docks," paper presented to the Society of Naval Architects, Philadelphia.
2. Cornick, H. F. (1958). *Dock and Harbor Engineering*, Vol. 1: *The Design of Docks*. Griffin, London.
3. Mazurkiewicz, B. K. (1980). *Design and Construction of Dry Docks*. Trans Tech Publications, Clausthal, Germany.
4. Crandall, P. S. (1987). *Dockmaster's Manual*. Crandall Dry Dock Engineers, Dedham, Mass.
5. Crandall, P. S. (1987). "Dry Dock Blocking Systems for Modern Ships," *Journal of Ship Production*, Vol. 3, No. 1, SNAME.
6. Salzer, J. R. (1986). "Factors in the Selection of Drydocking Systems for Shipyards," *Journal of Ship Production*, Vol. 2, No. 2, SNAME.
7. Crandall, P. S. (1976). "Large Floating Dry Docks for Large Ships," *Marine Technology*, Vol. 13, No. 2, SNAME.
8. Colson, C. (1894). *Notes on Docks and Dock Construction*. Longmans, Green, London.
9. PIANC (1988). "Report on Dry Docks," PIANC, Supp. to Bull. No. 63, Brussels.
10. NAVFAC (1969). *Drydocking Facilities*, DM-29. Dept. of the Navy, Alexandria, Va.
11. BSI (1985). *British Standard Code of Practice for Maritime Structures*, BS 6349, Part 3: "Design of Dry Docks, Locks, Slipways and Shipbuilding Berths." British Standards Institution, London.

12. Hassani, J. J. (Apr. 1973). "Very Large Graving Docks," ASCE National Structural Engineering Meeting.
13. Krauss, F. E. and Plude, G. H. (Nov. 1972). "1200-ft. Graving Dock," *Journal of the Waterways, Harbors and Coastal Engineering Division, Proc. ASCE.*
14. Millard, C. F. and Hassani, J. J. (1971). "Graving Dock for 300,000 Ton Ships," *Civil Engineering*, ASCE.
15. Tate, T. N. (Dec. 1961). "World's Largest Dry Dock," *Civil Engineering*, ASCE.
16. Stott, P. F. (Nov./Dec. 1957). "Modern Dry Docks: Design, Construction and Equipment. Siting Design and Construction," *The Dock and Harbour Authority.*
17. Morrison, R. A., Bird, G. R., and Waggoner, J. T. (1989). "Design and Performance of Hydraulic Relief System," Ports '89, Proc. of the Conf., Kenneth M. Childs, Jr., ed. American Society of Civil Engineers, New York.
18. Wu, A. H., Yachnis, M., and Brooks, E. W. (1988). "Structural Analysis of Graving Docks by the Finite Element Method," PIANC, Supp. to Bull. No. 63, Appendix B, Brussels.
19. Grieve, J. (Mar. 1960). "Pumping Plant for Dry Docks," *The Dock and Harbour Authority.*
20. Wauchope, G. A. (Mar./Apr. 1958). "Modern Dry Docks: Design Construction and Equipment. Dock Pumping Machinery," *The Dock and Harbour Authority.*
21. Gardener, H. (May 1958). "Modern Dry Docks: Design Construction and Equipment. Culvert Sluices and Valves," *The Dock and Harbour Authority.*
22. Hassani, J. J. (1987). "Shipyard Facilities—New and Old Closures for Dry Docks," *Journal of Ship Production*, Vol. 3, No. 3, SNAME.
23. Walker, P. J. (Jan. 1958). "Modern Dry Docks: Design Construction and Equipment. Caissons for Dock Entrances," *The Dock and Harbour Authority.*
24. Baily, S. C. (Sept. 1947). "Reinforced Concrete Sliding Caissons," *The Dock and Harbour Authority.*
25. Danks, J. Y. (Feb. 1958). "Modern Dry Docks: Design Construction and Equipment. Dock Gates," *The Dock and Harbour Authority.*
26. Lutts, C. G. (Mar. 1955). "Results of Hauling Load Measurement Tests on Marine Railway No. 11," Boston Naval Shipyard, Materials and Chemical Lab.
27. Crandall Dry Dock Engineers, Inc. (1980). "An Introduction to Railway Dry Docks and Transfer Systems." Crandall Dry Dock Engineers, Dedham, Mass.
28. Crandall Dry Dock Engineers, Inc. (1979). "Care and Maintenance of Marine Railway Dry Docks." Crandall Dry Dock Engineers, Dedham, Mass.
29. Mackie, K. P. (Apr. 1979). "Mechanical Systems of Drydocking, An Investigation in the Design and Construction of Slipways and Shiplifts," submitted to the University of Cape Town.
30. Crandall Dry Dock Engineers, Inc. (1989). "The Modern Marine Railway Dry Dock," *Shipbuilding Technology International 1989*. Sterling Publications, London.
31. Giuffre, A. and Pinto, P. A. (n.d.). "Genoa's Floating Prestressed Concrete Dock," Societá Generale per Lavori e Pubbliche Utilitá, Rome, Italy.
32. Cunningham, B. (1910). *A Treatise on the Principles and Practices of Dock Engineering.* Griffin, London.
33. Crandall Dry Dock Engineers, Inc. (1977). *Floating Dry Docks: Their History and Characteristics.* Crandall Dry Dock Engineers, Dedham, Mass.
34. Cook, V. E. (1957). "General Discussion of Floating Dry Docks," *SNAME Trans.*, Vol. 65.
35. Thatcher, K. C. (1978–79). "Floating Docks, Development and Modern Trends," Lloyd's Register Technical Association, Paper No. 4.
36. U.S. Navy, Naval Sea Systems Command (1987). *Military Standard—Safety Certification Program for Drydocking Facilities and Shipbuilding Ways for U.S. Navy Ships*, MIL-STD-1625C (SH).
37. Wasalaski, R. G. (Oct. 1982). "Safety of Floating Docks in Accordance with MIL-STD-1625A," *Naval Engineers Journal*, ASNE.

38. Jiang, I., Grubbs, K., Haith, R., and Santomartino, V. (1987). "DRYDOCK: An Interactive Computer Program for Predicting Dry Dock Block Reactions," *SNAME Trans.*, Vol. 95.
39. Palermo, P. M. and Brock, J. S. (Apr. 1956). "Investigation of Pressure of Keel Blocks during Drydocking of USS *Midway* (CVA 41)," David Taylor Model Basin Report No. 1003.
40. Potvin, A. B., Hartz, B. J., and Nickum, G. C. (1969). "Analysis of Stresses in a Floating Dry Dock Due to a Docked Ship," *SNAME Trans.*, Vol. 77.
41. Vaughan, H. (1966). "Elastic Analysis of a Ship in a Floating Dry Dock," *SNAME Trans.*, Vol. 74.
42. Yeh, G. C. K. and Ruby, W. J. (1952). "A New Method for Computing Keel Block Loads," *SNAME Trans.*, Vol. 60.
43. American Bureau of Shipping (1977). *Rules for Building and Classing Steel Floating Dry Docks.* American Bureau of Shipping, New York.
44. American Bureau of Shipping. *Rules for Building and Classing Steel Vessels.* American Bureau of Shipping, Paramus, NJ (published yearly).
45. Lloyd's Register of Shipping (1973). *Rules and Regulations for the Construction and Classification of Floating Docks.* Lloyd's Register of Shipping, London.
46. Amirikian, A. (1957). "Analysis and Design of Floating Dry Docks," *SNAME Trans.*, Vol. 65.
47. Andersson, T., Buczkowski, T., and Doerffer, J. W. (1976). "Some Aspects of the Design and Building of Large Floating Docks," *SNAME Trans.*, Vol. 84.
48. Crandall, P. S., (May 1975). "Floating Dry Dock Doubles as Launching Platform," *Civil Engineering*, ASCE.
49. Salzer, J. R. (1986). "Shipyard Transfer Systems," *Journal of Ship Production*, Vol. 2, No. 3, SNAME.
50. Chambers, H. B. (1976). "Heavy Load Moving Systems," *Marine Technology*, Vol. 13, No. 2, SNAME.
51. Seeholzer, H. (1988). "Economic Drydocking with the Lift-Dock," *Shipbuilding Technology International 1988.* Sterling Publications, London.
52. Pearlson Engineering Company. "Syncrolift—The World's Most Modern Drydocking & Transfer System," brochure and technical literature. Pearlson Engineering Company, Miami, Fla.
53. "Shiplift and Land Level Transfer System," a description. Todd Pacific Shipyards Corporation, Los Angeles Division, Los Angeles, Calif.
54. Vakharlovskiy, G. A. et al. (1970). "Modern Dock Structures for Large and Medium-Size Ships." Joint Publications Research Service, Washington, D.C.
55. Kramo Limited. "Kramo—Hydraulic Jack Lifting System," brochure and technical literature. Kramo Limited, South Humberside, England.
56. The Hy-Chain Organization. "Hy-Chain Lift," brochure and technical literature. The Hy-Chain Organization, a joint venture of AG Weser Seebeckweft, F.A.G. Kugelfischer, and Georg Schafer & Co., West Germany.
57. Lloyd's Register of Shipping (1984). *Rules and Regulations for Lifting Appliances in a Marine Environment.* Lloyd's Register of Shipping, London.
58. Tobiasson, B. O. (1989). "Vessel Considerations for Straddle Hoist Dry Docking," Ports '89, *Proc.* the Conf. Kenneth M. Childs, Jr., ed. American Society of Civil Engineers, New York.
59. Chamberlain, C. J. (1983). *Marinas: Recommendations for Design, Construction, and Management.* National Marine Manufacturers Association, New York.
60. Dunham, J. W. and Finn, A. A. (1974). *Small Craft Harbors: Design, Construction, and Operation*, SR-2, CERC, USACE, Fort Belvoir, Va.

11

Rehabilitation, Maintenance, and Repair

Marine structures require routine maintenance and periodic repairs because of the harsh environment and heavy usage to which they are subjected. In addition to regular maintenance and repairs, many facilities are upgraded and modernized to keep pace with developments in vessels and material handling technology. Others may be converted to new uses as a result of changing demands. In most developed countries, there are few available virgin sites for new marine terminals; hence there is a strong emphasis on restoring, upgrading, and/or converting existing facilities.

This chapter begins with a general discussion of important considerations in rehabilitating, reconstructing, and maintaining existing structures. The importance and the central elements of a proper material condition inspection are reviewed. Primary modes of deterioration are discussed. A general approach to restoration and structural repairs is presented, and a separate section is devoted to pile and bulkhead restoration measures, which are among the most ubiquitous of port structure problems. Finally, the uses of protective coatings, claddings, and cathodic protection systems are reviewed.

11.1 REHABILITATION, RECONSTRUCTION, AND MAINTENANCE

Marine structures are subjected to harsh environmental and service conditions as described in the following section. A proper maintenance program involves constant vigilance and repair of damage and deterioration as they occur. This is especially important for marine structures where structural failure or conditions that eventually render them unserviceable are the result of cumulative damage and gradual deterioration effects. Unfortunately, marine structures typically suffer from neglect, often reaching the stage where major repairs or reconstruction are necessary. Good design and detailing practice and the proper selection and application of materials and protective systems will greatly increase a structure's life and reduce the chances of catastrophic failure. Certain structures may be deliberately overdesigned in order to preclude maintenance requirements.

Some of the most common maintenance and repair problems in pier and wharf construction are as follows:

- Pile and bulkhead deterioration, especially within the tide zone.
- Fender system damage.
- Deck surface wear and erosion, including cracking and spalling of deck, beams, and pile caps.
- Crane rail attachment problems resulting in spalling of grout and misalignment of rails.
- Mooring hardware connections becoming loose or corroded.

According to UNCTAD (1), average annual maintenance costs as a percent of current new cost range from 0.3% for steel sheet pile structures, to 0.75% for concrete deck on concrete piles, to 1.0% for concrete deck on steel piles for quays and wharves. Rubber fender system maintenance can be estimated at 1.0% of current new (replacement) cost.

Subsequent sections of this chapter deal with identifying various modes of deterioration, general repair methods, and pile and bulkhead preservation and protective measures. This section presents a general review of considerations involved in upgrading and modernization, factors affecting the decision to rehabilitate versus reconstruct, and important aspects of inspections and design considerations. References (2) through (5) cover general port and waterfront structure maintenance. References (6) through (8) contain useful papers on case histories in marine structure rehabilitation.

Upgrading and Modernization

Upgrading and modernizing an existing facility whenever feasible often will produce cost savings over building a new facility or completely reconstructing the existing one. In addition, recycling an existing facility likely will avoid many regulatory delays and environmental objections. Modernization often can be carried out so as to minimize interference with ongoing activities. Also, savings in overall construction time often are realized. Upgrading usually is required as a result of changes in vessel and material handling technology, which typically mean larger, deeper-draft vessels and heavier, faster cargo handling equipment. In any event, upgrading, modernization, and/or conversion of an existing facility usually involves any or all of the following requirements:

- Structural strengthening for both vertical and lateral loads.
- Dredging for maintenance or increased water depths and maneuvering area.
- Change in the overall layout and an increase in backland storage areas.
- Safety and pollution control improvements.
- Provisions for new mechanical equipment.

The structural strengthening of an existing structure may prove to be a difficult and expensive task. If the existing structure at least can be restored to a

"like new" condition, then other approaches may be taken to deal with increased loads. The use of contemporary high-energy-absorption resilient fender units can greatly reduce vessel berthing reaction forces; in fact, fender systems often can be designed around the lateral load capacity of the pier (see Section 5.6). Increased mooring loads may be handled by providing supplemental mooring structures or providing local reinforcing and/or means of load distribution at mooring hardware. Marginal wharf structures may have additional mooring hardware located along the backshore perimeter if the lead of the larger vessels' hawsers permits. Strengthening of deck systems often involves the need to accommodate heavier wheel loads, which sometimes can be achieved by the addition of topping slabs to distribute the wheel loads, although this may result in the loss of some uniform live load capacity. Limiting the live load storage capacity may be an acceptable trade-off in many instances. Contemporary methods of analysis sometimes may reveal a certain reserve strength not counted upon in the original design. One should exercise caution in this regard, as certain building code requirements now are more conservative than earlier versions. More frequent application of dynamic analysis also has resulted in more conservative designs in some instances.

Increased dredge depths may result in foundation stability problems for wharves and bulkheads and in increased effective column lengths of piles.

Facility layout requirements vary, but contemporary uses usually require the addition of space. Backland storage areas, paving, and roadways often require maintenance or upgrading (9). Older-vintage general cargo terminals usually had transit sheds or warehouses built onto the wharves and piers themselves, whereas today's mobile equipment, larger cranes, and containerized cargo require larger and more open spaces than were once needed. Railroad spurs once were prevalent on piers and wharves, whereas today the number of tracks required, if any, generally is minimal.

Contemporary marine terminals must consider many safety-related and pollution control features that were not required in earlier times—most notably, fire protection and oil containment and collection systems.

Provision for larger and heavier cargo handling equipment is a frequent requirement. Older gantry cranes often suffer from metal fatigue and should be inspected carefully if being reconditioned for new use (10).

Rehabilitation versus Reconstruction

The decision to reconstruct versus repair and rehabilitate often is ultimately an economic one, but there are other important considerations, many of which should be accounted for in engineering economic terms. It is important to consider the design life of the restored facility versus the design life of a newly constructed one. The life cost then can be evaluated on a first cost/present worth

basis, provided periodic maintenance costs for both options are included, using rudimentary methods of economic analysis. If the structure is being upgraded in either dimensions or load capacity, and/or being converted to a new use, which is often the case, rehabilitating it to its former "like new" condition obviously will not suffice; additional engineering analysis is required. Cargo handling facilities, for example, usually need to be upgraded to meet the requirements of contemporary vessels and cargo handling equipment and methods. This almost invariably involves increases in the load carrying capacity of the structure. On the other hand, in utilizing a former cargo wharf or quay for residential or recreational purposes, one may find that the structure still possesses sufficient capacity in spite of material wastage; the problem then involves prevention of further deterioration and aesthetic considerations rather than increasing structural strength. The degree of deterioration and the nature of the current intended use thus become central to the decision making process.

Inspection and Design Considerations

In approaching the problem of rehabilitating an existing waterfront structure, the following information should be sought and carefully considered:

- Results of the material condition survey, including results of laboratory or nondestructive testing (NDT), such as cores and coupons, remaining material thickness measurements, etc.
- Original construction documents (as-built drawings), maintenance and repair records, pile driving logs, service history, etc.
- Results of soil borings and laboratory testing.
- Load test results (if warranted).
- Review of vintage material specifications and design standards.
- Review of products and methods available for repairs and construction.

A proper material condition survey is essential to the evaluation process. General inspection procedures are outlined in references (2) and (3), and additional information can be found in references (11) through (14). The degree of thoroughness required depends upon the structure type, past performance, and nature of the restoration being planned, and may range from a general, visual, preliminary type of inspection to a detailed mapping and charting of deteriorated areas. Material thickness and mechanical properties testing may be required at many locations or may be sufficient on a random sampling basis. Even if demolition is planned from the outset, a general condition survey will yield information on how the structure has stood up to the elements, which can be valuable input to the design of a new structure, and will also provide information on possible salvage value and the degree of difficulty of demolition

work. Underwater examination by divers also is an essential part of any condition survey. The divers conducting inspection work must be specially trained and experienced in such work. Underwater visibility typically is poor within most developed harbors, and the work proceeds more by feel than by sight.

Review of all available information on the past history of the structure, including as-built construction drawings, pile driving records, maintenance and repair records, and the service history, can prove invaluable in assessing rehabilitation. Unfortunately, this information generally is hard to find, and when available on a piecemeal basis, it may be difficult to interpret. If construction drawings are not available, then the principal dimensions and member sizes must be ascertained in the field during the material condition survey. Even when record drawings are available, general dimensions and soundings should be verified during the survey. In some cases, especially if dredging, filling, or an increase in load capacity is planned, new soil borings may be required if no record documents are available, and if there is any question as to the structure's stability and capacity.

In cases where borings may be unwarranted and/or inconclusive, and especially in structures where the substructure and foundation are not practically available to inspection, full-scale load testing may be the only way to verify capacity. The load test may consist of loading an entire area, as would be required in a retaining-type structure, and monitoring movements, pore pressures, and so forth; or it may be sufficient to selectively load-test individual piles. Load testing is the most conclusive of the various ways to verify the integrity of an existing structure, but is often costly and disruptive; so the engineer often must proceed with less definitive information than load test results.

In evaluating the capacity of an existing structure, the engineer must be aware of the materials, methods, codes, and standards that were employed at the time of the structure's original construction. For this purpose, maintaining a file of outdated codes and handbooks may prove very useful. The designer also may be confronted with a serious dilemma in analyzing an older structure under current standards, a problem that often is not easily resolved; then the designer's judgment must be applied on an individual-case basis as to whether the past service record constitutes an effective load test, or whether strengthening to current standards is warranted. A case history example is provided by the author in reference (15).

Finally, in approaching the rehabilitation problem, the designer must be fully familiar with the materials and methods currently available for restoration and marine construction in general. This may not be as simple as it sounds, as many products are being marketed as protective coatings and patch-and-fill repair-type compounds. Specifying particular ones requires a knowledge of engineering materials and, hopefully, of satisfactory performance records in similar applications in support of manufacturers' claims. Most repair methods are peculiar to the structural type and the material of construction.

11.2 GENERAL MODES OF DETERIORATION

Marine structures are subject to relatively rapid rates of deterioration due to such environmental forces as waves, currents, tides and extreme water levels, and ice, and mechanisms such as corrosion, physical and chemical attacks, and biodeterioration, as well as general wear, abrasion, and fatigue and accidental damage such as that caused by vessel impact and overloads. They also may become unserviceable because of the movement of bottom materials (i.e., through scour and siltation effects). Rapid deterioration rates also may be attributed to poor design, workmanship, and quality of materials. Environmental factors are discussed in Section 3.4 and their effects on material selection in Section 3.5. This section is intended to provide a general overview of deterioration mechanisms.

The site-specific physical and chemical properties of seawater often play an important role in deterioration rates. A more thorough overview of the effects of the marine environment on structural design has been provided by this writer (16). The general properties of various materials for marine use are described in references (17) and (18).

Principal modes of deterioration and examples of their primary effects are as follows:

- Corrosion of metals including concrete-reinforcing steel.
- Biodeterioration such as the decay of timber and attack by marine organisms.
- Physical–chemical processes such as freeze–thaw damage and sulfate attack of concrete.
- Mechanical damage such as abrasion and wear of timber and concrete and overloading, which affects all materials.

The overall long-term durability of a given structure depends upon local environmental conditions and the material of construction. The rates and severity of deterioration often exhibit a vertical zonation (see Figure 3.16), where tide range and extreme water levels are a major factor. Deterioration also may be localized to vulnerable and heavy-use areas, or to areas where incompatible materials are mixed or poor quality control was exercised. The true nature and mechanism of deterioration must be clearly understood before one prepares restoration plans. The basic modes of deterioration are discussed in turn in the following pages.

Corrosion

Corrosion in general is an electrochemical process whereby atoms of a given metal lose electrons, thus becoming positively charged (oxidation). The free

electrons then combine with atoms of an adjacent area or a surrounding substance (reduction). The electric current or migration of electrons from an anodic to a cathodic area may be caused or accelerated by various conditions that define the type and rate of corrosion found on a given structure. An explanation of corrosion processes is beyond the scope of this text. Further description of corrosion can be found in references (19) through (21).

Principal types of corrosion normally encountered on waterfront structures include *general and pitting corrosion*, manifested by scaling, pitting, or general wastage of large areas, due to reaction of the metal with its surrounding medium or with cathodic areas of the same member. *Crevice corrosion*, found in isolated areas, is due to localized oxygen depletion, such as that found under bolt heads and at discontinuous welds and joints. There also may be *galvanic corrosion*, caused by different electric potentials between dissimilar metals immersed in seawater, and less commonly *corrosion fatigue* and *stress corrosion* cracking, related to cyclic loading and high tensile stresses (particularly in high-strength steels), respectively.

The average rate of metal loss for mild steel in quiescent seawater is about .005 inch per year, or 5 mils per year (mpy). Localized pitting and crevice corrosion may proceed at several times this rate, however. Currents greatly increase corrosion, on the order of three times the average rate for a $1\frac{1}{2}$-knot current to nearly five times the rate for a 4-knot current. Turbulence and mixing increase corrosion rates by increasing the oxygen content and washing away corrosion by-products, thus exposing a new substrate.

Figure 11.1 illustrates the usual vertical distribution of corrosion on sheet pile bulkheads and piles. It shows highest areas of loss in the splash zone and immediately below MLW. Such typical profiles cannot necessarily be applied to any given structure, as the vertical distribution, which arises primarily from zones of differential aeration on the metal surface, may vary markedly under differing site conditions. For example, in some harbors with relatively large tide ranges, experience seems to indicate a much less pronounced spike in the splash zone with greatest attack just below MLW and frequently again just above the mudline, especially at shallow water sites.

Rates for sheet piling are approximately doubled for sheeting exposed to seawater on both sides versus one side filled. Studies of sheet pile bulkheads at several U.S. naval stations (22) revealed maximum corrosion rates on the order of 8 mpy in temperate climates up to 19 mpy in subtropical climates. Maximum rates were on the order of twice the average rates. Local perforations may render a bulkhead unserviceable because of leakage of fill long before general wastage reduces its structural strength to unacceptable levels (23). Splash zone rates on offshore structures of 55 mpy have been reported in the Gulf of Mexico (24). Maximum splash zone rates on offshore structures are more typically in the range of 25 to 40 mpy (25). Stray electric currents may be a source of very

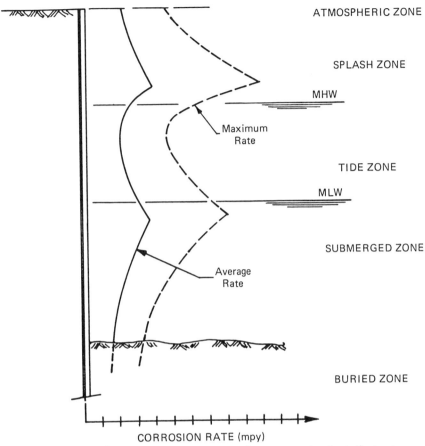

Figure 11.1. Vertical distribution of corrosion rates for representative sheet pile structure.

rapid corrosion rates, especially at shipyards where improper grounding of welding equipment or electric utility systems may result in localized electric currents and related corrosion. Faulty impressed-current cathodic protection systems, high voltage cable crossings, berthed vessels, or other extraneous sources also may contribute to high localized corrosion rates due to stray electric currents. Reference (21) is especially useful to the practicing waterfront engineer in providing further information on expected rates, steel selection and protection, and so forth.

An important point about corrosion that is especially relevant to repair work concerns the nature of galvanic action on new and corroded steel specimens. Because corroded steel is cathodic to new clean steel, a new patch welded to an existing pile, for example, will corrode at a much higher rate than the pile.

One way to offset this is to use a thicker patch piece to offset the differential. Mill scale and steel embedded in concrete are (usually) more cathodic than corroded steel. However, one should not jump to the conclusion that leaving mill scale on will be beneficial because once the mill scale is scraped off at any point, the base metal will suffer accelerated attack.

Another point worth mentioning involves selection of the member cross section itself. From a corrosion point of view, for example, pipe piles are superior to H-piles because pipe piles have less surface area, which is also easier to clean and inspect than the H-pile surface. Pipe piles retain coatings better because H-piles have sharp edges where coatings tend to thin out, and corrosion rates tend to be higher near the ends of the flanges. Also H-piles may suffer a more dramatic loss of capacity than pipe piles will, because of a reduced radius of gyration about the H-pile weak axis for a comparable average loss rate. This last point is significant in evaluating the remaining capacity of an existing H-pile structure.

Table 11.1 summarizes some practical design considerations for reducing corrosion problems.

Biodeterioration

Attack by marine organisms (borers) is an ever-present threat to timber-supported structures, even in temperate climates. Two types of borer are prominent: the woodgribble of the Limnoria family, which is a crustacean, and the teredo, which is a mollusk related to the clam, although it is wormlike in appearance. Figure 11.2 illustrates the characteristic pattern of damage inflicted by the two types of borers. The limnoria nibbles away shallow furrows at the surface and is responsible for the necked-down ''hourglass'' appearance of tim-

Table 11.1. Corrosion problem prevention checklist.

- Select proper materials; i.e., use compatible metals and corrosion- or fatigue-resistant ones when possible.
- Avoid using steel in harsh environments, i.e., tropical climates or beach bulkheads subject to erosion, corrosion, etc.
- Detail to avoid crevice and localized corrosion; i.e., avoid overlapping plates, use continuous welds, etc.
- Detail for easy maintenance; i.e., avoid using back-to-back angles and creating poorly drained areas, etc.
- Design members with extra thickness consistent with anticipated loss rate over design life. Use the extra thickness where most beneficial.
- Choose protective coatings and corrosion prevention systems judiciously and apply to maximum advantage; i.e., use resistant coating or cladding in tide zone with cathodic protection below low water.
- Consider effects of corrosion on fatigue life and potential for stress corrosion–cracking.

Figure 11.2. Marine borer damage patterns. (After Gaythwaite (15).)

ber piles often observed at low tide. Limnoria damage can reduce a pile diameter by as much as two inches per year (25). Toredo damage is less obvious because the animal grows inside the wood as it tunnels along, leaving the inside of the pile riddled with holes. The adult teredo actually may burrow to above MHW. All marine borers exhibit some degree of sensitivity to seawater: temperature, salinity, pH, currents, and the presence of pollutants. Optimum temperature ranges, depending upon the species, usually range from around 40° to 70°F although certain of the limnoria can survive in nearly freezing water. The pH generally must be within the range of 7.5 to 8.5 for all species. The teredo can tolerate salinities as low as eight parts per thousand (8‰), but the limnoria will become inactive at salinities below 16‰ to 12‰. Both the limnoria and the teredo are unlikely to attack when currents approach and exceed around two knots, and all species appear to have low resistance to general harbor pollution. Other crustacean borers include the sphaeroma and chelura families. The pholad family of molluscan borer has been known to penetrate poor-quality concrete. Means of classifying and quantifying marine borer damage can be found in reference (26).

Various forms of marine bacteria may play an important role in promoting biodeterioration, fouling, and corrosion. Sulfate-reducing anaerobic bacteria may produce hydrogen sulfide (H_2S) within bottom sediments and under the blisters of protective coatings or the bases of sessile fouling organisms. H_2S may affect the corrosion history of steel structures, causing sulfide stress cracking, for example. Other forms of bacteria may affect buried concrete or promote the breakdown of oil-based wood preservatives (16).

Timber structures often suffer from rot of the superstructure, deck, and stringers due largely to stagnant fresh water. Bacteria, various types of fungi, and insects, sometimes including wharf-boring beetles (*Nacerda*), often are evidenced in older, neglected timber structures. Rot is insidious and may be difficult to detect, especially in stringers and pile caps where the only evidence may be a surface coating of fuzz or slime or the characteristic brown or white powderlike residue of dry rot.

The most common and destructive type of decay organisms are fungi responsible for both brown and white dry rot and other forms of rot ranging from surface discoloration to total destruction of the wood. Such advanced stages usually are evidenced by the presence of fruiting bodies or a threadlike mycelium. Certain forms that cause only discoloration of the wood may not affect its static structural strength, but can affect its impact resistance dramatically, with obvious serious implications for waterfront structures.

Timber deck systems must be given careful scrutiny and provisions made for their future protection if they are being planned for reuse. A common sight along the waterfront is deteriorated top butt ends of fender piles that were not protected with some form of cladding.

Physical-Chemical Deterioration

Concrete is subject to physical (abrasion and freeze–thaw cycles) and chemical (chloride and sulfate attack, reactive aggregates, etc.) deterioration. In temperate and northern climates, concrete is especially vulnerable in the tidal zone where, with each tidal cycle, the concrete goes through a freeze–thaw cycle during the winter months. This effect may be exacerbated by the abrasive action of ice rising and falling with the tide or borne by currents. The best protection against this form of attack is to provide the most impermeable concrete possible, as described in Section 3.5. The use of air entrainment is essential in freeze–thaw climates.

Corrosion of reinforcing steel, associated with penetration of the concrete cover by chloride and oxygen ions, is a common cause of concrete degradation, especially within the tidal and splash zones. It actually is a complex electrochemical process (27) resulting in depassivation of the steel when the pH value of the concrete cover (normally around 13.0) is lowered below about 11.5. The

pH of seawater normally is within the range of 8.0 and 8.5. The use of dense impermeable concrete and an adequate cover is again the best protection against this type of degradation.

Sulfate attack of seawater constituents on the calcium hydroxide ($Ca(OH)_2$) and/or tricalcium aluminate (C_3A) of the hardened cement paste can result in softening and degradation of the concrete. The use of sulfate-resistant cements, such as Portland type II, with low to moderate C_3A contents, usually from 6% to 10% maximum, helps to reduce this form of concrete deterioration.

Improper design, application, and workmanship, as well as erosion and overloading damage, serve to accelerate the physical–chemical deterioration processes. Note the severe scaling and spalling of the concrete deck in Figure 11.3 adjacent to the expansion joint and rail pockets. The use of bituminous concrete for patchwork repairs probably has accelerated the degradation process through capture and condensation of moisture at the concrete surface. Concrete sometimes is employed as cladding to protect steel piles in the splash and tidal zones, although this is done where concrete itself suffers most in freezing climates. As can be seen in Figure 11.4, unless the concrete is properly installed

Figure 11.3. Degradation of concrete wharf deck at expansion joint and rail pockets. Bituminous patches likely have served to accelerate deterioration. (Photo courtesy of the Maguire Group, Inc.)

Figure 11.4. Deterioration of concrete pile shafts in tidal zone. Pile shafts to the left have been partially restored with the use of fabric forms. Installation of contemporary resilient fender units helps to reduce berthing impact loads. (Photo courtesy of the Maguire Group, Inc.)

and quality-controlled, the concrete jackets or shafts themselves may represent the biggest maintenance problem. References (28) through (31) contain important information on the performance of concrete in seawater.

Mechanical Deterioration

Waterfront structures, in particular fender systems (32), are subject to inadvertent mechanical damage due to accidental vessel impact and the cumulative effects of frequent normal berthings. Timber structures are especially susceptible to abrasion and wear deterioration due to ice and vehicle traffic as well as vessels. In localized areas, reinforcement or cladding sometimes can greatly alleviate wear problems, as in the provision of chafe plates at bollard locations and designing to avoid chafe conditions as far as practicable.

Fatigue is not usually taken into account in the design of most waterfront structures. However, it should be considered in the design of active ferry terminal and cargo handling facility structures, especially with the current trend toward more slender structural elements. Fatigue due to wave loading becomes a factor at exposed sites and in deep water.

Although wave action on an inner-harbor structure may not exert significant forces on the structure, over time cumulative erosion effects due to entrained sand, debris, and so on, can have a pronounced effect at exposed sites. During severe storms and when extreme water levels allow waves to penetrate where they normally would not reach, mechanical damage due to thrown debris and plucking of riprap slope protection may result.

Waterfront structures often are subjected to overloads due to operations. This is especially true for an older structure and may be a major factor in its requiring rehabilitation, as current trends toward higher-capacity cargo handling equipment, larger vessels, and stacking of palletized and containerized cargo may subject the structure to loads well beyond the original design capacity. Frequent breakage of fender piles, for example, indicates the need to upgrade the fender system. Signs of overload usually are readily detected; permanent deformations, settlements, localized damage such as from dropping heavy weights on deck, misalignment of members or the entire structure, failure of bracing or secondary members, and so on, all indicate overload conditions.

11.3 GENERAL REPAIR METHODS

Any proposed method of structural repair should consider amelioration of the conditions that led to the original deterioration. For example, if improper deck drainage led to the decay of timber framing elements, then the drainage situation should be remedied before the rotted timber is replaced. If a fender system is suffering continual damage, then it should be replaced with a more resilient system with a higher energy-absorption capability. If the deterioration was a result of faulty workmanship or materials, replacement in kind with properly installed good-quality materials should be sufficient. Repair methods also must consider relative cost, down time and disruption of operations, longevity of the repairs, access of personnel and equipment, and structural integrity of the completed work. Repair methods will vary with material and member types, as well as location with respect to tide water levels. This section deals primarily with repairs to deck systems and superstructures, generally above the MHW level although access to underdeck areas may be limited to certain stages of the tide.

Concrete problems generally manifest themselves in cracks, spalls, and surface erosion. Cracks may be due to structural problems such as overloading or understrength concrete or to physical/chemical problems such as shrinkage, reactive aggregates, and so forth. The presence of a whitish precipitate, which often forms "stalagtites" hanging down from the underside of deck slabs and beams, indicates cracks that penetrate the full depth of the slab or beam. The white precipitate or leachate usually is calcium hydroxide ($Ca(OH)_2$), formed by moisture penetrating through the cracked concrete. It sometimes is observed as a thin white coating on the undersides of slabs, termed effluorense. Cracks that are of a structural nature and other cracks that are not subject to further

movements such as those caused by initial support settlements or volume change of the concrete during curing may be sealed by pressure injection with epoxy grout, provided that the in-place concrete is sound. Epoxy compounds have excellent bond strength and essentially can restore the full strength of the concrete across the crack. Cracks as small as 5 mils (.005 inch) to as large as 250 mils have been sealed by the pressure injection process. Crack widths of less than approximately .01 inch generally may be considered small enough not to allow moisture to penetrate to the reinforcing steel. Working cracks, which are subject to movements such as those due to temperature change or structure deflections, may be sealed with elastomeric fillers, provided that the cracks do not affect the overall structural integrity. The use of epoxy injection to repair working cracks will only cause new cracks to appear or the injected crack to reopen.

Spalls or localized pockets where the concrete has been chipped or eroded away often are patched with epoxy- or polymer-modified mortars and cements. The choice of the appropriate mortar type and mix depends upon the extent and depth of the spall and whether it is on a horizontal, vertical, or overhead surface. Where reinforcing steel is exposed, it must be properly cleaned and a bonding agent applied. In general, all loose concrete must be chipped away and cleaned so that the mortar or patching compound adheres to a clean and sound substrate. Edges of spalls generally should be saw-cut to ensure the adherence of patching material. Beams and pile caps with extensive spalling on all surfaces may need to be formed and have the mortar mix pressure-injected into the form surrounding the beam. Another common method of restoring the underside of pier and wharf decks with extensive surface spalling employs shotcreting. Shotcrete or gunite is basically a mix of cement, sand, fine aggregate, and water applied by air hose under pressure. Shotcrete also may be polymer- or epoxy-modified. It has excellent bond and generally is applied over wire fabric to a depth of two inches or more, in several passes (33).

Large voids or extensive areas of unsound concrete must be cleaned back to sound concrete and filled with a durable mix of cast-in-place (C.I.P.) concrete, which often requires extensive formwork (see Figure 11.5). Replacement concrete should be as similar as possible to the cast-in-place concrete in regard to the maximum aggregate size and the water–cement ratio. Fresh concrete is not chloride ionized as the older in-place concrete is, and may cause accelerated corrosion of existing reinforcing steel. This can be prevented by coating the exposed steel with a suitable insulation compound such as an epoxy. Contact surfaces may be moistened, but no standing water should be present prior to placement. Prepacked concrete also may be used for general repairs of larger voids. Its application consists of packing the form with coarse aggregate and then filling the voids with an intrusion grout consisting of Portland cement, sand, water, and a pozzolan and intrusion admixture. The dry-pack method

Figure 11.5. Concrete beam repairs—schematic example.

consists of ramming a very stiff mortar mix into small openings in thin layers, and is suitable only for small holes such as form tie holes and narrow deep slots.

Deck surfaces are subject to surface erosion, generally known as scaling, due to deicing salts or other chemicals, and to wear or damage from traffic and heavy equipment, as well as general weathering. Decks with little and shallow scaling and with only small hairline cracks, or cracks that have been sealed by pressure injection, may be coated with a sealant to prevent moisture from penetrating deeper into the slab. Typical sealants range from linseed oil compositions to epoxy systems, polyurethanes, and other synthetic resins. The use and performance records of the various deck sealant systems are described in detail in reference (34). Where the deck is more severely eroded, an overlay system of C.I.P. concrete or polymer-modified concrete (PMC) may be used. The overlay acts both to seal the underlying structure and to provide a more durable wearing surface. If C.I.P. concrete toppings are used, they usually are on the order of 2.5 to 4 inches thick and must be reinforced with wire fabric. Much thinner overlays, resulting in much lighter dead loads, can be attained by using PMC. Of the many possible PMC mixes, the use of latex-modified concrete is perhaps the most widespread. Suitable latex formulations result in greatly improved bond, tensile, and flexural strengths of cements and mortars. They can be feather-edged to minimum thicknesses of 0.5 inch or less, although the durability of such thin applications under wheel loads is questionable. Occasionally, attempts are made to patch or overlay concrete decks with asphaltic or

bituminous concrete. This is not recommended practice because asphalt can be penetrated by moisture, and moisture and vapor will be retained at the concrete surface, thus allowing degradation of the concrete to continue, especially if reinforcing steel is not covered.

As is apparent from the foregoing discussion, a wide variety of polymeric materials and cement admixtures are available for repairs. A detailed discussion of their chemistry and application is beyond the scope of this text. General guidance for the repair of concrete deck systems can be found in reference (35), and for the use of epoxy- and polymer-modified concrete in references (36) and (37). A more detailed general treatment of the various concrete repair materials and methods and of the placement of concrete under water is provided by Thoresen (38). It is most important that the design engineer understand the properties, application requirements, and past performance history of a given product. This requires close scrutiny of product manufacturers' literature and documentation. All materials must be applied in strict accordance with the manufacturers' recommendations, and surface preparation prior to application must be clearly specified and carefully inspected.

Steel members generally are repaired by replacement, plating, or encasement. (Encasements are discussed in the following section.) In replacing or restoring existing steel members with new plating, careful consideration must be given to achieving a suitable weld or connection that is both sound and resistant to corrosion. New metal, even if of the grade of the existing metal, typically is more anodic than corroded metal, so the new plating may suffer an accelerated rate of corrosion. The use of thicker material for plating will somewhat offset this effect. Representative steel member repairs are illustrated in Figure 11.6. To prevent accelerated corrosion of the faying surfaces, the surfaces should be properly cleaned to near-white metal and continuous watertight welds used. Underwater welded repairs and welding under wet conditions are described in reference (39). If analysis of field measurements indicates that enough metal remains to resist the design loadings, then engineers can employ protective measures as discussed in the following sections, without resorting to strengthening-type repairs.

The usual way to repair timber structures is to replace the affected members in kind, but sometimes individual members may be repaired or strengthened by the addition of splice plates or shoring members. Figure 11.7 illustrates some typical timber splice details that can be used to join replacement sections of affected members. Longitudinal checks or splits sometimes are held together by the use of stitch bolts. Often it will be found that metal fastenings within the tide zone and splash zones will deteriorate more rapidly than the timber itself. Where decay or rot is present, the source of trapped moisture causing the rot should be eliminated if possible before the placing of new timbers. Marine borer attack usually is treated by wrapping or encasement, as discussed in the follow-

REHABILITATION, MAINTENANCE, AND REPAIR 501

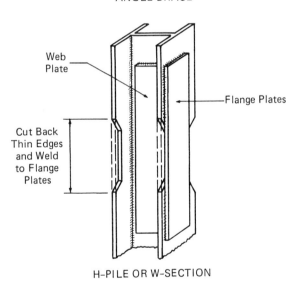

Figure 11.6. Repair of steel members—schematic.

ing section. Fire-damaged members should be carefully inspected. If there is sufficient sound material below the surface char, then the member may be allowed to remain as is. In general, all timber used in marine construction should be treated with preservatives, with the exception of some naturally resistant

Figure 11.7. Timber splice and stitch bolt details—schematic.

dense tropical hardwoods. Holes and cutoffs should be field-treated. Section 3.5 describes material and treatment requirements for marine construction. The general repair of wood structures is covered in reference (40).

Stone masonry work, which typically is made from dense and durable cut granite blocks, seldom exhibits material deterioration problems. Settlements due to foundation problems and washing out of material through open joints are the usual problems. Open joints may be filled with small stone and mortared. Epoxy or other polymeric grouts may enhance bond and durability. In the sealing of open joints with any kind of mortar, termed pointing, extreme caution must be exercised to leave adequate openings or other provision for drainage of water from behind the wall. Riprap stone may be displaced by wave or ice action; and when the displaced stone is replaced, consideration should be given to

whether or not the original stone is of adequate weight for its purpose. Occasionally, sedimentary rocks or rubble that is not resistant to weathering may be found in existing riprap slope protection. Such stone or materials should be replaced with more resistant granitic rock.

Rubber fender units suffer from gradual wear, overload damage, and exposure to ultraviolet light (ozone attack). Preformed rubber units may crack along their juncture with the metal base plates to which they are bonded; fenders usually are replaced in kind when such cracks occur. Fenders that are subject to frequent overload should be replaced with more resilient units. The softness and other properties of the rubber used in fender units can be widely varied to suit particular applications. The fender manufacturer's technical staff should be consulted before replacement units are specified. As previously discussed, fender systems can be designed to reduce berthing loads and hence damage to other structural elements.

Other common waterfront materials requiring frequent inspection and replacement include wire rope, chain, connection hardware, elastomeric bearings, and utility piping and systems. Routine inspection and reduction of exposure to deteriorating agents will greatly reduce long-term maintenance and replacement costs.

11.4 PILE AND BULKHEAD RESTORATION METHODS

In approaching the problem of pile foundation or bulkhead deterioration, one first must determine whether there is sufficient sound material remaining to meet strength requirements, or whether structural strengthening also is required. In the first instance, preservation methods such as protective coating and claddings, cathodic protection of steel, plastic or corrosion-resistant metallic wrappings, and encasement in concrete jackets may be applied. Where strengthening is required, general remedies include encasements, posting and shoring, and, in extreme cases, underpinning, which generally requires the driving of additional new piles. For most pier and wharf structures, it is likely that reconstruction of the entire structure will be more cost-effective than an attempt to underpin the entire foundation.

Posting sometimes is carried out where severe deterioration is confined to a small area, such as may be the case with timber piles that are rotted at their top butt ends. The decayed area is cut out one pile at a time, and a new piece of timber wedged into place and secured with splice plates (see Figure 11.8). Shoring may be feasible when impact damage is confined to one or only a few piles. When deterioration is more extensive, such as severe corrosion of steel piles within the tide zone or marine borer attack of timber in the immersed zone, then encasement in concrete jackets often is employed. Before using concrete jackets the engineer should consider the structure's overall rigidity, as the con-

504 DESIGN OF MARINE FACILITIES

Figure 11.8. Timber pile posting detail.

crete jackets may be subject to cracking if large structure deflections are possible.

Concrete jackets may be formed by using removable steel forms or more commonly today by the use of leave-in-place fiberglass (GRP), plastic, or fabric forms. Figure 11.4 shows concrete pile shafts that have been reconstructed by this process, and Figure 11.9 schematically illustrates the application of concrete jackets. The forms are suspended from the deck above or from falsework,

Figure 11.9. Jacketing of steel pile with concrete—schematic.

and are tightly secured at their base with a base plate. Concrete then is pumped into the form using the tremie process, where it must be placed under water. Extreme care must be exercised to keep the concrete from falling through the water. Concrete for pile encasements should be of a rich cement, using an aluminous cement with low C_3A content, such as Portland Type II cement. Aggregates should be well graded and durable, and mixing water should be free of chlorides. Air entrainment is recommended and is essential in areas subject to freeze–thaw. Pozzolanic admixtures also may be used as fluidizers. Pile shafts, which are applied only within the tide zone, should extend at least two to three feet below MLW, and the portion of the pile that is exposed at the

bottom of the concrete jacket should be coated to avoid creation of a local galvanic corrosion cell. The jacket may be extended to the bottom of the cap or not, depending upon the prevailing conditions and height of the cap above MHW. A minimum of four inches cover to the steel pile with at least 2.5 inches clear to the added reinforcing steel normally is required. Welded wire fabric (WWF) often is used as reinforcing steel for convenience in placing and for its bidirectional properties. Encasements for timber piles must extend to at least one foot or more below the mudline to prevent limnoria attack. Concrete piles also may be restored by jacketing or by shotcreting.

Steel bulkheads too may be restored by concrete encasements (see Figure 11.10). Localized corrosion of sheet pile webs sometimes is restored by plating. Corroded or damaged interlocks may be restored by welding an inverted angle section to the adjoining sheets and filling the V-shaped void with grout. Additional description of typical steel bulkhead repairs is provided by Kray (41). Bulkheads most commonly fail because of anchor rod failure and toe kick-out due to scour or overdredging (42).

Where piles possess sufficient structural strength but must be protected from further deterioration, the use of wrapping materials often is an economical solution. Wrapping materials for waterfront structures are usually of polyvinyl chloride (PVC) or polyethylene sheeting from 30 to 60 mils thick. The pile surface must be thoroughly cleaned and smooth so as not to tear the wrapping material, which is tightly wound around the pile and secured with corrosion-resistant metallic bands. With proper installation, this method denies free oxygen to the pile surface, thus arresting both corrosion and marine borer activity. The application of flexible pile wraps on timber piles is described in reference (43), and on steel piles in reference (44). Offshore structures have been very successfully protected by the application of sheet metal, Monel® Alloy 400, wrapping (21). The application of protective coatings and mastic claddings and of cathodic protection systems for piles and bulkheads is discussed in the following sections. Additional description of pile protective methods can be found in references (5), (17), (21), (22), (25), (45), and (46).

11.5 PROTECTIVE COATINGS AND CLADDINGS

Protective coatings serve as a barrier to isolate the steel substrate from the corrosive action of the environment. As such, a suitable coating system must be impermeable and must possess good adhesion and resistance to the marine environment and to fouling and abrasion. Other considerations include resistance to sunlight and hydrocarbons and compatability with cathodic protection systems. The ideal requirements for a coating system will vary with its class of service (i.e., immersion, tidal, and splash zone and atmospheric). It is essential that any coating system be applied over a properly cleaned and prepared sur-

Figure 11.10. Steel sheet pile bulkhead repair—schematic.

face. In general, the performance of a protective coating depends upon the surface preparation, application, film thickness, and environmental conditions. The following discussion is intended to review some of the more common coating systems specified for marine structures. Generic coating systems are specified by the Steel Structures Painting Council (SSPC) (47). Recommended practice for the corrosion protection of submerged steel pipe is provided by the National Association of Corrosion Engineers (NACE) (48).

Coal-tar epoxies are among the most common systems used for immersion and splash zone service. They consist of a two-component catalyzed thermosetting resin, usually either amine or polyamid cured. The epoxy has good adhesion characteristics, is tough and resistant to chemicals and solvents, and combined with coal-tar has excellent water resistance. Coal-tar epoxy coating systems should be applied over a near-white sandblasted surface in accordance with SSPC-10 (47), to a minimum dry film thickness (dft) of approximately 16 mils. Up to 20 to 24 mils dft may be specified for splash zone and severe exposures. The lifetime of an epoxy coal-tar system is directly proportional to the film thickness, and a properly applied coating should give 10 to 15 years of service. Straight coal-tar coatings are inexpensive and water-resistant but lack the toughness, durability, and solvent resistance of epoxy- or polyurethane-modified systems. Other drawbacks of coal-tar systems are their easy penetration by fouling organisms and their poor performance when used in conjunction with cathodic protection systems. Straight epoxy systems give good service, and some polyamid cured epoxies have been developed that will cure under moist conditions and even under water. These coatings, which generally are more expensive than coal-tar epoxies, also must be applied over a near-white sandblasted surface.

Fusion bonded epoxy coatings are one-part heat-cured systems, as used in epoxy-coated reinforcing bars. The process also can be applied to structural shapes, and the finished product offers superior durability and uniformity of coating to epoxy coal-tar systems. As the process involves heating the member to high temperatures and applying a powder under controlled conditions, it is not amenable to field application.

Polyethylene coatings applied by a heat process constitute a relatively recent and promising coating system (49). Polyethylene coatings applied to approximately two to three mils thickness have been used on offshore pipe pile structures, and long lifetimes, of 40 to 50 years, are predicted. Pipe pile ends that are field-spliced during driving are fitted with field-applied shrink-fit polyethylene collars. Like fusion bonded epoxy systems, however, this coating system is not yet practical for in-place applications.

Other coatings suitable for immersion service include phenolic mastics, chlorinated rubber compounds (often used inside ballast tanks), and top-coated metallized zinc or aluminum systems. Metallizing is a process of spraying molten metal particles onto a prepared surface. Metallized surfaces must be top-coated, usually with a vinyl, epoxy, saran, or furan top coat, to keep the zinc surface film from eroding rapidly. Such systems must be applied under controlled or shop conditions and are not frequently employed on waterfront structures.

Zinc-rich coatings often are used as prime coats for epoxy, vinyl, or acrylic coatings, and also may be used by themselves in atmospheric exposures. They offer good splash zone and freshwater protection as well. Inorganic zinc and

organic zinc coatings are so named by the nature of the material used as a binder. They are usually applied in relatively thin coats of 2 to 5 mils dft. Usually 1 mil dft should be allowed for each year of life required. Zinc-rich paints provide cathodic protection to the metal substrate when the paint film is scratched. Organic zincs have better film-forming properties than inorganic zincs and are generally more suitable for primers.

Vinyl coatings have excellent resistance to high humidity and salt atmospheres and sometimes are used for immersion service as well. Polyurethanes are known for their toughness and abrasion and impact resistance, and they are very flexible; but their adhesion properties are only fair, and aromatic urethanes, in particular, have poor resistance to weathering.

There are many other coating systems with specific advantages and disadvantages for a given application and a myriad of trade name products for the engineer to consider. It is important to review the product literature of different manufacturers and finally to consult the proposed manufacturer's technical staff prior to specifying a coating system for a specific project. A comprehensive treatment of protective coatings for marine structures can be found in references (17), (19), and (21).

Claddings are extra-thick coatings designed to be field-applied by spraying, troweling, or a gloved hand, usually for splash zone protection. They can cure under moist conditions, and some splash zone mastics can be applied under water by a diver's hands. The underwater curing materials typically are two-component, 100% solids, nonshrinking polyamide epoxy compounds. They normally are applied to 125 to 250 mils dft, with underwater applications often limited to patching or narrow zones. They should be applied over a near-white sandblasted surface and have fair to good impact and abrasion resistance. These coatings also will bond to concrete, wood, or fiberglass surfaces.

11.6 CATHODIC PROTECTION SYSTEMS

Cathodic protection systems provide corrosion protection to immersed steel surfaces by imposing a reverse electric current, in essence making the pile act as a cathode rather than an anode. As seawater acts as the electrolyte, the system works only when the metal surface is submerged. However, partial protection within the tide zone is provided during high water. Cathodic protection systems are of two basic types: the galvanic (sacrificial) anode system and the impressed current system. Figure 11.11 schematically illustrates cathodic protection system installations.

In the galvanic anode system, sacrificial anodes of aluminum, magnesium, or zinc are attached directly to the piles, thus inducing a flow of electrons away from the anodic material and toward the cathodic pile surface. Anodes must be

Figure 11.11. Cathodic protection systems—schematic.

properly sized and located to ensure a uniform distribution of current density over a planned structure life or anode replacement time.

In an impressed current system, a DC electric current is applied to all the piles through electric bonding cables. Anodes of graphite, silver/lead, or plat-

inized metals are either suspended from the deck, pile-mounted, or bottom-supported in wells. As a rule of thumb, an electric current of one ampere flowing for one year will corrode away 20 pounds of steel. Usual current densities range from .005 to .06 amps per square foot of protected surface. The electric current raises the pH at the metal surface, causing precipitation of calcareous (chalk) deposits. High current densities result in a hardened chalk, and low densities in a soft chalk that is easily washed away. It is essential that all concrete reinforcing steel be bonded to the circuit. This must be done during the original construction, and it is recommended that any new piers with concrete decks on steel piles be completely bonded even if there is no intention to install rectifiers and activate an impressed current system initially. Compatibility of cathodic protection systems, especially impressed current systems, is an important consideration. Deepwater structures may be protected solely by cathodic protection below MLW and with splash zone coatings from the tide zone to the deck.

Other important factors affecting the design of cathodic protection system include the resistivity of the water as measured by its salinity or dissolved solids content, water temperature range, dissolved oxygen content, pH, tide range and currents, pollutants, and bottom soil characteristics. The required current density is further affected by turbulence such as that caused by structure vibrations, prop wash, storms, and so on, which increases the oxygen content at the pile surface and thus the required current density. Aged structures with corroded pitted metal surfaces are more difficult to protect than new structures and generally require higher current densities. Electrical interference from stray currents also can have an important effect on system design and performance.

As can be seen from the preceding discussion, cathodic protection systems must be designed by specialists with experience in marine structures. References (21) and (50) give a detailed description of the design of cathodic protection for marine structures, and reference (51) presents an instructive case history of retrofitting a deepwater wharf with an impressed current system.

REFERENCES

1. UNCTAD (1985). *Port Development: A Handbook for Planners in Developing Countries*. Geneva, Switzerland.
2. AAPA (1970). *Port Maintenance*. American Assoc. of Port Authorities, Washington, D.C.
3. DOD (June 1978). *Maintenance of Waterfront Facilities*, MC-104. Dept. of Defense, U.S. Navy, Washington, D.C.
4. ICE (1978). *Maintenance of Maritime Structures*. The Institute of Civil Engineers, London.
5. Johnson, S. (1965). *Deterioration, Maintenance and Repair of Structures*. McGraw-Hill, New York.
6. Wong, K., ed. (1983). *Proc. ASCE*, Ports '83, Spec. Conf. on Port Modernization, Upgrading, and Repairs, New Orleans, La.

7. ICE (1986). *Maritime and Offshore Structure Maintenance*, Proc. of 2nd Int. Conf., Institute of Civil Engineers. Thomas Telford, London.
8. BSCE (1984). "Rehabilitation of Waterfront Structures," seminar notes, sponsored by the Waterways Group of the Boston Society of Civil Engineers Section, ASCE, Boston, Mass.
9. PIANC (1986). "Port Maintenance Handbook: Chapter 3—Roads and Storage Areas," Supp. to Bull. No. 54, Brussels, Belgium.
10. PIANC, (1985). "Port Maintenance Handbook: Chapter 5—Mechanical Equipment," Supp. to Bull. No. 50, Brussels, Belgium.
11. ACI (1984). "Guide for Making a Condition Survey of Concrete in Service," ACI 201.1R-68, American Concrete Institute, Detroit, Mich.
12. DOT (1979). *Bridge Inspectors Training Manual*. U.S. Dept. of Transportation/FHA, Washington, D.C.
13. Fiorato, A. E. and Johal, L. S. (1981). *Inspection Guide for Reinforced Concrete Vessels*, Report No. CG-M-11-81, Vols. I and II. U.S. Dept. of Transportation, USCG, Washington, D.C.
14. Agi, J. J. (1983). "Structural Evaluation of Marine Structures," *Proc. ASCE*, Ports '83, Spec. Conf., New Orleans, La.
15. Gaythwaite, J. W. and Carchedi, D. R. (Mar. 1984). "Rehabilitation of Waterfront Structures—Design Criteria and General Considerations," Seminar on Rehabilitation of Waterfront Structures, Boston Society of Civil Engineers, Section ASCE, Boston, Mass.
16. Gaythwaite, J. W. (1981). *The Marine Environment and Structural Design*. Van Nostrand Reinhold, New York.
17. Whiteneck, L. L. and Hockney, L. A. (1989). *Structural Materials for Harbor and Coastal Construction*. McGraw-Hill, New York.
18. USACE (Feb. 1983). *Construction Materials for Coastal Structures*, SR-010, CERC, Ft. Belvoir, Va.
19. Laque, F. L. (1975). *Marine Corrosion—Causes and Prevention*. Wiley, New York.
20. Uhlig, H. H. (1948). *The Corrosion Handbook*. Wiley, New York.
21. Dismuke, T. D. et al. eds. (1981). *Handbook of Corrosion Protection for Steel Pile Structures in Marine Environments*. American Iron and Steel Inst., Washington, D.C.
22. Ayers, J. R. and Stokes, R. C. (Aug. 1961). "Corrosion of Steel Piles in Salt Water," *Proc. ASCE*, WW-3, Vol. 87.
23. Buslov, V. (Aug. 1983). "Corrosion of Steel Sheet Piles in Port Structures," *Proc. ASCE*, JWPCOE, Vol. 109, No. 3.
24. Creamer, E. V. (1970). "Splash Zone Protection of Marine Structures," Offshore Technology Conf., OTC Paper No. 1274, Houston, Tex.
25. Chellis, R. D. (1961). *Pile Foundations*. McGraw-Hill, New York.
26. NAVDOCKS (1965). *Marine Biology Operational Handbook*, MO-311. Dept. of the Navy, Washington, D.C.
27. Bazant, Z. P. (June 1979). "Physical Model for Steel Corrosion in Concrete Sea Structures—Theory and Application," *Proc. ASCE*, ST-6, Vol. 105.
28. ACI (1980). *Performance of Concrete in Marine Environment*, SP-109, American Concrete Institute, Detroit, Mich.
29. ACI (1988). *Concrete in the Marine Environment*, Proc. of the 2nd Int. Conf., SP-109, American Concrete Institute, Detroit, Mich.
30. ACI (Mar. 1982). "Concrete in a Marine Environment," Spec. Issue *Concrete International*, Vol. 4, No. 3, ACI.
31. Buslov, V. (1979). "Durability of Various Types of Wharves," *Proc. ASCE*, Coastal Structures '79, Spec. Conf.
32. Padron, D. and Han, H. Y. (Aug. 1983). "Fender System Problems in U.S. Ports," *Proc. ASCE*, JWPCOE, Vol. 109, No. 3.

33. ACI (1985). "Guide to Shotcrete," ACI 506R-85, American Concrete Institute, Detroit, Mich.
34. Pfeifer, D. W. and Scali, M. J. (Dec. 1981). "Concrete Sealers for Protection of Bridge Structures," National Cooperative Highway Research Program Report 244, Trans. Res. Board, NRC, Washington, D.C.
35. ACI (1980). "Guide for Repair of Concrete Bridge Super-structures," ACI 546-1R-80, American Concrete Institute, Detroit, Mich.
36. ACI (1984). "Use of Epoxy Compounds with Concrete," ACI 503R-80, American Concrete Institute, Detroit, Mich.
37. ACI (1986). "Guide for the Use of Polymers in Concrete," ACI 548.1R-86, American Concrete Institute, Detroit, Mich.
38. Thoresen, C. A. (1988). *Port Design: Guidelines and Recommendations*. Tapir Publishers, Trondhiem, Norway.
39. Fulton, R. N. (1983). "Underwater Welded Dock Repairs," *Proc. ASCE*, Ports '83, Spec. Conf., New Orleans, La.
40. ASCE (1982). *Evaluation, Maintenance and Upgrading of Wood Structures—A Guide and Commentary*, by the Comm. on Eval., Maint., and Upgrading of Timber Structures of the Comm. on Wood of the Struct. Div., ASCE, New York.
41. Kray, C. J. (1983). "Rehabilitation of Steel-Sheet Pile Bulkhead Walls," *Proc. ASCE*, Ports '83, Spec. Conf., New Orleans, La.
42. Horvath, J. F. and Dette, J. T. (1983). "Rehabilitation of Failed Steel-Sheet Pile Bulkheads," *Proc. ASCE*, Ports '83, Spec. Conf., New Orleans, La.
43. NAVFAC (1975). *Specification 75M-810a for Installation of Flexible Barriers on Marine Borer Damaged Wood Bearing Piles*. Dept. of the Navy, Alexandria, Va.
44. Wakeman, C. M. et al. (Dec. 1973). "179 Years in Sea Water?," *Materials Protection and Performance*, Vol. 12 No. 1, NACE.
45. Heinz, R. (Dec. 1975). "Protection of Piles: Wood, Concrete, Steel," *Civil Engineering*, ASCE, New York.
46. Watkins, L. L. (May 1969). "Corrosion and Protection of Steel Piling in Seawater," U.S. Army COE, Tech. Memo No. 27.
47. SSPC (1975). *Steel Structures Painting Manual*, Vol. 2: *Systems and Specifications*. Steel Structures Painting Council, Pittsburgh, PA.
48. NACE (1976). *Recommended Practice: Control of External Corrosion on Underground or Submerged Metallic Piping Systems*, Standard RP-01-76. Nat. Assoc. of Corrosion Engineers, Houston, Tex.
49. Tominaga, M. and Heidengren, C. R. (Mar. 1988). "Steel Pipe Foundations for Deep Water," *Civil Engineering*, ASCE, New York.
50. Lehmann, J. A. (Oct. 1979). "Cathodic Protection for Marine Structures," *Proc.* NACE, Western Regional Conf., San Francisco, Calif.
51. Wagner, J. and Fitzgerald, J. H. (Mar. 1979). "Cathodic Protection for Wharf Foundation Piles at the Port of Anchorage, Alaska," *Proc.* NACE, Corrosion '79, Paper No. 254, Atlanta, Ga.

Appendix 1

Conversion Factors

U.S. Customary		SI Metric
1 inch (in.)	=	2.54 centimeters (cm)
1 foot (ft)	=	.3048 meter (m)
1 yard (yd)	=	.9144 meter (m)
1 statute mile	=	1.609 kilometers (km)
1 square inch (sq. in)	=	6.45 square centimeters (sq cm)
1 square foot (sq ft)	=	.093 square meter (sq m^2)
1 cubic inch (cu in.)	=	16.39 cubic centimeters (cc)
1 cubic foot (cu ft)	=	.0283 cubic meter (m^3)
1 pound (lb)	=	.453 kilogram (kg)
1 ton	=	.907 metric ton (m.t.)
1 pound (force)	=	4.448 Newton (N)
1 kilopound (kip) (force)	=	4.448 kilo Newtons (kN)
1 kip per foot (k/ft)	=	14.59 kilo Newtons per meter (kN/m)
1 pound per square inch (psi)	=	6.89 kilo Newtons per square meter (kN/m^2)
1 pound per square foot (psf)	=	47.88 kilo Newtons per square meter (kN/m^2)
1 pound per cubic foot (pcf)	=	16.02 kilograms per cubic meter* (kg/m^3)

* (not an S.I. unit)

Nautical Units

1 nautical mile = 1.151 statute miles = 6076.1 feet = 1.852 kilometers
1 knot = 1.151 miles per hour = 1.688 feet per second = .515 meter per second
1 fathom = 6.0 feet = 1.829 meters
1 long ton = 1.12 short tons = 2240 pounds = 1.016 metric tons

Appendix 2
Selected Information Sources: Journals, Periodicals, and Conference Proceedings

TECHNICAL JOURNALS

Journal of Waterway, Port, Coastal and Ocean Engineering, American Society of Civil Engineers, New York (bimonthly).

Coastal Engineering, An International Journal for Coastal, Harbour and Offshore Engineers, Elsevier Scientific Publishing Co., Amsterdam, The Netherlands (quarterly).

Bulletin, of the Permanent International Association of Navigation Congresses, Brussels, Belgium (quarterly).

Terra et Aqua, International Association of Dredging Companies, The Hague, The Netherlands (quarterly).

PERIODICALS

The Dock and Harbour Authority, Foxlow Publishing Co., London (monthly).

Port Construction and Ocean Technology, General Publishing Ltd., Surrey, England (bimonthly).

Port Development International, London.

Ocean Industry, Gulf Publishing Co., Houston, Tex. (monthly).

Offshore, Penn Well Publishing Co., Tulsa, Okla. (monthly).

World Dredging and Marine Construction, Symcon Publishing Co., San Pedro, Calif.

Cargo Systems International, Journal of the International Cargo Handling and Coordination Association, London, England (monthly).

World Wide Shipping, WWS, Blauvelt, New York (eight issues per year).

World Port and Harbour Abstracts, British Harbour Research Association, London (quarterly).

CONFERENCE PROCEEDINGS

"Ports '89," American Society of Civil Engineers (ASCE); held every three years since 1977.

"Civil Engineering in the Oceans," ASCE; four conferences held, in 1968, 1969, 1975, and 1979.

"Conference on Coastal Engineering," ASCE; first conference held in 1950, with twenty-second conference planned for 1990.

"International Navigation Congress," PIANC; held intermittently since 1885, with twenty-seventh congress planned for 1990.

"World Ports Conference," of the International Association of Ports and Harbours (IAPH); held intermittently with sixteenth conference in 1989.

"Conference on Ocean Engineering under Arctic Conditions" (POAC); first conference in 1971, seventh conference in 1983.

"Offshore Technology Conference"; held annually in Houston, Tex. since 1969. Co-sponsored.

"World Dredging Conference," World Dredging Association; held approximately every 18 months since 1967.

"Coastal Zone '89," ASCE; first conference in 1978, sixth conference in 1989.

SOURCES OF VESSEL DATA

Marine Technology, The Society of Naval Architects and Marine Engineers (SNAME), Jersey City N.J. (quarterly).

Marine Log, Simmons-Boardman Publishing Corp., New York (monthly).

Maritime Reporter and Engineering News, Maritime Activity Reports, Inc., New York (monthly).

Shipping World and Shipbuilder, Marine Publications International, Ltd., London (monthly).

Ship and Boat International, Metal Bulletin Journals, Ltd., London (ten issues per year).

The Motor Ship, Gavin Howe, Surrey, England (monthly).

Index

Abrasion, 80, 330, 449, 496, 506
Access bridge, 386, 387
Added mass, 142-144, 215, 221, 370-372
Air
 density, 197
 entrainment, 386, 388
Airy wave, 66
Anchor
 bolts, 284
 burial, 396, 397
 chain, 394, 395
 drag, 396
 embedment, 397
 gravity, 395, 396
Anode, 509-511
Apron, 56, 57
Artificial habor, 49
Astronomical tide, 75, 77
Atmospheric
 corrosion, 83
 pressure, 77
 zone, 92
Azobe, 88

Back filling, 319
Backing bar, 284
Ballast
 control, 382, 462
 tanks, 382-386, 457, 462, 463
 vessel, 18, 200, 207
Barge carrier, 33, 35, 58
Barges, 37, 361
Basin
 anchorage, 50
 coastal, 49
 dry dock, 342, 415-438
 impounded, 49
 turning, 50
Batter piles, 234, 249, 250, 252, 256, 332, 240, 341
Beaufort Scale, 62

Berth
 arrangements, 50-56
 definitions, 50
Berthing
 beam, 158
 coefficients, 135, 136
 dolphin, 50, 181, 269-276
 energy, 134-141, 145-152
 impact, 134-141, 145-152
 reaction force, 134, 135, 140, 154, 155
 velocity, 145-147
Bent (*see* Pile)
Biodeterioration, 81, 82, 92, 489, 492-494
Bitt, 177, 193
Block coefficient, 20, 138, 366
Block walls, 235-237
Blocking (dry dock), 411, 412, 495, 462
Bollard, 177, 193
Borers, 82, 492, 493
Borings (*see* Subsurface Explorations)
Breaking wave, 119, 123
Breakwater, 3, 300 (*see also* Floating breakwater)
Breasting dolphin (*see* Berthing)
Breast line, 180
Brow (*see* Gangways)
Bulk
 cargo, 58, 59, 101, 112
 carriers, 30
 terminals, 8, 58, 59
Bulkhead
 anchored, 234, 317, 319-325
 cantilevered, 234, 319, 320
 definitions, 234, 317
 line, 50
 repairs, 503-506
 sheet pile (*see* Sheet pile)
 watertight, 369, 380
Bull rail, 283
Buoys, 9, 364, 400
Buried Zone, 92
Butyl Rubber, 173

519

Cableway, 113
Caisson
 concrete, 234, 239, 267, 268, 274, 325
 pneumatic, 269
Camel, 156, 362, 363
Capping beam, 269
Capstan, 195
Cargo
 bulk, 58, 59, 101, 112
 general, 56, 97–102
 handling, 56
 loads, 97–102
 ships, 30, 32
Catamaran, 41, 358, 360
Catenary, 390–393
Cathodic protection, 83, 454, 509–511
Catwalk, 181, 233, 276, 277
Celerity, 67
Cell
 construction, 234, 273, 323–328
 sheet pile, 234
Chain
 anchor, 394, 395
 fender, 168
 hauling, 446, 447
 wheel, 446
Channel (definitions), 49, 50
Chock, 193
Claddings, 509
Clapotis, 121
Classification Society, 376, 466
Cleat, 193
Closed construction, 233, 234, 239 (*see also* Solid fill construction)
Coastal
 basin, 49
 engineering, 3
 flooding, 314, 319
Collier, 58
Composite construction, 261, 262, 456
Concrete
 admixtures, 86
 cover, 86, 261, 382
 deck system, 259–261
 deterioration, 494–496
 hull, 379, 382
 mix, 86–88
 piles (*see* Pile)
 precast, 260, 261, 267–269
 prestressed, 260, 261
 reinforcement, 86, 88, 261
 repairs, 497–500
Combination bulk carriers, 30
Container
 cranes, 57, 110, 112
 dimensions, 100, 101
 ships, 32
 terminals, 57
Containment boom, 363, 364
Conveyor, 8, 112
Corrosion
 fatigue, 490
 galvanic, 490
 general, 88, 489–492
 protection, 83, 86
 rates, 490
Cradle (marine railway), 443
Crane
 beams, 277–280
 container, 110, 112
 crawler, 109
 fixed, 110
 loads, 109–112
 portal, 109
 rails, 277, 278
 track, 277–280
 truck, 103–106
Crib construction, 234
Cruise ship, 35–37
Curb, 283
Current
 forces, 77–79, 116, 118, 119
 general, 77, 79
 prop-wash, 129, 130
 rotary, 79
 tidal, 77–79
 velocity, 77–79

Damping coefficient, 221, 370–372
Davit, 480
Deadweight (definitions), 18
Design
 criteria, 12, 13, 44, 45
 life, 47
 vessel, 9, 45
 water level, 77
 water line, 17
Dewatering, 313, 427, 462
Diaphragm cells, 234
Diffraction theory, 116, 120, 213, 228

Directional spectra, 70, 373
Displacement (definitions), 18
Dock (definitions), 50, 233
Dolphin
 berthing, 50, 181, 230, 269
 breasting, 269
 cantilever, 271, 272
 cell, 273
 cluster, 269, 270
 general design, 269-276
 mooring, 50, 181, 238, 269, 389
Draft (definitions), 17-19
Drag force, due to
 current, 118-120, 204-209
 wave, 116-119
 wind, 196-203
Drainage
 deck, 281, 282
 soil, 349, 350
Dredging, 2, 5, 48, 313
Drift force, 211, 213, 215, 216
Dry bulk
 cargo, 58, 59
 carriers, 30
 terminals, 58, 59
Dry docks
 basin, 415-438
 floating, 452-469
 geotechnical design, 412, 342-344
 graving, 415-438
 railway, 438-452
 vertical lift, 469-476
Dynamic
 analysis, 128, 262-264, 398
 compaction, 347, 348
 load factor, 128, 262, 263
 response, 128, 262-264, 359-375, 398

Earthquake (*see* Seismic design)
Eccentricity
 coefficient of, 135
 pile, 245, 251
Effective stress, 302-304
Ekki, 88
Elastomeric
 bearings, 276
 materials, 173
Erosion, 129, 314, 319 (*see also* Scour)
Euler equation, 264
Expansion joints, 280, 281

Fabric
 filter, 82, 316, 350
 jackets, 82, 504
 reinforcement, 350, 351
Factor of safety, 14, 45, 128, 154, 165, 186, 192, 242, 314, 394
Fair lead, 193, 394
Fairway, 49
Fatigue, 83, 128, 266, 267, 382, 489, 496
Fender
 design, 160-173, 193
 selection, 152-156
 types, 156-160
Ferry
 boats, 35
 terminals, 57, 58, 129, 130
Fetch, 64
Fire protection, 285
Fishing vessels, 40, 41
Floating
 breakwaters, 364-366
 docks, 401-402
 dry docks, 452-469
 link spans, 361, 362
 piers, 358-361
Flood valve, 384, 385, 428, 462, 463
Fork Lift Truck, 97, 104
Fouling, 81, 82, 494
Foundations (marine), 311, 312
Freeboard, 18, 369
Free surface effect, 368
Frequency domain, 221, 222, 263
Friction
 coefficient of, 169
 surface, 207, 208
Froude number, 144, 226

Gangways, 101, 115, 388 (*see also* Brows)
General cargo
 terminals, 5, 56
 vessels, 30, 32
Graving dock, 415-438
Gravity
 fenders, 159
 walls, 234, 235, 317, 326
Greenheart, 88, 243, 331
Gross tonnage (definition), 17, 19
Grout, 278, 498
Gunite, 86, 498
Gust (wind), 62, 63, 203

Handrail, 115
Harbor (definition), 49
Hauling
 chain, 446, 447
 machine, 446
Hawser, 188
HDPE, 173
Heave
 motion, 178, 180, 216, 370, 372–375
 period, 178, 180, 372
Heeling moment, 140, 200, 201, 204, 207, 366–368
Hoist, 480, 481
Hose tower, 114
Horsepower
 brake, 37, 383, 384
 pump, 383, 384
 shaft, 37
Hull
 construction, 23, 24, 375–382
 form, 20–23
 types, 358–360
Hurricanes, 61, 184
Hydrodynamic
 coefficients, 221–223
 mass (see Added mass)
Hydraulic
 filling, 319
 jacks, 467, 472
 lifts, 474, 475

Ice
 abrasion, 80
 accumulation, 80, 124
 expansion, 124
 forces, 124–127, 176, 177
 impact, 126, 127
 information, 80–81
 jacking, 124, 125
 thickness, 79
 thrust, 80
 uplift, 80, 124, 125
Impressed current, 491, 509–511
Impulse response function, 223
Inertial force coefficient, 117, 223
Irregular waves, 213, 222, 393

Jacket
 pile, 503–506
 template, 234
Jack-up barge, 233, 234

Jets, 61
Jetty, 233

Keel blocks, 411, 445
Kinematic viscosity, 228
Kuelegan-Carpenter number, 119

Lamella wall, 234, 257
LASH
 terminals, 58
 vessels, 33–36
Launching, 59, 375, 409, 450, 452
Lift dock, 469–477
Lift force, 119
Lighting, 286
Limnoria, 492, 493
Linear loader, 113
Lines
 mooring, 177, 181, 186–193
 vessels, 20
Linkages, 159
Link-spans, 57, 361, 362
Liquefaction, 308, 319
Liquid bulk
 cargoes, 59, 114
 carriers, 26
 terminals, 59
LNG
 terminals, 59
 vessels, 28, 29
Loading
 arms, 113–115
 platforms, 59, 274
LPG
 terminals, 59
 vessels, 28, 29

Marginal wharves, 50, 56
Marinas, 401–405
Marine railway, 438–452
Matrix methods, 264
Mean
 high water, 75
 low water, 75
 sea level, 75
Metacentric height, 140, 367, 368, 371, 325, 459
Merchant fleet, 25, 26
Military
 facilities, 60
 vessels, 41

INDEX 523

Models
 mathematical, 61, 152, 221–223, 264, 398
 physical, 226–228, 398
Mole, 238
Moment
 to list, 368, 369
 to trim, 368, 369
Mooring
 arrangements, 180–183, 388–390
 buoys, 400
 chains, 394, 395
 design criteria, 183–185, 393, 394
 dolphin (see Dolphin)
 forces, 176–180, 196–221, 365, 394, 398
 free swinging, 390, 397–400
 hardware, 193–196, 284
 single point, 364, 397–400
 small craft, 402–405
Morison equation, 116–118
Multiplication effect, 459, 460

Navigation
 aids, 284, 285
 channels, 49, 50
 lights, 284, 285
Naval vessels (see Military vessels)
Neap tide, 75
Net tonnage (definition), 19

OBO vessel, 30
Offshore
 engineering, 3
 ports, 9, 49
 structures, 242
Open construction, 233, 234, 238
Ore carriers, 30

Pad eye, 193
Paint (see Protective coatings)
Pallets, 56, 99, 100
Panamax (vessel), 23
Passenger
 ships, 35–37
 terminals, 58
Pier (definitions), 233, 234
Pierhead line, 50
Pile
 bents, 250–, 255, 256
 design considerations, 328–331
 prestressed concrete cylinder piles, 243, 320, 331–332
 prestressed concrete piles, 243, 331–332
 steel H-piles, 243, 333
 structural design, 242–257
 timber piles, 243, 331
Pile analysis:
 axial load capacity, 334–336
 depth to fixity, 339–340
 downdrag, 341
 lateral load capacity, 336–339
 uplift resistance, 340
Pipelines, 114, 115
Pitch
 motion, 178, 216, 369–371
 of fenders, 170
 period, 178–180, 371
Plimsol mark, 17
Pneumatic
 caisson, 269
 fender, 156, 157
Polymer modified concrete, 88, 500
Pontoon, 258–360, 372–376, 452
Port (definition), 49
Preloading, 348, 349
Probability of encounter, 47
Prop-wash, 129, 130, 249
Protective coatings, 506–509
Pumping station, 428–431
Pumping systems, 383, 384, 427–432, 462, 463, 466
Pumps, 383, 384

Quick condition, 318
Quick release hook, 194, 195
Quay, 234, 300, 311

Radial loader, 113
Radius of gyration, 138, 249, 371, 492
Rails (see Crane)
Railway loadings, 109, 110
Ranging (harbor), 73, 177, 211 (see also Seiche)
Raker pile, 234
Random loads, 262
Rayleigh distribution, 65, 68
Reflection (see Wave)
Refraction (see Wave)
Regular waves, 68, 213, 393
Reinforced Earth, 350
Relieving platform, 234, 238
Rennie dock, 454

Retaining structures
 bulkheads, 300, 311–312, 317, 319–325, 351
 concrete caissons, 312, 317, 325, 327
 design considerations, 317, 319, 321
 gravity quay walls, 312, 317, 326
 steel cofferdams, 312, 317, 325, 326
Response amplitude operator (RAO), 216–218, 373
Return period (definition), 47
Reynolds number, 117, 196, 204, 207, 228
Rigging, 197, 200
Risk analysis, 47
Roll
 motion, 178, 216, 369–372
 period, 178, 180
Rollers, 449, 467
Roll-on/roll-off
 ships, 33
 terminals, 57, 129, 137
Rope, 188–190 (*see also* Wire rope)
Rubber, 160, 173

Salinity, 61, 81, 92, 493
Scantlings, 25
Scope, 394, 397, 402
Scour, 58, 77, 79, 489 (*see also* Erosion)
Sea state, 65
Seiche, 73, 177, 211, 212
Seismic design, 81, 127, 128, 262, 264–266, 308
Semi-submersible, 358, 360
Sewage, 286
Sheet file
 bulkheads, 234, 503, 506 (*see also* Retaining structures)
 steel, 320–323, 326, 328, 351, 352, 490
Shipyards, 59, 60
Shotcrete (*see* Gunite)
Side haul dry dock, 449, 450
Significant wave (definition), 65
Sinker, 393
Site investigation, 292, 293
Slam load, 120
Slip (*see* Berth)
Slipway (*see* Marine railway)
Slopes
 analysis, 313
 protection, 312, 313, 315
 stability, 311, 313, 314

Soil properties
 consolidation, 305
 permeability, 307
 shear strength, 304
Soil testing-in situ
 constructive monitoring, 295, 298, 347–348
 soil properties, 295, 298, 307
Soil testing: laboratory, 307
Soil types
 bedrock, 307
 clay, 306, 309
 fill, 310
 inorganic silt, 309
 organic silt and peat, 310
 sand and gravel, 308
Solid fill structures:
 design considerations, 312, 328
 dolphins, 300, 311–312, 326–328
 piers, 300, 311–312, 326–329
Solitary wave, 123
Sonar: docking, 146
Spectrum
 responses, 216–218, 372
 wave (*see* Wave spectra)
Splash zone, 92, 490, 493, 506, 511
Spring
 constant, 156, 221, 263, 264, 272, 372
 lines, 180
 tide, 75
Spur pile, 234
Standoff
 distance, 172
 force, 186, 209–211
Steam lines, 286
Steel
 structural, 82–86
 repairs, 500
Stillwater Level, 121
Storm conditions, 62, 182, 184
Stowage Factor, 19
Straddle
 carrier, 104
 lift, 477
Stringers, 23, 257
Strouhal number, 119
Submerged zone, 92
Subsurface explorations
 drilled, 293, 296
 geophysical methods, 295, 299, 300
 planning, 300–301
 test pits, 295, 300

INDEX 525

Surge
 harbor, (see Seiche and Ranging)
 motion, 178, 210, 212
Sway, 178, 210
Swell (ocean), 69, 73, 211

Tanker
 terminals, 59, 181
 vessels, 26-28
Teredo, 492, 493
Tidal
 currents, 77, 79
 datums, 75-77
 zone, 92, 493, 506, 511
Timber
 construction, 242, 246, 257-259
 dolphins, 269-271
 fenders, 156
 piers, 242, 257-259
 properties, 88-90
 repairs, 500-503
 treatment, 90
Time domain, 221, 222, 264, 398
Tonnage (vessel), 18, 19
Transfer
 bridge, 58, 361, 386-388
 systems, 450-452, 461, 467-469, 476
Transit shed, 115
Transmission coefficient, 71, 365
Transporter, 107, 108
Traveling staging, 410
Tremic concrete, 86, 257, 505
Trestle, 238, 276, 277
Truck
 crane, 103-106
 loading, 103-105
Tsunami, 74
Tugboats, 37-40
Turbidity, 81
Turning basin, 55
Typhoons, 61, 184

ULCC (vessel), 28
Underdrain system, 422
Under keel clearance, 53, 143, 145, 210, 213
Utility systems, 285-287

Vehicle
 carrier, 33
 loads, 102-109

Ventilation, 385, 410
Vertical wick drains, 349
Vibrating pile compaction, 346, 347
Vibro-compaction, 344, 345
Vibro-replacement (stone columns), 345, 346
Virtual mass (see Added Mass)
Visibility, 81, 145
VLCC (vessel), 28
Volumetric coefficient, 22
Vortex shedding, 266

Wave
 action, 319
 attenuation, 365, 366
 crest elevation, 123
 data, 63, 64
 definitions, 66
 diffraction, 70, 71
 energy, 67
 forces, 116-123, 211-228
 number, 66
 period, 66-68, 70
 reflection, 71, 72
 refraction, 70
 shoaling, 70
 spectra, 68-70, 216
 steepness, 68
 uplift, 123
Wetted surface, 207, 208
Wharf (definitions), 50, 233, 234
Winches, 195, 196
Wind
 areas, 201
 data, 62, 63, 203
 forces, 115, 116, 196-203, 278, 279
 gusts, 62, 203
 rose, 62
 spectra, 394
 velocity, 62, 203
Wing wall, 452
Wire rope, 190, 192, 472
Work boats, 37-40

Yaw
 moment, 197, 200, 211
 motion, 178, 210, 398